DIGITAL MICROWAVE TRANSMISSION

ML

STUDIES IN ELECTRICAL AND
ELECTRONIC ENGINEERING 38

DIGITAL MICROWAVE TRANSMISSION

I. FRIGYES

Technical University of Budapest

Z. SZABÓ
P. VÁNYAI

Research Institute for Telecommunication
Budapest, Hungary

AMSTERDAM—OXFORD—NEW YORK—TOKYO 1989

This work is the enlarged and revised version of
"Digitális mikrohullámú átviteltechnika"
published in Hungarian by Műszaki Könyvkiadó, Budapest
The distribution of this book is being handled by the following publishers

for the USA and Canada
Elsevier Science Publishing Company, Inc.
655 Avenue of the Americas,
New York, NY 10010, USA

for the East European countries, Democratic People's Republic of Korea,
People's Republic of Mongolia, Republic of Cuba, and Socialist Republic of Vietnam
Kultura Hungarian Foreign Trading Company, P.O.B. 149, H-1389 Budapest 62, Hungary

for all remaining areas
Elsevier Science Publishers
Sara Burgerhartstraat 25
P.O. Box 211, 1000 AE Amsterdam, The Netherlands

Library of Congress Cataloging–in–Publication Data
Frigyes, István.
    [Digitális mikrohullámú átviteltechnika. English]
    Digital microwave transmission / I. Frigyes, Z. Szabó, P. Ványai.
        p. cm. — (Studies in electrical and electronic engineering; 38)
    Rev. translation of : Digitális mikrohullámú átviteltechnika.
    Includes bibliographical references.
    ISBN 0-444-98858-0 (U.S.)
    1. Microwave communication systems. 2. Digital communications.
    I. Szabó, Zoltán, okl. villamosmérnök. II. Ványai, Péter.
    III. Title. IV. Series.
    TK7876.F7413 1989   89-17216
    621.382—dc20   CIP

ISBN 0-444-98858-0 (vol. 38)
ISBN 0-444-41713-3 (series)

© I. Frigyes – Z. Szabó – P. Ványai, 1989

© English translation — T. Sárkány, 1989

Joint edition published by
Elsevier Scientific Publishing Co., Amsterdam, The Netherlands
and Akadémiai Kiadó, Budapest, Hungary

Printed in Hungary by Szegedi Nyomda, Szeged

STUDIES IN ELECTRICAL AND ELECTRONIC ENGINEERING

PREFACE

The purpose of this book is to present theoretical and practical problems encountered in digital microwave transmission. The authors intend to provide information useful to those wishing to become acquainted with this topic, including experts in other fields aiming to learn the fundamentals, and readers interested in details.

In the first four chapters, theoretical background information is given. Following the Introduction, we present problems of digital transmission and synchronization problems of microwave transmission, and properties of microwave channels are outlined. In the chapters which follow, specific equipment components such as modulators, demodulators and carrier recovery circuits are dealt with, together with baseband signal processing and auxiliary equipment. Of course, microwave circuits and assemblies are not considered, in spite of the fact that some of these, *e.g.* antennas in satellite transmission systems, have a decisive effect on transmission performance.

We have not endeavoured to give accurate mathematical details as this has been considered to be beyond our scope. However several methods of probability calculus and stochastic processes have been applied, together with methods needed for describing signals, such as the Fourier transform, signal representation by complex envelope, and others. In some places, reference has been made to a few theorems of circuit theory, assumed to be known by the reader.

Owing to the restricted space available, some important topics have not been dealt with: error correcting encoding, synchronization problems encountered outside the strictly considered transmission path, special network and network organization problems such as multiple access methods, rural networks, and others.

This English edition is a substantially revised form of the Hungarian first edition published in 1980. Digital microwave transmission has changed immensely since this date, since the intervening years have witnessed the digitalization of the majority or at least a very large part of long haul networks worldwide. In microwave transmission, this has resulted in a widespread use of high speed digital systems, focusing attention on the modulation methods suitable for high speed transmission, together with propagation problems arising in high speed transmission. This technical evolution has necessitated revision and enlargement of the original book. Furthermore, large scale integrated circuits are now generally available, rendering unnecessary the presentation of detailed block diagrams, or even circuit diagrams. This again required revision of the original book.

Continuous phase modulation, together with signal space encoding, and spread spectrum systems are topics which, though falling within the scope of this book, have not been covered in detail because of the limited space available, and also because they can currently be considered as special topics of mobile and military communication. However, the authors are convinced that this is only a temporary situation.

Finally, the authors wish to express their thanks to their translator, T. Sárkány, for his helpful editorial remarks which have contributed to a more comprehensible text.

The authors

CONTENTS

Contents   xi

Contents

CHAPTER 1

INTRODUCTION

It is a common statement that a worldwide trend towards digital transmission is evolving in an ever increasing number of communication networks. The main reason for this process which has been taking place since the early sixties is economics: according to several investigations, the solution of telecommunication tasks by digital means is much less expensive than by analog means. In more detail, the practical utilization of the PCM principle has been rendered possible by the development of computers since this resulted in a price reduction of digital integrated circuits so that even relatively small-scale production for the telecommunication market proved to be economical. On the other hand, computer development has also created a demand for the transmission of digital signals, notably data signals. Consequently, the development and widespread use of PCM transmission equipment, first of all primary multiplex equipment, began in the early sixties.

Parallel to this development, and partly as a consequence of it, a rapid evolution of digital transmission was set in motion. The most important phase of this development was the evolution of digital switching techniques which has resulted in the general use of stored program controlled digital exchanges. These have outperformed any of their competitors from all aspects, including economy. It is probably true that this was the most decisive event in the evolution of digital networks.

The introduction of digital switching had different impacts, depending on whether the country concerned is small or medium sized, or large. In the former, the distances between inhabited localities are relatively small, bringing the exchanges relatively close to each other. Economical considerations in these countries have resulted in early change-over to digital transmission, thus avoiding the expense of too frequent analog-to-digital and digital-to-analog conversions. On the other hand, the transmission routes of larger, sparsely populated countries are longer, resulting in larger distances between exchanges. Consequently, the complete change-over to digital transmission has been delayed in these countries until digital transmission became less expensive than analog transmission — primarily as a consequence of fibre optic links. In these countries, the advantages of network integration, which is a present-day trend, have also contributed to the spreading of digital transmission.

Our book deals with a particular kind of digital transmission, the digital microwave transmission. For better characterization of this field, we present a few definitions. Throughout, a signal sequence is said to be digital if the following conditions are met:

(*a*) the sequence is made up of elements of a known finite set;
(*b*) all elements have finite and equal time durations (generally denoted by *T*);
(*c*) the starting time instants of all time durations are known.

By the above definitions, the more general conventional definition is somewhat restricted as it applies only to so-called synchronous signal sequences. In practice, an even more restricted class of signals is used as the digital signals are binary (*i. e.*, there are only two possible levels), realized by rectangular pulses of duration *T*. The time intervals *T* are consecutive: the time instant $t=T$ of the ith signal element coincides with the time instant $t=0$ of the $(i+1)$th signal element. This kind of signal is called an NRZ (Non-Return-to-Zero) signal.

In view of these definitions, the relatively few equipment types which make up a digital channel can be determined: these are the signal multiplexing/demultiplexing equipment and the encoding/decoding equipment. Accordingly, the structure of a digital transmission channel is shown in Fig. 1.1. For simplicity, a one-way (simplex) channel is shown but actually, two-way (duplex) channels are used in most cases. The speed of the bit sequence, or in other words, its clock frequency

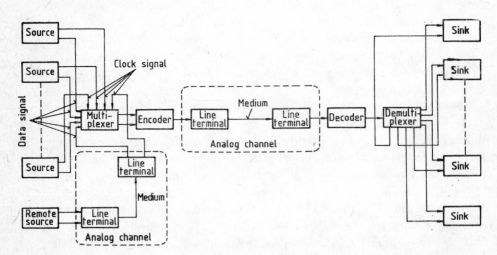

Fig. 1.1. Digital transmission channel.

$1/T$, is necessarily changed by the multiplex equipment: the speed of the multiplexed signal may be the sum of the input signal speeds, but in some cases it may be somewhat higher.

The encoding process results in another signal which is also digital. This encoding of digital signals may be advantageous for several reasons, perhaps the most significant being the error correction required in some kinds of digital transmission. This kind of encoding is accompanied by an increase in the transmission speed.

The signals supplied by the sources and received by the sinks, as shown in Fig. 1.1, are digital signals, corresponding to the above definitions (*a*), (*b*) and (*c*). Equipment converting analog into digital and digital into analog signals is therefore excluded from our investigations. This figure shows, in addition to the digital sources and sinks, the "muldex" (multiplex–demultiplex) equipment and the "codec" (encoder–decoder) equipment, and also the line terminal equipment and the transmission medium. These have the function of analog rather than digital processing, involving distortion and noise addition. These effects cannot be·interpreted by the above defined channel and its source or sink. Consequently, the line terminal equipment is simply regarded as a source or a sink.

The digital channel itself has only two quality parameters, the error probability, *i.e.* the relative number of errored signals reaching the sink, and the jitter, *i.e.* the timing error. The definition of these characteristics is obvious: the signal is in error if the *j*th element of the bit stream is regarded by the sink as the $(k \neq j)$th element. The timing is in error if the time duration of the signals reaching the sink differs from the time duration at the source. The permissible values of these quality parameters depend on the specific field of application.

The description of real transmission media is more appropriate by analog than by digital channels. The digital signals will be attenuated and distorted by the transmission medium which will add noise to them even if the signal sequence is directly coupled to the medium.

In Figure 1.1, the part of the digital channel in the dotted frame can thus be regarded as an analog channel, characterized by the signal-to-noise ratio, the linear distortion, the nonlinear distortion, etc. The above defined parameters of the digital channel depend obviously on the characteristics of the analog channel. In Fig. 1.1, all equipment of the analog channel is represented by the line terminal equipment which has to be designed so that the digital parameters should not exceed the specified limits. Our book deals primarily with a special analog channel, the microwave analog channel, intended for the transmission of digital signals.

Figure 1.2 is a simplified, rather general, representation of such an analog channel. It is a matter of choice whether the line codec is part of the analog channel or

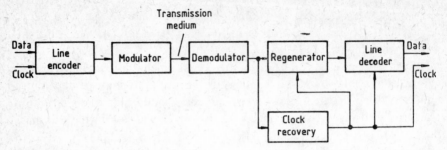

Fig. 1.2. Analog channel for the transmission of digital signals.

not. However, it is justified to regard it as part of the analog channel since, in contrast with the error correcting encoding, it is intended exclusively to simplify the realization of other parts such as the modulator and clock recovery, or to achieve a more efficient utilization of the transmission medium. The modulator has the function of converting the digital signals into signals matching optimally the properties of the transmission medium. Similarly to the digital signals, the line signals generated by the modulator are elements of a finite known set. However, the number of elements in the line signal set is not necessarily equal to the number of elements in the digital signal sequence to be transmitted. (We have seen that the latter is two in most cases; the number of elements in the line signal set is frequently higher, *e.g.* 3, 4, 8, 16, *etc.*) The line signals may be either baseband signals — this is generally the case in wirebound transmission — or modulated carrier signals applied for microwave or fibre optical transmission, but also for wirebound transmission if medium or high speed data have to be transmitted.

The demodulator has the function of recovering the original line signals. However, this can only partly be accomplished owing to the channel noise and distortion: some of the signals are erroneously received (there is no possibility for correction within the analog channel), and the time duration of the signals is altered because of the distortion. The regenerator has the function of restoring the time duration of the signals (within the allowable jitter). In most variants, this operation requires the clock signal, supplied by the clock recovery circuit, defining the starting instants of the individual signal elements. However, so-called direct regenerators require only the modulated signal. In most cases, the clock is recovered from the demodulated signal as input signal (according to Fig. 1.2), but sometimes, the clock recovery circuit is driven directly by the modulated carrier. The former and latter are called series and parallel clock recovery, respectively.

In the previous considerations, the transmission medium has not yet been mentioned. Most present-day digital links utilize a cable as the transmission medium, primarily for transmission of 1.5 or 2 Mbit/s bit streams over short distances. However, in many cases, a microwave channel is better suited, *e.g.* in local net-works it may substitute trunk cables. Furthermore, long distance analog FM radio relay systems can also be substituted by digital systems — the latter are frequently more suitable alternatives. Even low capacity short links are frequently less expensive when realized by digital radio if there are no established cable ducts, or the existing cable capacity cannot be increased. For very large distance transmission, submarine cables can advantageously be substituted by digital satellite links. If the area between terminal stations, separated by medium dis-tances, is not accessible, the only practical choice for limited traffic requirements is satellite or trans-horizon digital radio transmission. The best examples for this kind of application are the telecommunication links between the mainland and off-shore oil rigs. Finally, it should be mentioned that the circular waveguide propagating the TE_{01} mode has the highest transmission capacity of all presently known transmission media (though the above list does not cover all application fields).

Figure 1.3 shows in somewhat more detail the characteristic parts of the micro-wave radio channel used for the transmission of digital signals. It should be noted that this system is frequently called the "digital radio relay system" or "digital satellite system". These terms will be applied also in the following though this is

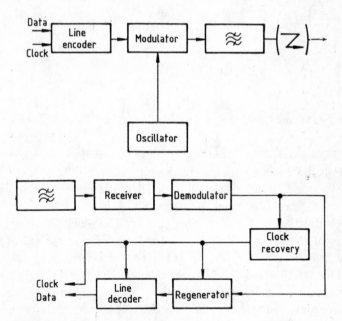

Fig. 1.3. Microwave channel for the transmission of digital signals.

not quite correct because it is the analog channel, according to our definition, which is made up of this equipment. In Fig. 1.3, the modem section of Fig. 1.2 is shown in somewhat more detail. Again a simplex channel is presented, without showing auxiliary equipment such as the service channel, alarm and stand-by switch-over equipment. In the case of a waveguide connection, the antennas shown in Fig. 1.3 are substituted by the devices transmitting and receiving the TE_{01} mode.

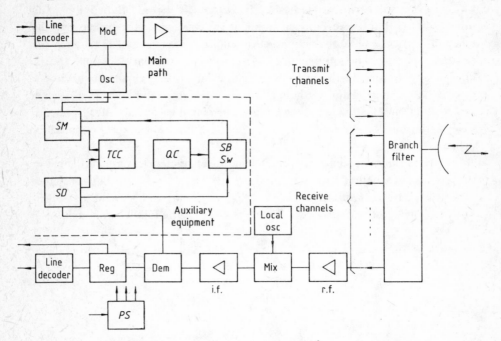

Fig. 1.4. Microwave terminal station — main path and auxiliary equipment.

Finally, Figure 1.4 is a more detailed representation of a terminal station, including also the auxiliary equipment. Blocks *SM* and *SD* have the functions of multiplexing and demultiplexing the service channels. According to the version shown, the service channel multiplex signal is transmitted by additional frequency modulation of the transmit oscillator. The block *TCC* serves for the transmission of the telecontrol and occasional telecommand signals. (These are needed because the repeater stations are normally unattended.) The *SB Sw* block, used for stand-by switching, receives the switch control signal from the *QC* quality control unit. Finally, *PS* denotes the power supply of the station.

In the following, the relevant characteristics of digital radio equipment will be outlined. Practically, the complete microwave frequency range is suitable for establishing digital radio links. The lower limit of this range can be regarded as being 300 MHz while the upper limit is well inside the millimeter band — 60 to 80 GHz are the highest frequencies at which radio relay experiments have been published. The classical frequency range of radio relay equipment is about 2 to 11 GHz. Practically all analog systems are operated in this range so that in countries with extended analog radio relay network, this range can be regarded as saturated. This is why digital radio is not readily authorized in this range, even although it is perfectly suited for digital transmission. In these countries, the range 10 to 20 GHz is the range usually used for digital systems. On the other hand, the propagation properties of this range are not well suited for high capacity analog transmission which is another argument for digital exploitation. In spite of this, several digital radio relay systems of different capacities are operated in the frequency range below 10 GHz, and their application in this range will probably become more widespread. The range above 20 GHz is, for the time being, only occasionally used because the range below 20 GHz is not yet saturated, so there are few arguments for increasing the frequency.

At present, telecommunication satellites are operated in the 4 and 6 GHz band, and also in the 11 and 14 GHz band, both for analog and digital transmission. There are also experiments under way at higher frequencies up to about 30 GHz. The trans-horizon radio relay systems will not be dealt with because of their peripherical significance. These are operated in the lower part of the microwave frequency range at 1 to 1.5 GHz, though trans-horizon systems at higher frequencies are also known. Finally, the frequency range of waveguide transmission lies in the millimeter wave band, and falls between 30 to 100 (possibly 120) GHz.

The transmission capacity of digital microwave systems varies between wide limits. According to a CCIR Recommendation, systems with transmission speeds below 10 Mbit/s, between 10 and 100 Mbit/s, and above 100 Mbit/s, are classified as low, medium, and high capacity systems, respectively. The highest capacity digital radio relay systems with a transmission speed of about 600 Mbit/s are suitable for the transmission of about 7600 telephone channels which is substantially higher than the 2700 channels transmitted by the highest capacity analog FM systems. Lower speed equipment is normally designed for signal speeds corresponding to one of the hierarchical PCM levels. The so-called "European hierarchy", recommended by the CCITT, has the following transmission speeds: 2048 kbit/s (30 channels), 8448 kbit/s (120 channels), 34 368 kbit/s (480 channels), and 139 264 kbit/s (1920 channels). These telephone channel numbers are derived from the 30 channel primary PCM multiplex by digital multiplexing. Obviously, other signals of appropriate speeds can also be transmitted over these systems, *e.g.* 6 high quality sound program channels can be transmitted at 2 Mbit/s, one or two videophone channels at 8 Mbit/s, and by suitable encoding, one television channel at 34 Mbit/s. The speed of the satellite channels is usually 60 to 65 Mbit/s though the next generation of satellites will probably have much higher speed (*e.g.*, up to 500 or 600 Mbit/s).

In conclusion of this chapter, let us investigate the performance requirements. Obviously, the majority of digital microwave equipment, similarly to other transmission equipment, is used for the transmission of telephone channels; it is thus feasible to design the microwave channel to meet voice transmission requirements. This means that services requiring higher transmission performance are required to provide this by themselves. This can actually be accomplished as we have seen that a digital channel has two performance parameters, the error probability and the jitter, and both can be substantially reduced: the error probability by error correction encoding, and the jitter by jitter reducing circuits. It is thus feasible to include the error correcting codecs and de-jitterizers within the analog-to-digital and digital-to-analog converters of these special services.

The following error probability values, specified for high quality transmission, are applicable to a long connection; for shorter links, only a fraction of these is allowed (*e.g.*, according to proportional reduction). According to a CCIR Recommendation, the length of a long connection is 2500 km or its multiple.

For voice transmission, the upper limit allowed for error probability is 10^{-6} or 10^{-7}. For data transmission, the requirement is 10^{-5} or in other cases, much less. For television transmission, the permitted error probability is highly dependent on the encoding scheme. In the case of a simply encoded video signal, the picture quality does not depend essentially on the error probability — there is no need for better than voice transmission quality — but the required transmission speed is very high. Television encoding methods, used in practice, though reducing the redundancy, are highly sensitive to errors, and require an error probability of less than 10^{-9}, possibly 10^{-12}. Similarly strict requirements have also to be met by sound program transmission.

The permissible jitter also depends on the service provided. Voice signals are relatively less sensitive to jitter though some basic noise is generated in the decoded voice signal by jitter. Television and sound program transmission is much more sensitive to jitter and exceptionally high jitter requirements have to be met by the PCM transmission of FDM signals.

CHAPTER 2

TRANSMISSION OF DIGITAL SIGNALS OVER ANALOG CHANNELS

The analog channel has been defined in Chapter 1. In a real (non-ideal) channel, the transmitted signal is affected in several ways so the shape of the received signal which has to be evaluated may differ significantly from the original shape. In the case of digital transmission, this may result in faulty evaluation of the received signal. In a favourable case, the distortion will still allow a practically correct evaluation of the message which is thus received without loss. However, in some cases, a higher distortion introduced by the channel may result in a false message. (In this concept, "distortion" is interpreted in a somewhat generalized sense, including not only the actual distortion of the filters and nonlinear circuits but also the additional thermal noise and interference noise originating from other channels, etc.)

This property of digital transmission allows the precise determination or measurement of the transmission channel quality. (On the other hand, a more-or-less arbitrary parameter is used to characterize the transmission of analog signals such as r.m.s. error, peak error, the average value of the received signal with zero input, etc.). The exact determination of digital transmission quality allows a unique formulation of the transmission task: the equipment should be designed to obtain the highest number of correctly received signals, or in other words, the error probability should be as low as possible. In this chapter, the main disturbing effects will be investigated, and based on this investigation, the system design relations will be presented. In accordance with the scope of this book, the disturbances of microwave channels will primarily be investigated, but without making explicit references to this fact. Figure 2.1 shows the representation of an analog channel including these disturbing sources. The signal sequence transmitting the useful message is represented by the modulated carrier $s(t)$, and this is distorted by the transmitter filter $T(\omega)$ — centered at the carrier frequency f_c —, and by the nonlinearity of the following amplifier. In the transmission path, the signals of other channels are added to the desired signal resulting in disturbance. The frequency of channel $z_0(t)$ is identical to that of the desired channel (or,

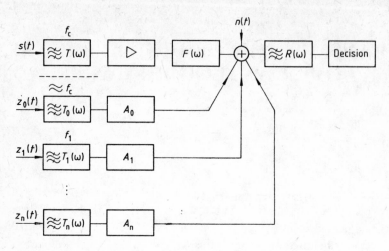

Fig. 2.1. Corrupted analog channel.

more generally, is within the pass-band of the receiver filter) while the frequencies of the remaining channels are different.

All channels have been assumed to have identical powers, *i.e.* $P = P_0 = P_1 = \ldots$ $\ldots = P_n$, but the interfering powers will be modified by the amplifiers or attenuators $A_0, A_1, \ldots, A_n$. The distortion introduced by the transmission channel itself is represented in Fig. 2.1 by $F(\omega)$. In terrestrial radio channels, this is caused by selective fading which is significant in the microwave range, at bit rates of 70 Mbit/s or higher. Onto this distorted useful signal, loaded by interfering channels, the additive noise $n(t)$ is superimposed, which is then reduced by the receiver filter $R(\omega)$, while introducing further distortion.

The effect of the disturbances shown in Fig. 2.1 may differ in various cases. When single digital signals are transmitted then the effect of additive noise only has to be considered, *i.e.* the receiver structure resulting in optimum transmission while additive noise is superimposed has to be found, together with an optimum waveform of the transmitted signal $s(t)$. However, when digital signal sequences are transmitted, the distortions may also be significant in additon to the additive noise. The distortions result in an overlapping of neighbouring signals, permitting only a smaller noise level for error-free reception (this effect is called intersymbol interference). Finally, interchannel interference may affect transmission (as shown in Fig. 2.1) if more than one channel is transmitted over a single path.

2.1 Transmission of single signals

This transmission, illustrated in Fig. 2.2, results in problems which are most fundamentally treated in the literature of telecommunication (see Refs [1, 2, 3, 4, 5, 6] at the end of this chapter). One of a discrete set of M messages (M is a finite number), the ith message m_i, is generated by the source. For each message, a different signal waveform is transmitted by the transmitter, the signal $s_i(t)$

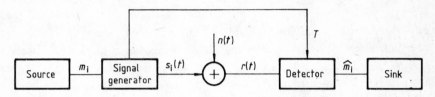

Fig. 2.2. Digital link corrupted by Gaussian noise.

corresponding to m_i. The transmission over the channel results in noise super-position, the sum of the signal and the noise being

$$r(t) = s_i(t) + n(t) \tag{2.1}$$

where $n(t)$ is a sample function of the Gaussian white noise process with a two-sided spectral density of $N_0/2$. From the received signal $r(t)$ the receiver has to produce an estimate of which signal $s_j(t)$ has been transmitted by the transmitter and as a result of this estimate, message $\hat{m}_i$ is transmitted to the sink.

In digital telecommunications where M is finite, there is a theoretical possibility of storing in the receiver all possible signal waveforms $s_j(t)$. The receiver, having to select one of the known waveforms, will evidently choose that waveform which best fits the waveform $r(t)$. In synchronous digital transmission which is exclusively dealt with in this book all signals have the same duration T. In the receiver, not only the received signal $r(t)$ but also the signal duration and the timing of the waveform $s_i(t)$ are available, the latter being transmitted to the receiver over a separate path (the practical realization of this transmission will be dealt with in Chapter 3). This separate path is denoted by T in Fig. 2.2. Note that the signals s_i may have parameters which are either unknown or not accurately known to the receiver such as the carrier frequency or phase. This question will also be dealt with in Chapter 3.

2.1.1 Vector representation of digital signals

In the following, the descriptive geometrical method introduced by Ref. [7] will be applied. Let the signal corresponding to message m_i be one of the set of elements $s_j(t)$ ($j \leq M$ and $0 \leq t \leq T$). Let the energy of the individual signals be finite, *i.e.*

$$\int_0^T [s_j(t)]^2 \, dt = E_j < \infty; \quad j = 1, 2, ..., M. \tag{2.2}$$

The energy of the individual signals is not necessarily identical.

For such a set of signals comprising M elements, an orthonormal base can always be found which has elements $\varphi_k(t)$; $k = 0, 1, ..., D$ where $D \leq M$ and which can be used to express the signals $s_j(t)$ as follows:

$$s_j(t) = \sum_{k=1}^{D} a_{jk} \varphi_k(t) \tag{2.3}$$

where

$$a_{jk} = \int_0^T s_j(t) \varphi_k(t) \, dt. \tag{2.3a}$$

For the functions φ_k we have

$$\int_0^T \varphi_k(t) \varphi_j(t) \, dt = \delta_{jk} \tag{2.4}$$

where

$$\delta_{jk} = \begin{cases} 1; & j = k \\ 0; & j \neq k. \end{cases} \tag{2.4a}$$

The most important part of the above statement is the fact that the series expressed by (2.3) has a finite number of terms. This seems to be plausible by common sense reasoning: M different signals can evidently be expressed as a sum of other signals, and the number of terms in this sum cannot be higher than M; a formal proof is based on the Gram–Schmidt orthogonalization procedure [1]. This procedure also yields the number D of base elements and the functions $\varphi_k(t)$.

The individual elements of the set are uniquely expressed by the D-tuple of coefficients if (2.3) is taken into account. These numbers can be regarded as the orthogonal components of a vector so the signals of the set are represented by individual points of a D-dimensional vector space, called signal space. The time function $s_j(t)$ is uniquely characterized by the signal vector s_j. From the knowledge of the time function and the base, the vector components can be determined from expression (2.3a), and from these vector components, the time function can be calculated from expression (2.3).

In the following, we shall need two vector relations. According to the first relation, the squared absolute value of the vectors equals the energy of the time function in question because

$$\int_0^T [s_j(t)]^2 \, dt = \int_0^T [\sum_k a_{jk}\varphi_k(t)]^2 \, dt = \int_0^T \sum_k a_{jk}^2 \varphi_k^2 \, dt = \sum_k a_{jk}^2 \qquad (2.5)$$

where the last equation follows from (2.4). According to the second relation, the scalar product of two vectors yields the integral of the product of the corresponding time functions because it again follows from (2.4) that

$$\int_0^T s_j(t)s_l(t) \, dt = \int_0^T \sum a_{jk}\varphi_k \sum a_{lk}\varphi_k \, dt = \sum a_{jk}a_{lk}. \qquad (2.6)$$

Our last task is to derive the vector representation of the noise voltage in (2.1). It will be assumed that the noise voltage has a stationary Gaussian distribution and represents a white noise, with a two-sided spectral density function $N_0/2$ and zero expected value. Similar to the signal vector components, noise vector components can also be expressed:

$$n_k = \int_0^T n(t)\varphi_k(t) \, dt \qquad (2.7)$$

where $n_1, n_2, ..., n_k, ..., n_D$ are components of a noise vector $\mathbf{n}$. The noise components n_k are random variables with Gaussian distribution and zero expected value which can be shown to have a variance $N_0/2$ and which are independent of each other.

The noise vector $\mathbf{n}$ does not comprise the complete noise process $n(t)$ since the set of functions $\varphi_k(t)$, comprising a finite number of functions, cannot be complete. One sample function of the noise process can be expressed in the form

$$n(t) = \sum n_k \varphi_k(t) + h(t) \qquad (2.7a)$$

where the process $h(t)$ is orthogonal to the signal space, *i.e.* to all base functions φ_k. It can be shown that in the case of a suitable receiver structure, $h(t)$ has no effect on the reception performance.

Summarizing, relation (2.1) can be expressed in vector form:

$$\mathbf{r} = \mathbf{s}_i + \mathbf{n} \qquad (2.8)$$

i.e., the process $r(t)$ has been expressed in the form of a D-dimensional random vector variable. Because of the above properties, the probability density function of the noise vector **n** is given by

$$p(\mathbf{n}) = \frac{1}{(\pi N_0)^{D/2}} \, e^{-|\mathbf{n}|^2/N_0}. \tag{2.9}$$

2.1.2 Optimum receiver structures

We have seen that in digital transmission, the error probability should be kept as low as possible by utilizing an optimum receiver structure. This structure will minimize the average of the incorrectly received signals; averaging should be performed for all vectors $\mathbf{s}_1, \mathbf{s}_2, ..., \mathbf{s}_M$ of the signal space. In this definition, it is implicitly stated that all signals are regarded to have the same weight so that the loss of any of the signals results in equal cost. In order to reach an optimum decision, the receiver has to know the individual vectors $\mathbf{s}_i$ (this statement will subsequently be extenuated); the probability of their transmission which shall be denoted by $P(\mathbf{s}_i)$ and termed *a priori* probability; and also the vector **r** of the noisy received signal.

The decision process is defined as follows. The signal space is divided into regions in which one and only one vector $\mathbf{s}_i$ is assigned to each region. Should the received vector **r** fall into the kth region of the signal space then the decision should be for a transmitted signal vector $\mathbf{s}_k$. It can be shown (see, for instance, Ref. [1]) that the lowest error probability is obtained if the regions are chosen so that in the kth region,

$$P[\mathbf{s}_k|\mathbf{r}] > P[\mathbf{s}_i|\mathbf{r}]; \quad i = 1, 2, ..., M \neq k \tag{2.10}$$

where P is the probability of the bracketed event.

The quantities in (2.10) are called *a posteriori* probabilities, and the receiver performing the operations (2.10) is designated as the maximum *a posteriori* probability receiver. In Ref. [1], among others it is shown that this kind of receiver has to maximize the following decision function for $i=k$ cases:

$$P(\mathbf{s}_i)p(\mathbf{r}|\mathbf{s} = \mathbf{s}_i) \tag{2.11}$$

where the second factor is the conditional probability density of the vector **r**.

Expressions (2.8) and (2.9) can be used to find an explicit form for (2.11): during the transmission interval of signal vector $\mathbf{s}_i$, **r** is a random vector variable with Gaussian distribution having an expected "value" $\mathbf{s}_i$, the decision function will be

$$P(\mathbf{s}_i) \frac{1}{(\pi N_0)^{D/2}} \, e^{-(|\mathbf{r}-\mathbf{s}_i|^2/N_0)}. \tag{2.11a}$$

The optimum receiver has thus to decide on the signal $i=k$ maximizing the above expression or, which is equivalent, to evaluate the following expression:

$$|\mathbf{r}-\mathbf{s}_i|^2-N_0 \ln P(\mathbf{s}_i) = \text{minimum.} \qquad (2.12)$$

We perform the squaring operation in (2.12) and noting that $\mathbf{r}^2$ does not depend on i, the function of the optimum receiver is to maximize the following quantity:

$$\mathbf{r}\mathbf{s}_i+\frac{1}{2}[N_0 \ln P(\mathbf{s}_i)-|\mathbf{s}_i|^2]. \qquad (2.13)$$

It is seen from (2.12) that if all signals $\mathbf{s}_i$ have identical *a priori* probabilities (*i.e.*, when $P(\mathbf{s}_i)=1/M$) then the decision should be for that signal which has the smallest distance from $\mathbf{r}$. In other cases, the boundaries of the decision regions will depend on the variance N_0.

Figure 2.3 shows the vector realization of the optimum receiver while the actual realization with suitable waveforms is given in Fig. 2.4; this latter is based on the previously explained interpretation of the scalar product.

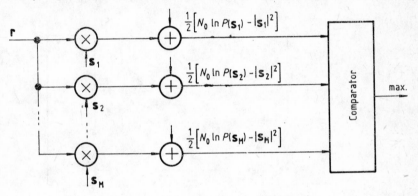

Fig. 2.3. Optimum receiver (vectorial form).

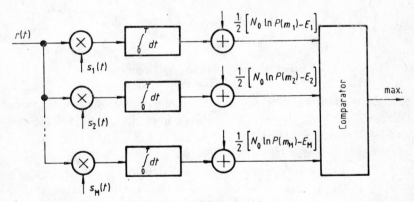

Fig. 2.4. Correlation receiver.

The circuit in Fig. 2.4 is called a correlation receiver. However, an interesting, somewhat different realization of this receiver can also be devised by taking into account the fact that the functions $s_i(t)$ differ from zero only during the time interval 0 to T. The following relation then holds:

$$\int_0^T r(t)s_i(t)\,dt = \int_{-\infty}^{\infty} r(\tau)s_i(\tau)\,d\tau = \int_{-\infty}^{\infty} r(\tau)s_i(T-t+\tau)\,d\tau\big|_{t=T}. \qquad (2.14)$$

In the last equation, we recognize that the response at the instant $t=T$ is expressed for a filter driven by $r(t)$ and having an impulse response function

$$h(t) = s_i(T-t). \qquad (2.15)$$

The latter filter is called a filter matched to $s_i(t)$, and (2.14) allows us to devise a matched filter receiver as shown in Fig. 2.5 which is a variant of the optimum receiver; $s_i(t)=0$ for $t>T$ and thus the matched filter can always be realized (as it has an impulse response which is identically zero for negative t values).

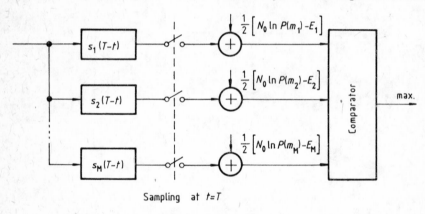

Sampling at $t=T$

Fig. 2.5. Matched filter receiver.

2.1.3 Error probability of the optimum receiver

The calculation of the error probability at the output of the optimum receiver is a simple task, at least theoretically, as shown in the following. Assume that the received signal vector **r** falls in the ith region of the signal space; the decision is then for signal s_i or message m_i, *i.e.* $\hat{m}=m_i$.

The decision is correct if the given signal was actually the ith signal or $m=m_i$. It thus follows that during transmission of the ith signal, the probability of an incorrect decision is equal to the probability of the **r**   vector falling into some region $j \neq i$. Taking into account the probability density of the noise vector given in (2.9), the probability of the correct decision when the ith signal is transmitted is given by

$$P_{Ci} = \int_{V_i} p(\mathbf{r}|\mathbf{s} = \mathbf{s}_i) \, dv \qquad (2.16)$$

where the integration should be performed for that region V_i of the D-dimensional space which pertains to the received message m_i. Thus the "average" probability of the correct decision is

$$P_C = \sum_{i=1}^{M} P(\mathbf{s}_i) \int_{V_i} p(\mathbf{r}|\mathbf{s} = \mathbf{s}_i) \, dv \qquad (2.16a)$$

and the average error probability is

$$P_E = 1 - \sum_{i=1}^{M} P(\mathbf{s}_i) \int_{V_i} p(\mathbf{r}|\mathbf{s} = \mathbf{s}_i) \, dv. \qquad (2.17)$$

As a reminder,

$$p(\mathbf{r}|\mathbf{s} = \mathbf{s}_i) = \frac{1}{(\pi N_0)^{D/2}} \exp\left[-|\mathbf{r} - \mathbf{s}_i|^2 / N_0\right]. \qquad (2.18)$$

It follows from the above results and particularly from expressions (2.13), (2.17) and (2.18) that the structure and error probability of the optimum receiver are independent of the actual waveform of the transmitted signals and depend exclusively on the vector configuration in signal space. This is an important property of digital signals transmitted with additive Gaussian noise, allowing the design of the transmission system (for lowest error probability) in the signal space, and the choice of the signal waveform to be based on additional requirements. We shall come back to this question in Section 2.2.1.

Let us now investigate a few specific cases in order to illustrate the previously obtained somewhat general results. For simplicity, only the case $P(\mathbf{s}_i) = 1/M$ will be investigated which has a great practical significance.

The simplest set of signals is the binary ($M = 2$) set of NRZ (Non-Return-to-Zero) signals. Figure 2.6 shows the waveform, the (single) base function and vector configuration for this set, together with the boundary regions resulting in an optimum partition of the signal space. From (2.17) and (2.18), the error probability is given by

$$P_E = \frac{1}{2} \operatorname{erfc}\left[\frac{1}{2}\sqrt{\frac{E}{N_0}}\right] = \frac{1}{2} \operatorname{erfc}\left[\frac{\sqrt{R_s}}{2}\right] \qquad (2.19)$$

where the notation $E/N_0 \triangleq R_s$ has been introduced. The quantity R_s will frequently be called signal-to-noise ratio.

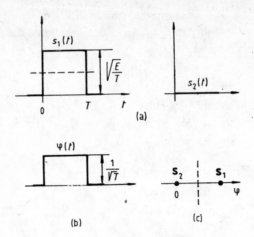

Fig. 2.6. (a) NRZ signal set; (b) base function of the set; (c) vector diagram pertaining to the set.

The next example will be the binary phase modulation (PSK: Phase Shift Keying):

$$s_1(t) = \sqrt{\frac{2E}{T}} \cos \omega_c t$$

$$s_2(t) = \sqrt{\frac{2E}{T}} \cos (\omega_c t + \Phi).$$ (2.20)

This set of signals can be represented in a two-dimensional space with the following base functions:

$$\varphi_1 = \sqrt{\frac{2}{T}} \cos \omega_c t$$

$$\varphi_2 = \sqrt{\frac{2}{T}} \sin \omega_c t.$$ (2.21)

It is seen from (2.21) that the orthogonality condition for functions φ_1 and φ_2 is met if $2\omega_c T$ is an integral multiple of 2π as

$$\int_0^T \varphi_1 \varphi_2 \, dt = \frac{1 - \cos 2\omega_c T}{2\omega_c T}.$$ (2.22)

This is not a serious constraint for microwave links as $2\omega_c T \gg 1$, so that the value of the integral in (2.22) is always negligible.

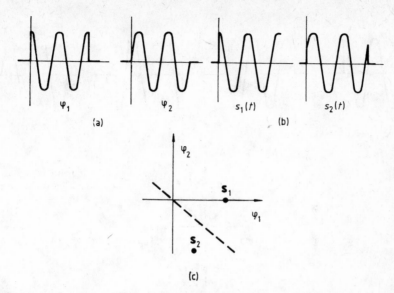

Fig. 2.7. (a) Base functions of the PSK signal set; (b) PSK waveforms; (c) vector diagram
pertaining to the set.

The signal waveforms, signal vectors and signal space boundaries corresponding
to (2.20) and (2.21) are shown in Fig. 2.7. For the vectors of Fig. 2.7c, the error
probability is given, with good approximation, by

$$P_{\mathrm{E}} = \frac{1}{2}\operatorname{erfc}\left[\sqrt{R_{\mathrm{s}}}\sin \Phi/2\right]. \qquad (2.23)$$

It can be seen from Fig. 2.7c and (2.23) that the condition for a highest distance
between signals and thus the smallest error probability is given by $\Phi = \pi$; in
this case, $\mathbf{s}_2 = -\mathbf{s}_1$. The general binary phase modulation is two-dimensional
but the antipodal signal set — similar to the NRZ signal set — is one-dimensional
(the expression (2.23) then yields the exact value of the error probability).

Let us now investigate the four-level phase modulation (QPSK: Quaternary
Phase Shift Keying) which is also characterized by a two-dimensional signal set.
The signal waveforms are the following:

$$s_1(t) = \sqrt{\frac{2E}{T}}\cos \omega_{\mathrm{c}}t; \qquad s_3(t) = \sqrt{\frac{2E}{T}}\cos(\omega_{\mathrm{c}}t + \Phi_2)$$

$$\qquad (2.23a)$$

$$s_2(t) = \sqrt{\frac{2E}{T}}\cos(\omega_{\mathrm{c}}t + \Phi_1); \quad s_4(t) = \sqrt{\frac{2E}{T}}\cos(\omega_{\mathrm{c}}t + \Phi_3)$$

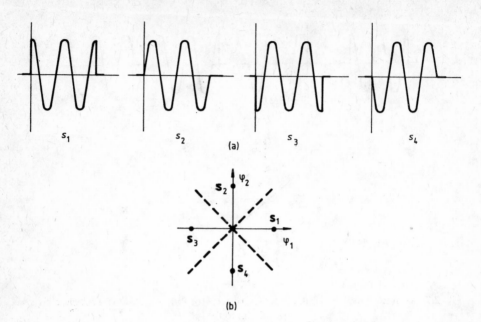

(a)

(b)

Fig. 2.8. (a) QPSK signal set; (b) vector diagram pertaining to the set.

and the base functions are given by

$$\varphi_1 = \sqrt{\frac{2}{T}} \cos \omega_c t; \quad \varphi_2 = \sqrt{\frac{2}{T}} \sin \omega_c t. \tag{2.23b}$$

The distance between the signals will be highest if $\varphi_k = k \dfrac{\pi}{2}$. Figure 2.8 shows this case but the base functions, being identical to those of Fig. 2.7b, are not shown. For the illustrated special case, the error probability is given by

$$P_E = \text{erfc}\left[\sqrt{\frac{R_s}{2}}\right] - \frac{1}{4}\left\{\text{erfc}\left[\sqrt{\frac{R_s}{2}}\right]\right\}^2. \tag{2.24}$$

For a relatively high signal-to-noise ratio R_s, the second term can be neglected resulting in a signal-to-noise ratio requirement for a given error probability P_E which is 3 dB higher than for 2PSK. *M*-ary *M*-PSK signals can also be defined, similarly to the QPSK signals.

Another kind of two-dimensional signal set is given by the QAM class (Quadrature Amplitude Modulation), involving the amplitude modulation of the sine and cosine components of a carrier frequency signal. The vector representation and the boundaries of the signal space are shown in Fig. 2.9 for the case when four

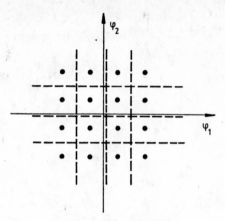

Fig. 2.9. 16QAM signal set.

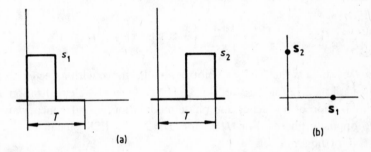

Fig. 2.10. (a) Signals of an orthogonal biphase signal set; (b) vector diagram pertaining to the set.

different amplitudes are allowed for both carrier components. As the next example, consider the so-called biphase signal set which has two states. Signal waveforms are shown in Fig. 2.10a. The signal set is orthogonal (as the product of the two signals is identically zero), with the vector diagram shown in Fig. 2.10b.

Let us now investigate the following four-state signal set:

$$s_1 = a \cos \omega_c t; \qquad\qquad s_3 = a \cos [(\omega_c + 2\delta\omega)t + \vartheta_2]$$
$$s_2 = a \cos [(\omega_c + \delta\omega)t + \vartheta_1]; \quad s_4 = a \cos [(\omega_c + 3\delta\omega)t + \vartheta_3] \tag{2.25}$$

where $\delta\omega = 2\pi/T$ and ϑ_i are arbitrary initial phases. These signals are mutually orthogonal to each other and thus the dimensionality of this set is given by $D = M = 4$. This signal set can be regarded as the digital version of frequency modulation, usually denoted by FSK (Frequency Shift Keying). Figure 2.11 shows the vector configuration with the aid of the symbolic illustration of the four-dimensional space. The boundary hyper-planes are not shown as they would

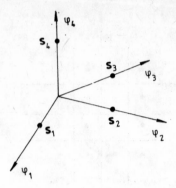

Fig. 2.11. Four-dimensional orthogonal signal set.

not be illustrative. The error probability for the orthogonal signal set is given by

$$P_E = 1 - \int_{-\infty}^{\infty} \frac{\exp(-x^2)}{\sqrt{\pi}} \left\{ \frac{1}{2} \operatorname{erfc}\left[-x - \sqrt{R_s}\right] \right\}^{M-1} dx. \qquad (2.26)$$

The last example is the family of biorthogonal signals, with a vector configuration shown in Fig. 2.12 for the case $D=3$, $M=6$. One possible realization method is the combination of frequency and phase modulation, each frequency appearing with two opposite phases. For $D=1$ or $D=2$, the PSK signal set is simultaneously a biorthogonal set.

The definition and asymptotic expression of the function erfc x appearing frequently in the error probability expressions is the following:

$$\operatorname{erfc}[x] \triangleq \frac{2}{\sqrt{\pi}} \int_{x}^{\infty} e^{-u^2} du \sim \frac{e^{-x^2}}{x\sqrt{\pi}}. \qquad (2.27)$$

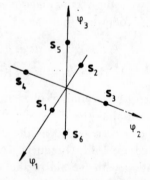

Fig. 2.12. Three-dimensional biorthogonal signal set.

2.1.4 Carrier frequency transmission

It has been assumed in the foregoing that the M possible signals of the set which may be transmitted are exactly known by the receiver, and the implications of this statement have not been detailed. From the examples presented, the NRZ and biphase signal sets are baseband ones — comprising baseband pulses — so the individual signal waveforms can be stored relatively easily in the receiver. However, in the rest of the examples, the situation is different, the individual signals being distinguished by different modulation states of the carrier. The unmodulated carrier frequency is given by

$$y(t) = a \cos(\omega_c t + \Phi) \qquad (2.28)$$

which comprises, in additon to its energy, two further parameters, the frequency and phase. In the correlation receiver shown in Fig. 2.4, the frequency ω_c and phase φ have to be known. Neglecting the frequency instability of the oscillators (assuming that this is much less than the bandwidth required) ω_c can be assumed to be known. However, the carrier phase is not known to the receiver, and this information has to be transmitted to the receiver by separate means. Transmission systems in which the transmitter phase is made known to the receiver are called coherent systems. A coherent transmission system, with indication of the separate phase transmission path, is shown in Fig. 2.13; it is seen that this system differs somewhat from the simple system shown in Fig. 2.2. Everything said in previous sections can be applied to such coherent systems. In the following, a few coherent transmission systems and their performance will be investigated.

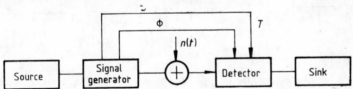

Fig. 2.13. Coherent link corrupted by Gaussian noise.

A binary coherent phase modulation receiver can be realized somewhat more easily than the optimum receiver shown in Fig. 2.4 as the two antipodal signals allow the application of a single multiplier as phase detector and a single zero-comparator. The modulator and demodulator are shown in Fig. 2.14a. The QPSK modulator and demodulator are shown in Fig. 2.14b; here two multipliers and two zero-comparators are sufficient instead of four. The figures refer to the (optimum) case in which the possible phase states are 0 and π (in Fig. 2.14a), and 0, $\pi/2$, π and $3\pi/2$ (in Fig. 2.14b). The figures also show the decisions based on the

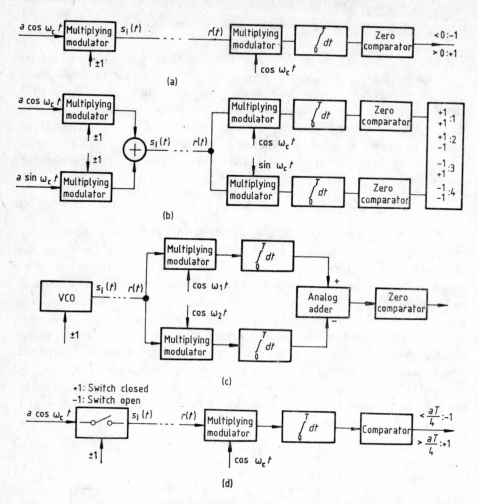

Fig. 2.14. Principle of a few modulators and coherent demodulators: (a) binary PSK; (b) QPSK; (c) binary FSK; (d) binary ASK.

comparator states. Figure 2.14c shows a binary frequency modulator and coherent demodulator while in Fig. 2.14d, a binary amplitude modulator and coherent demodulator are given. The error probability of the latter (so-called ASK) system is given by

$$P_E = \frac{1}{2} \operatorname{erfc} \left[\frac{\sqrt{R_s}}{2} \right]. \tag{2.29}$$

In Figure 2.15, the error probability is plotted as a function of the R_s signal energy/noise spectral density in the case of coherent demodulation, for PSK,

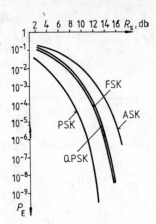

Fig. 2.15. Error probability of a few modulation systems as a function of $R_s = E/N_0$.

FSK and ASK systems. It is seen from the comparison of the two-state systems that best results are achieved by the PSK system, the (orthogonal) FSK system being 3 dB worse, and the ASK system being additionally 3 dB worse. This means that this is the additional energy requirement for achieving the given error. The optimum nature of PSK systems is explained by the fact that they allow the realization of an antipodal signal set, *i.e.* $s_1(t) = -s_2(t)$. In the case of a single transmitted signal, this is only possible with phase modulation as the frequencies of antipodal signals are necessarily equal. (Note that in the FSK system, the investigated orthogonal signal set is not the best, and can be somewhat improved by increasing the frequency deviation denoted by $\delta\omega$ in (2.25). However, the improvement obtainable is not significant. On the other hand, a substantial increase in the frequency deviation will again result in orthogonal signals; this means that with a large frequency deviation, the performance will theoretically be equal to that pertaining to the deviation $\delta\omega = 2\pi/T$.)

It is worth mentioning here a special case of FSK, called MSK (Minimum Shift Keying); this has advantages which will be treated later together with multichannel transmission. In an MSK system, the phase varies continuously (without phase jumps). If this can be realized then half of the frequency deviation given in (2.25), $\delta\omega = \pi/T$ is sufficient to result in orthogonal signals. In the MSK system, this (minimum) frequency deviation is applied so that during time interval T, the signal phase changes by $+\pi/2$ or $-\pi/2$ with respect to the initial phase. The MSK signal set is thus made up of the two following orthogonal signals:

$$s_1 = a\cos\left[\left(\omega_c + \frac{\pi}{2T}\right)t\right]; \quad s_2 = a\cos\left[\left(\omega_c - \frac{\pi}{2T}\right)t\right]. \tag{2.30}$$

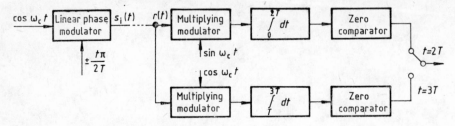

Fig. 2.16. Principle of an MSK modulator and demodulator.

The block diagram of the MSK modulator and demodulator is shown in Fig. 2.16.

Let us now mention in advance a property of actual digital transmission systems according to which a continuous sequence of binary signals is transmitted. The second bit interval will again be one of the signals given in (2.30), substituting t by $t-T$. This means that the phase of $s(t)$ *versus* time will change as shown by one of the straight lines in Fig. 2.17, assuming zero phase at $t=0$. Figure 2.17 shows that the input signal to the integrator circuit in the demodulator will be a positive half sine wave if s_2 is transmitted in the first bit interval, and a negative half sine wave if s_1 is transmitted, independently of the second bit information (this applies to the 0 to $2T$ time interval). The energy of the second bit can thus be utilized for the decision which means a 3 dB increase in R_s. Consequently, a certain increase in circuit complexity will have the effect of rendering the MSK signal performance equal to the antipodal PSK signal performance. The detection of the second bit will of course require another detector operating in the T to $3T$ time interval, as shown in the lower branch of Fig. 2.16 [9, 10].

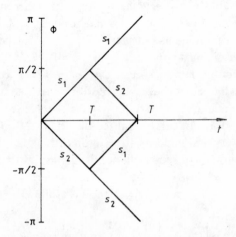

Fig. 2.17. Time-phase tree of an MSK system.

Let us now investigate the case in which there are no means of realizing the extra link denoted by Φ in Fig. 2.13, *i.e.* the transmitter phase is not known to the receiver. In the following, the optimum non-coherent receiver structure will be determined for a transmission medium with additive Gaussian noise. Again the $P(m_i) = 1/M$ case will be considered (the M *a priori* probabilities are identical).

It can first be observed that if the ith signal of the given signal set is expressed by

$$s_{it} = a(t) \cos \left[\omega_c t + \vartheta(t) \right] \tag{2.30a}$$

then the received signal will be given by the following expression:

$$s_{ir} = a(t) \cos \left[\omega_c t + \vartheta(t) + \Phi \right] =$$
$$= a(t) \cos \Phi \, \cos \left[\omega_c t + \vartheta(t) \right] - a(t) \sin \Phi \, \sin \left[\omega_c t + \vartheta(t) \right]. \tag{2.31}$$

In other words, each signal vector corresponds to a point in a two-dimensional space. The two components of vector $\mathbf{s}_{ir}$ can be denoted by s_{ci} and s_{si}, and the dimensionality of the originally D-dimensional signal space has become $2D$. As the noise process $n(t)$ also has sine and cosine components, noise will be added to both terms of (2.31) while traversing the channel. The two components n_c and n_s of the noise vector are nearly statistically independent of each other.

The optimum non-coherent receiver should optimize the probability density

$$p(\mathbf{r}_c, \mathbf{r}_s | \mathbf{s}_i) \tag{2.31a}$$

where p now denotes the probability density

$$\int_0^{2\pi} p(\mathbf{r}_c, \mathbf{r}_s | \mathbf{s}_i, \Phi) p(\Phi) \, d\Phi \tag{2.32}$$

and

$$\mathbf{r}_c = \mathbf{s}_c + \mathbf{n}_c; \quad \mathbf{r}_s = \mathbf{s}_s + \mathbf{n}_s. \tag{2.32a}$$

Φ is assumed to have a uniform distribution between 0 and 2π. Utilizing (2.31) and (2.9) we have

$$p(\mathbf{r}_c, \mathbf{r}_s | \mathbf{s}_i, \Phi) = \frac{1}{(\pi N_0)^D} \exp \left[\frac{-1}{N_0} (|\mathbf{r}_c - \mathbf{s}_i \cos \Phi|^2 + |\mathbf{r}_s - \mathbf{s}_i \sin \Phi|^2) \right]. \tag{2.33}$$

Calculating the expected value denoted in (2.32) and rearranging, we note that the optimum receiver will decide on m_i where

$$I_0 \left(\frac{2X_i}{N_0} \right) e^{-N_i/E_0} = \text{maximum} \tag{2.34}$$

and

$$X_i \overset{\triangle}{=} \sqrt{|\mathbf{r}_c \mathbf{s}_i|^2 + |\mathbf{r}_s \mathbf{s}_i|^2} \tag{2.35}$$

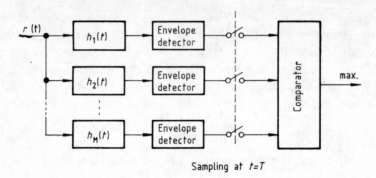

Fig. 2.18. Optimal non-coherent receiver realized by envelope detectors.

where $I_0(.)$ is the zero-order modified Bessel function of the first kind. If all s_i signals have the same energy then the extreme value will not be influenced by the second factor in (2.34). Both I_0 and the root operation being monotonous functions the optimum receiver will maximize X_i^2 (when the transmitted signals have equal probabilities and energies). The general structure of a non-coherent receiver will not be presented but a specific variant, the envelope detector receiver, is given in Fig. 2.18. The denoted filters have the impulse response of

$$h_i(t) = s_i(T-t) \cos \omega_c t, \qquad (2.35a)$$

while the envelope detectors have the function of producing the envelope of the incoming signal.

We now come to the investigation of specific non-coherent structures. The receiver of a binary orthogonal FSK system is shown in Fig. 2.19a, and has an error probability

$$P_E = \frac{1}{2} e^{-R_s/2}. \qquad (2.36)$$

The error probability of an M-ary non-coherent FSK system is given by

$$P_E = 1 - \int_0^\infty \exp\left[-(x+R_s)\right] I_0\left[\sqrt{4R_s x}\right] G(x) \, dx;$$

$$G(x) = (1-e^x)^{M-1}. \qquad (2.37)$$

A special kind of non-coherent demodulator for FSK signals is the limiter–discriminator, well known from analog FM systems, measuring the instantaneous frequency of the received signal. This method can be regarded as special because its operating principle differs substantially from those of the optimum receivers treated earlier: the concept of "instantaneous frequency" does not appear when looking for the *a posteriori* probability, so this kind of demodulator cannot be

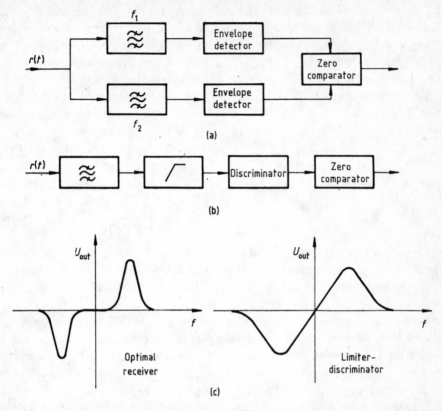

Fig. 2.19. Non-coherent FSK demodulators: (a) utilizing envelope detectors; (b) utilizing a limiter–discriminator; (c) detector characteristics.

deduced from the theory. In spite of this, it is surprisingly shown, for example by Ref. [4], that the error probability obtainable with a limiter–discriminator is exactly equal to the error probability of the optimum receiver as given in (2.36).

The structure of the limiter–discriminator, shown in Fig. 2.19b, is fairly similar to the binary non-coherent demodulator shown in Fig. 2.19a. An essential difference, however, is the presence of the limiter and primarily, that the bandpass filter preceding the limiter passes the M different frequencies, and furthermore the output voltage of the discriminator is proportional to the input frequency. On the other hand, the bandwidth of the matched filters covers only the vicinity of the given frequencies and not even for binary FSK is the output voltage proportional to the input frequency.

The two-state non-coherent ASK system is theoretically somewhat more complicated than the FSK system. According to (2.34), we have $E_1 \neq E_2$, *i.e.*, the energies of the two signals are not equal, resulting in a more complex optimum receiver than shown in Fig. 2.18 for the two-state variant. It can be seen from (2.34) that the optimum threshold position (*i.e.*, the optimum boundary of the signal space between the two signals) depends on the noise magnitude or the signal-to-noise ratio R_s. The optimum placement of the threshold can be determined from the probability densities $p(r)$ during the transmission of the two signals. For a signal $s_1 = 0$, $r(t)$ is Gaussian noise which has an envelope of Rayleigh distribution:

$$p(r|s_1) = \frac{r}{\sigma^2} e^{-r^2/2\sigma^2}; \quad 0 \leqq r < \infty \tag{2.37a}$$

while for a signal $s_2 = a \cos \omega_c t$, $r(t)$ has an envelope with Rice distribution:

$$p(r|s_2) = \frac{r}{\sigma^2} \exp\left[-\frac{r^2 + a^2/2}{2\sigma^2}\right] I_0\left(\frac{ra}{\sqrt{2}\sigma}\right). \tag{2.37b}$$

These two functions are plotted in Fig. 2.20, their intersection point yielding the optimum threshold value. For high signal-to-noise ratios, this optimum is at half amplitude, but the optimal threshold is higher for lower values of R_s; in the above expression, $R_s = a^2/2\sigma^2$.

Let us now investigate the possibility of non-coherent PSK transmission. There is evidently no such possibility when applying the preceding concept according to which the transmission of a single signal is evaluated leading to the optimum receiver structure shown in Fig. 2.18. This is explained by the fact that the phase of the carrier carrying the information to be transmitted would be lost by the non-coherent transmission. On the other hand, if two consecutive signals are investigated in the decision process their phase difference can be determined (similarly to the MSK transmission), and utilized for the decision. This transmission (or more precisely demodulation) method based on the phase coherence of consecutive signals is called differentially coherent transmission.

In 2PSK transmission, the transmitter generates antipodal PSK signals, and the receiver has to decide whether two consecutive bits have been equal or different. We thus have an orthogonal signal set lasting for an interval $2T$ because the signals $[s(t), -s(t)]$ and $[s(t), s(t)]$ are mutually orthogonal in the $(0, 2T)$

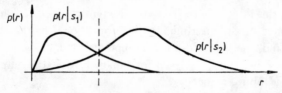

Fig. 2.20. Probability density functions of non-coherent ASK transmission, and the optimal comparator level.

time interval. This means that the bit error probability of the binary, differentially coherent system is obtained by taking into account the fact that the resultant energy of two consecutive signals is $2E$. By applying expression (2.36), the error probability *versus* signal-to-noise ratio will be given by

$$P_E = \frac{1}{2} e^{-R_s} \tag{2.38}$$

where R_s is now the ratio of the energy of one bit and the noise spectral density.

For PSK transmission with more than two states, the calculation of the error probability is somewhat more difficult, and the structure of the optimum receiver has also to be correspondingly modified. The receiver structure, based on Ref. [11], is shown in Fig. 2.21. This is essentially the structure of Fig. 2.19a, amended by circuits giving phase difference. The error probability can be calculated from a knowledge of the probability density of the difference $y_i - y_{i-1}$, taking into account the fact that an error is generated when the received $y_i - y_{i-1}$ differs from the actual phase difference by more than π/M. The probability density has been determined for example in Refs [12] or [13]. Approximate expressions for the bit error ratio are given in Refs [7] and [14]. The latter is extremely accurate for $R_s > 5$ dB:

$$P_E = \text{erfc} \left[\frac{R_s}{\sqrt{1+2R_s}} \sin \frac{\pi}{M} \right]. \tag{2.39}$$

Another method for the demodulation of 2PSK signals is the use of a discriminator [15] as for FSK signals, but this method does not yield the lowest error probability. This method is based on the fact that the instantaneous frequency is the derivative of the instantaneous phase, resulting in a zero or positive pulse for a transmitted s_1 vector, and a zero or negative pulse for a transmitted s_2 vector at the discriminator output. The regeneration of the transmitted signal could theoretically be handled by an integrating circuit but we see from [15] that this would increase the error probability to an unacceptable value. This is why a relaxation

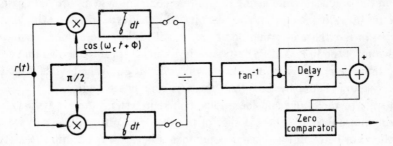

Fig. 2.21. Principle of a differentially coherent demodulator.

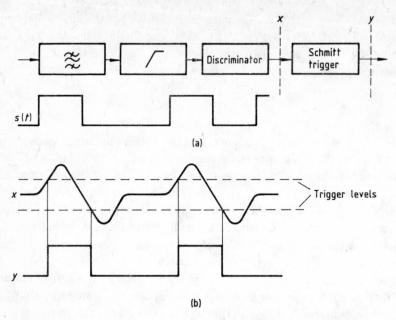

Fig. 2.22. (a) Principle and (b) typical waveforms of a relaxation demodulator.

circuit (*e.g.*, a Schmitt-trigger) having hysteresis is normally used to transfer the three-state signal of the discriminator back to a two-state signal. This kind of PSK demodulator is known as a relaxation demodulator, having structure and waveforms as shown in Fig. 2.22; in this figure, a bit stream is illustrated instead of a single bit.

To conclude this section, the advantages, if there are any, of the coherent systems as compared with the non-coherent systems will be investigated. Let us first look at the two-state case by plotting bit error ratios as a function of the signal-to-noise ratio for the coherent and non-coherent FSK system, and further for the coherent and differentially coherent PSK system (see Fig. 2.23). It is seen that in the practically interesting range, the coherent systems are only marginally better, and so it seems that the technical difficulties associated with the transmission of the phase information are not justified. For more than two states, the behaviour of the FSK and PSK systems is different. It can be shown that coherent and non-coherent FSK systems are asymptotically equivalent while the coherent PSK system is better by 3 dB than the non-coherent system.

In spite of the above considerations, coherent systems are always better when offering more freedom in the choice of the vectors. Coherent transmission allows a distinction between two carrier components in phase quadrature so signal sets comprising quadrature components can be applied, offering transmission with

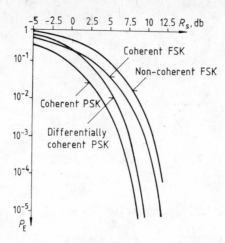

Fig. 2.23. Error probability of coherent and non-coherent FSK, also coherent and differentially coherent PSK.

good bandwidth efficiency (an example of this is given in Fig. 2.9). Essentially only FSK can be applied for non-coherent transmission, but this has poor bandwidth efficiency.

The differentially coherent system is a transition between coherent and non-coherent systems. Though offering distinction between quadrature carrier components by non-coherent methods, it is nearly as complex as the coherent system, and requires a 3 dB higher signal-to-noise ratio.

Summarizing these considerations, it can be said that in most cases, practical requirements are more easily met by coherent or differentially coherent systems, especially when the available transmission bandwidth is limited. Consequently, coherent systems are utilized in the vast majority of cases because they require a smaller signal-to-noise ratio. Non-coherent systems are justified in only two cases: *1* — if low equipment cost is a primary requirement; ASK, FSK or relaxation PSK may then be applied, being realizable by extremely simple circuits; *2* — if the coherence cannot be implemented even within a bit interval T because of the instability of oscillators. This may be the case primarily in low bit-rate transmission systems [11].

2.1.5 Optimum choice of signal set

In Section 2.1.1 to 2.1.4, the optimum receiver structure for a given signal set and the performance of the receiver have been investigated. This section deals with the optimization of the signal set, primarily for coherent transmission. It will be assumed that all messages have equal $P(m_i)$ probabilities.

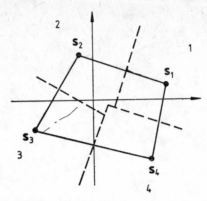

Fig. 2.24. An arbitrary two-dimensional signal set.

It has been shown in Sections 2.1.1 and 2.1.2 that the probability of the correct reception of the ith signal is equal to the integral of the $p(\mathbf{r}|\mathbf{s}_i)$ conditional probability, taken to the ith decision region of the signal space, or, in other words, to the probability of the vector $\mathbf{r}$ falling into this region. It thus follows that the error probability will be smaller by applying higher signal energy. As an example, Fig. 2.24 shows an arbitrary four-level, two-dimensional signal set. It is seen that the distance between any signal and the region boundary can be made as high as required by multiplying all vectors by sufficiently high numbers. The only condition to be fulfilled is that none of the signal vectors should be parallel to any of its region boundaries.

Evidently, such a trivial optimization of the signal set has no practical significance because of the infinite energy required. In order to find a meaningful optimum, the signal energies should be restricted, and the task should be formulated in the following way: we look for a signal vector set having M elements and minimizing the average error probability while the signal energies should not be higher than E_{max}. This means that the vectors should be inside or on the surface of an M-dimensional sphere, but the actual dimensionality can be smaller than M.

This problem can be treated by variational calculus and is solved in Ref. [16] as follows. A global optimum for the transmission of M-ary digital signals is the so-called "regular simplex" $D=M-1$ dimensional signal set which can be defined by products

$$|\mathbf{s}_i|^2 = E_{max}$$

$$\mathbf{s}_i \cdot \mathbf{s}_k = \frac{E_{max}}{M-1}.$$

(2.40)

It is seen that for large M values, the simplex signals are approximately equal to the orthogonal signals for which $\mathbf{s}_i \cdot \mathbf{s}_{k \neq i} = 0$. It is also seen that in the binary case when $M=2$, (2.40) defines the antipodal signal set while for $M=3$, the vector end-points will coincide with the vertices of an equilateral triangle.

It will be shown in the next section that the occupied bandwidth is proportional to the dimensionality D. This is why the application of $M-1$ dimensional signals is frequently forbidden, and the transmission can only be realized by a sub-optimum signal set. A procedure to optimize the signal set with various restrictions for D, e.g. $D < M-2$, is detailed in the literature (e.g., Ref. [5] and the references given therein). These problems have not much practical significance and will thus not be dealt with in more detail. Normally, several regular simplex signal sets of smaller dimension are applied, with suitable relative positions.

Most of the digital transmission systems, both for radio and cable transmission, utilize two-dimensional signal sets for efficient spectrum utilization which thus have great practical significance. For $D=2$ and equal signal energies, the best results are obtained by an M-ary phase modulation system, with the following signal vector polar coordinates:

$$\varrho_{si} = \sqrt{E}; \quad \Phi_{si} = \frac{2\pi}{M} i, \quad i = 1, 2, ..., M. \tag{2.41}$$

Figure 2.25 presents an example for $M=6$, showing the signal vectors and the signal space. Note that all signals have two neighbouring signals much closer than the rest. This means that for large signal-to-noise ratios, there is a negligible probability to decide for a distant message instead of a neighbouring one. This results in a tight upper bound for the error probability of a two-dimensional signal set which can easily be calculated, as shown in the following. Let us note that passing the nearest boundary has the greatest probability. During the transmission of signal s_1; for example, the probability of a correct decision can thus be approximated by the probability of the received vector r falling within the circle

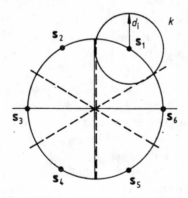

Fig. 2.25. 6PSK signal set for explaining expression (2.43).

shown in the figure. (The radius d_i of the circle is the distance of the signal vector from the nearest boundary line.) So for a coherent detection of a two-dimensional signal set we have

$$P_E < 1 - \frac{1}{M} \sum_{i=1}^{M} \int_0^d \int_0^{2\pi} p(\mathbf{r}|\mathbf{s}_i) \varrho \, d\varrho \, d\varphi. \tag{2.42}$$

Denoting the coordinates of vector $\mathbf{s}_i$ by (x_i, y_i) and those of vector $\mathbf{r}$ by (x, y) yields

$$p(\mathbf{r}|\mathbf{s}_i) = \frac{1}{\pi N_0} \exp\left[-\frac{(x-x_i)^2 + (y-y_i)^2}{N_0}\right] = \frac{1}{\pi N_0} e^{-\varrho^2/N_0}. \tag{2.42a}$$

Substituting this into (2.42) we have

$$P_E < e^{-d^2/N_0} \approx P_E. \tag{2.43}$$

Let us now return to the investigation of the two-dimensional signal sets. We have seen that assuming equal energy signals, the M-ary phase modulation will give the best results. However, better configurations can be obtained by discarding the equal energy concept. For instance, let us amend the six-state phase modulation vectors shown in Fig. 2.25 by a seventh vector $\mathbf{s}=0$. The minimum d_i has thus not been decreased compared to the 6PSK, and thus the error probability has not been changed practically either. It may be concluded generally that the lowest error probability is obtained by using PSK for $M \leq 6$ but this is no longer true for $M > 6$.

For a high number of states and $D=2$, the signal set will be optimum if the vector end-points coincide with the points of an equilateral triangular grid. An example for such an arrangement is shown in Fig. 2.26. Assuming the sine and cosine components of the carrier as base functions as before, it is seen that neither the amplitude nor the phase of the signal vectors is constant, *i.e.* the optimum two-dimensional signal set is obtained by a suitable combination of the amplitude modulation and phase modulation. In this case, the signal energy is far from constant.

For various reasons, the triangular grid shown in Fig. 2.26 is not realized in practice. The greatest shortcoming of this set is the extreme difficulty of implementing a corresponding modulator and especially a coherent demodulator. However, 8- and 16-state sets approximating the optimum set are extensively used both for cable and microwave transmission. The quadrature amplitude modulated system for $M=16$ states as shown in Fig. 2.9 is widely applied and for high bit rates, even 64 state QAM systems are used. For $M=8$, another combination of amplitude and phase modulation can be configured (see Ref. [17]).

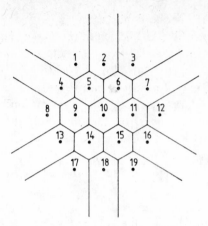

Fig. 2.26. PSK signal set forming a triangular grid, and domain limits.

PSK and QAM systems can be compared relatively easily. The distance between neighbouring signal vectors is given for MPSK systems by

$$d_p = 2\sqrt{E}\sin \pi/M \tag{2.44a}$$

and for QAM systems by

$$d_{QA} = \sqrt{2E}\,\frac{1}{\sqrt{M}-1}. \tag{2.44b}$$

Taking into account (2.43), the energy required to obtain a given error probability is less for a QAM system than for a PSK system, the difference being 1.64 dB for 16QAM (not a significant improvement) and 6.27 dB for 64QAM. It should be noted that in (2.44b), E denotes the energy of the largest signal. For a comparison based on average energies rather than peak energies, the difference between 16QAM and 16PSK systems is 4.19 dB.

2.1.6 Multilevel encoding of binary signals

In practice, most digital sources generate binary signals. However, it is frequently rewarding to apply code converters and thus obtain digital signals having more than two levels. It will be shown that multilevel modulation systems allow a (theoretically) arbitrary reduction of the occupied bandwidth by increasing the power or *vice versa,* the power can be reduced (down to a certain limit) by occupying more bandwidth. Before going into more detail about these procedures, the methods of obtaining more than two-level signals from binary sequences will be explained.

The principle of code conversion is rather self-explanatory: a single symbol known as dibit, tribit, tetrabit, *etc.* is formed from $n=2, 3, 4,$ *etc.* bits. A symbol made up from n bits has $M=2^n$ possible states, thus obtaining 4, 8, 16, ..., state systems.

When transmitting, over a channel loaded by Gaussian noise, M-ary signals derived by encoding binary signals and the transmission properties will depend basically on the dimensionality D of the applied signal set. Let us first of all determine the frequency range occupied by a single element of a digital signal set. This depends on the definition of the occupied band. Theoretically, this is infinite because the interval is finite. There is thus no frequency range in which the Fourier transform of the signal waveform is identically zero (see Ref. [1]). The practically occupied band is taken as the band not available for the transmission of other signals; outside this band, channels transmitting other information can be allocated.

The practically occupied band will be investigated in Section 2.3. As preliminary information we note that the noise bandwidth of the matched filter is $1/T_s$ where T_s is the symbol time (assuming double sideband modulation). This means that the band occupied by the given signal is certainly not less than $1/T_s$; it is practically 10 to 20 per cent higher which will not be taken into account at the moment. If the properties of the signals allow the transmission over a single filter (which is the practical case when all signals have the same frequency) then the complete signal set will occupy a frequency band of $1/T_s$. If, taking the other extreme, the frequencies of the individual signals are all different and the signals are even orthogonal (e.g., M-ary FSK signals are transmitted) then the frequency difference needed is $1/T_s$ and the occupied bandwidth is M/T_s. We have thus shown for this special case that the occupied bandwidth is proportional to the dimensionality. This statement can be expressed by the generally valid so-called dimensionality theorem (see Ref. [1, 18]).

Let W be the available band covering at least 92 per cent of the signal energies; then, for large TW products, the highest number of dimensions falling into the band W is $1.2 TW$. We shall not prove this theorem but instead apply it to investigate the numerous possibilities of multilevel encoding. It will be assumed that the "cost" of a digital transmission system has two components, the necessary transmitter power and the occupied bandwidth. (A third component is the complexity of the system which will not be treated here.) For a fixed transmitter power denoted by P, the energy of a symbol comprising n bits is given by

$$E = PnT. \tag{2.45}$$

Let our first task be to reduce the occupied band to as low a value as possible. According to our preceding considerations, this may be achieved by choosing a relatively high value of n and applying a two-dimensional signal set. For a phase modulation with 2^n states the occupied band will be

$$W = \frac{1}{T_s} = \frac{1}{nT} \tag{2.46}$$

i.e., nth of the value for binary PSK. On the other hand, the distance between neighbouring signals is much less than for the binary case: to retain a given error probability the power has to be increased by the factor

$$P/P_1 = \frac{1}{n \sin^2 \pi/2^n} \approx \frac{2^{2n}}{n\pi^2} \tag{2.47}$$

when taking into account expressions (2.44a) and (2.45). Here P_1 denotes the power for the binary case; the approximation holds for high values of n.

Applying QAM instead of PSK, expression (2.44b) yields the following power factor:

$$P/P_1 = \frac{2(2^{n/2}-1)^2}{n} \approx \frac{2^{n+1}}{n}. \tag{2.48}$$

Table 2.1. Power and occupied band of PSK and QAM signal sets,
for different values of n

n	$M = 2^n$	TW_n	$P_n/P_1^{PSK, dB}$	$P_n/P_1^{QAM, dB}$
1	2	1	0	0
2	4	0.5	0	0
3	8	0.33	3.57	—
4	16	0.25	8.17	6.53
5	32	0.2	13.2	—
6	64	0.17	18.4	12.10
7	128	0.14	23.7	—
8	256	0.12	29.2	17.5
9	512	0.11	34.7	—
10	1024	0.10	40.3	22.8

In Table 2.1, the relative bandwidth and relative power for different values of n are summarized, $n=1$ denoting the two-state values as reference. It is seen from this table and Eqs (2.46) to (2.48) that the required bandwidth can be reduced arbitrarily at the cost of transmitter power increase by using a multilevel signal set. However, in practice, the bandwidth cannot be reduced by an order of magnitude as a linear bandwidth reduction would require a nearly exponential increase in power. The table also shows that the choice of $n=2$, i.e. four-level modulation, is an extremely good compromise: the occupied band is halved with unchanged transmitter power. This is why the 4PSK or the equivalent 4QAM modulation is used extensively for digital signal transmission. Other methods of modulation are probably justified only by important requirements such as bandwidth efficiency; this is why 8, 16 or 64 level modulations are applied in microwave systems used for high bit rate transmission. In conclusion, note that expressions (2.47) and (2.48), giving the necessary power, are valid for low error probabilities only as long as expression (2.43) is valid. For small signal-to-noise ratios, the higher error probability will depend not only on the smallest signal vector distance. In such cases, the probability of deciding on a non-neighbouring signal vector cannot be neglected. Therefore, the applied approximation is not very good.

Let us now investigate the opposite task of power reduction by bandwidth increase. It should be noted that the distance between signal vectors is not changed by extending the signal set by an additional vector (*i.e.*, increasing M to $M+1$) which is perpendicular to the already existing vector space. In other words, this means that the dimension D has been increased to $D+1$ by the introduction of the $(M+1)$th vector.

As a first approximation, it may thus be stated generally that the signal energy needed for a given error probability is independent of the number of states if orthogonal signals are applied, *i.e.* the signal power needed according to expression (2.45) is inversely proportional to the number of bits n making up the symbols. On the other hand, the occupied bandwidth increase is nearly proportional to the number of states:

$$W = M/T_s = 2^n/nT. \tag{2.48a}$$

The preceding statement on the power requirement needs some correction as it would contradict the known fact that bit rates higher than given by the channel capacity cannot be transmitted over the channel ([1]). Without going into detail, the following statement seems to be reasonable. For a high distance d between signal vectors, *i.e.* for high signal-to-noise ratios, this distance will determine the error probability. However, with reduced signal distance, the probability of an incorrect decision is higher if several signals are at a distance d from the given signal. In other words, the error probability is primarily determined by the distance d at high signal-to-noise ratios, and by the number M of signals for low signal-to-noise ratios. So for an orthogonal modulation system, the necessary power is given by

$$P/P_1 = 1/n, \quad \text{if} \quad PT/N_0 \gg -1.6 \, \text{dB}. \tag{2.49}$$

The "much higher" notation in the above expression means only a few dB (according to detailed studies).

Table 2.2 presents the relative bandwidth and the relative power for different values of n in the case of orthogonal signal sets, assuming that the condition given in (2.49) is met. The data of the bi-orthogonal system are also given; for higher values of n, the error probability of the bi-orthogonal system is well approximated by that of the orthogonal system. In the binary case of $n=1$, the power required by orthogonal and bi-orthogonal signals is not the same; in the table, the former is taken as reference and denoted by P_0.

It is seen from Table 2.2 that the application of an orthogonal or similar signal

Table 2.2. Power and occupied band of orthogonal and biorthogonal signal sets, for different values of n

n	$M = 2^n$	$P_n/P_0^{\text{Ort, dB}}$	$W_n T^{\text{Ort}}$	$P_n/P_0^{\text{Biort, dB}}$	$W_n T^{\text{Biort}}$
1	2	0	2	-3	1
2	4	-3	2	-3	0.5
3	8	-4.8	2.7	-4.8	1.3
4	16	-6	4	-6	2
5	32	-7	6.4	-7	3.2
6	64	-7.8	10.7	-7.8	5.3
7	128	-8.5	18.3	-8.5	9.1
8	256	-9	32	-9	16
9	512	-9.5	56.9	-9.5	28.4
10	1024	-10	102.4	-10	51.2

set allows an almost arbitrary reduction in the power by increasing the occupied frequency range. Comparison of Tables 2.1 and 2.2 shows an interesting duality between bandwidth and power: within the limitation given by expression (2.49), power may be traded for bandwidth and *vice versa*. In practice, however, this trade-off is limited as the linear reduction of one of the quantities involved requires a nearly exponential increase of the other quantity. It is interesting to note that the time needed for the transmission is also increased together with the increase in bandwidth or power (because the values of the first, second, *etc.* bits in the receiver are only available after a time interval nT).

It should be noted in conclusion that in digital transmission, there are much more effective methods of reducing the error probability by increasing the bandwidth, these being the methods of error correcting coding. These methods are treated, among others, in Refs [6, 9, 11, 19].

2.1.7 Bit error probability — symbol error probability

In Sections 2.1.3 to 2.1.5 error probability signified the probability of detecting a signal element $j\neq i$ when the element i of the signal set is transmitted. This quantity can be designated as symbol error probability. Assuming that the M-ary signal set has been generated by coding a binary bit sequence, we shall now investigate the procedure by which the bit error probability of the original binary bit sequence can be derived from the symbol error probability. It is evident that the knowledge of the bit error probability is more important from the practical point of view.

The following estimate can be made without giving a general statement. Assume that a symbol comprising n bits is in error. In this case, the highest number of bits in error is n — there are no more bits in the symbol — and the lowest number of bits in error is 1, otherwise the symbol would not be in error. Assuming again that the number of bits is n times the number of symbols, we thus have

$$P_E/n \leqq P_{Eb} \leqq P_E \tag{2.50}$$

where P_E denotes the symbol error probability and P_{Eb} denotes the bit error probability. In forthcoming sections of this book, this distinction of notation will only be applied if necessary. If the case is unambiguous, the bit error probability will also be denoted by P_E.

A more definite relation can be derived for specific modulation systems. For instance, in the case of an orthogonal signal set, there is an identical probability of receiving any vector instead of the transmitted one. It thus follows from elementary considerations of probability calculus and from Bayes theorem that

$$P_{Eb} = \frac{2^{n-1}}{2^n - 1} P_E. \tag{2.51}$$

For a phase modulation having more than two states it has to be taken into account that in the case of an error, there is an overwhelming probability of receiving one of the neighbouring vectors of the transmitted one (see Fig. 2.26). Assuming again symbols having n bits, it is thus feasible to perform a code conversion according to which the symbols corresponding to neighbouring phase positions differ in a single bit only. This is called Gray encoding. It can be seen directly that the digits of a number generated by Gray encoding of an n-digit binary number are given by the following relation:

$$g_i = b_{i-1} \oplus b_i; \quad i = 1, ..., n; \quad b_0 = 0. \tag{2.52}$$

Here b_i is the ith digit of the binary number, g_i is the ith digit of the Gray encoded number, and increasing subscript numbers denote decreasing binary positions.

For M-ary phase modulation, small error probabilities and Gray encoding, we have

$$P_{Eb} = \frac{1}{n} P_E. \tag{2.53}$$

A similar coding according to which the neighbouring symbols differ in just one bit can also be performed for a QAM signal set.

2.2 Transmission of signal sequences

It was shown in the preceding section that if a single signal is transmitted then the only disturbance is due to the additive noise which is well approximated by Gaussian noise for microwave channels. However, in practice, the transmission of signal sequences is always performed. As already mentioned at the beginning of this chapter, intersymbol interference can then be generated between consecutive bits as a result of the distortion caused by the linear filters. Let us now investigate from this point-of-view the matched filter receiver which has been found as optimum receiver in Section 2.1.2. For simplicity, the analysis will be carried out for binary baseband NRZ signals.

For antipodal NRZ signals, $s_1(t) = -s_2(t) = \sqrt{E}$. The impulse response of the corresponding matched filter is given by

$$h(t) = s_1(T-t) = s_1(t). \tag{2.54}$$

If such a filter is driven by a single signal, e.g. $s_1(t)$, then the output signal waveform will be

$$y(t) = \int_{-\infty}^{\infty} s_1(t-u)h(u)\,du = \begin{cases} 1 - \left| 1 - \dfrac{t}{T} \right|; & 0 \le t \le 2T \\ 0; & t < 0, \quad t > 2T. \end{cases} \tag{2.54a}$$

On the other hand, if this filter is driven by a signal sequence then the output will evidently be the sum of the pulse responses generated by the individual input signals. Figure 2.27 shows an arbitrary binary signal sequence and the output waveform of the matched filter. It is seen that this hardly resembles the input waveform but at instant $t=kT$, y is solely determined by the actual kth signal, i.e. there is no harmful interference. (It should be noted that this statement holds only if the instants kT are precisely known. Should the sampling be performed with a timing error ΔT then the intersymbol interference will either increase or decrease the error.)

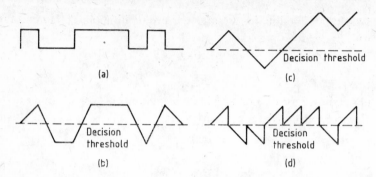

Fig. 2.27. Transmission of signal sequences: (a) NRZ signal sequence of the transmitter; (b) output signal of the matched filter; (c) output signal of an ideal integrating circuit; (d) output signal of an integrate-and-dump filter.

The matched filter is thus also suitable for the reception of signal sequences without intersymbol interference. However, the filter is difficult to realize as the transfer function of the matched filter, *i.e.* the Fourier transform of the pulse response (2.54) is given by

$$H(f) = e^{j\pi fT} \frac{\sin \pi fT}{\pi fT} \tag{2.55}$$

and this cannot be realized by an RLC network comprising a finite number of time invariant elements. However, for transmission of a single signal, it can be approximated perfectly by an ideal integrating circuit taking into account the fact that the sampling gate is opened only at the instant $t = T$ (see Fig. 2.5). It can be shown (see Ref. [4]) that if the ideal integrator is substituted by a simple RC section having a time constant of $4T$ then the loss will be less than 1 dB. However, such an ideal or nearly ideal integrator will not be at all suitable for the transmission of signal sequences. This can be proved directly by plotting the response of such an "approximately matched" filter to the bit stream of Fig. 2.27a (see Fig. 2.27c). On the other hand, the transfer function given in (2.55) can be realized by a time variant network: if an ideal integrating circuit or an RC network is discharged immediately following the sampling instant at $t = T$, then the wanted signal is obtained. This kind of filter is called integrate-and-dump circuit, and its response to the bit stream mentioned earlier is shown in Fig. 2.27d.

This kind of integrate-and-dump filter, and moreover, even its carrier frequency variants, can be realized in principle. However, the practical realization presents difficulties so it is feasible to determine the signal waveforms and transfer functions providing a transmission without intersymbol interference, with special emphasis on time invariant realizations. This problem is investigated by several books, *e.g.* [19–22], so in the following section, a short treatment with the main results only is presented. Following this, the effect of non-ideal filters on the error probability will be surveyed for transmission with phase modulation. The possibility of equalizing the filter distortion will also be dealt with. Finally, signal sets will be investigated for which the intersymbol interference has an advantageous rather than a harmful effect.

2.2.1 Filter characteristics for transmission without intersymbol interference. Nyquist criteria

Baseband signals will first be investigated, and the results will be generalized for QAM systems. No other modulation systems will be dealt with but we note that PSK can be regarded as a special case of QAM. This is proved by expressing one element of the PSK signal set in the form $e^{j\Phi} = \cos \Phi + j \sin \Phi$, which gives the amplitudes of the in-phase and quadrature components at the transmission time of the given signal vector.

Let us now investigate the model shown in Fig. 2.28. The source signal has again M elements but we now restrict our investigations to the case when the signal set elements $s_i(t)$ differ in amplitude only, *i.e.* a pulse amplitude modulation system (PAM) is investigated. This system can practically be visualized by assigning to every message m_k a constant a_k by which the amplitude of the generator output pulse is multiplied. The optimum values of amplitudes a_k are chosen according to the principles given in Section 2.1.5.

Let us for the time being assume that the pulse generator gives Dirac impulses at instants $t = kT$. (This restriction will not be used later.) The Dirac impulses having varying amplitudes according to the changing m_k values, reach the transmission channel via transmit filter $A(\omega)$. These are the output signals which have been expressed in Section 2.1 in vector form and which have been subjected to various investigations.

The receiver part in Fig. 2.28 is principally identical to the receivers treated in Section 2.1. The overall transfer function of the system is the product of the two filter characteristics. We have now to find the class of characteristics $C(\omega) =$

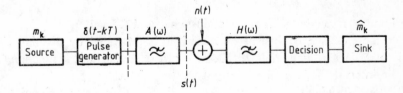

Fig. 2.28. Model of a baseband transmission system with unit impulse signals.

$= A(\omega) \cdot H(\omega)$ which results in a transmission without intersymbol interference. The transmission is considered to have no intersymbol interference if at the instant of the kth decision, it is only the response to the kth signal which differs from zero, and all other responses are zero.

Let us first consider the case of ideal low-pass filters, *i.e.* let the filter phase response be linear, and the $C(\omega)$ amplitude response be expressed by

$$C(\omega) = \begin{cases} 1; & 0 \leq |\omega| \leq \omega_0 \\ 0; & |\omega| > \omega_0. \end{cases} \tag{2.56}$$

This filter has an impulse response of

$$c(t) = \frac{\sin \omega_0 t}{\omega_0 t} \tag{2.57}$$

which has a maximum value of $c=1$ at $t=0$; and it has a value of zero at $t=k\pi/\omega_0$.

Assume now that this ideal low-pass filter is driven by the signals generated by the pulse generator shown in Fig. 2.29, *i.e.* the input to the filter $A(\omega)$ is given by

$$s'(t) = \sum_{k=-\infty}^{\infty} a_k \delta(t-kT). \tag{2.58}$$

In this case, the desired response will be free from intersymbol interference if the filter cut-off frequency is $f_0 = \omega_0/2\pi = 1/2T$.

For baseband signals, the bandwidth $B = 1/2T$ is the smallest one resulting in a transmission free from intersymbol interference. "Bandwidth" in this case only means the band outside which the transfer function value is identically zero, in contrast to the noise bandwidth treated earlier. (The frequency $1/2T$ is called the Nyquist frequency.)

Let us now investigate a class of filters having bandwidths wider than $1/2T$ and impulse responses with zeros at the instants $t=kT$. However, these filters will also be assumed to have a finite bandwidth in the above sense. It follows from the sampling theorem [19, 23] that all filters satisfying the relation

$$C_{eq} \triangleq \sum_k C\left(\omega_0 + \frac{2k\pi}{T}\right) = \text{const}; \quad |\omega| \leq \frac{\pi}{T} \tag{2.59}$$

have impulse responses with zeros at instants $t=kT$.

The above defined equivalent characteristic is generated by dividing the ω axis into sections of magnitude $2\pi/T$, shifting these sections of the function $C(\omega)$ into the interval $\pm \pi/t$, and adding all these sections.

All filters satisfying the relation (2.59) are said to satisfy the first Nyquist criterion. By this condition is explicitly expressed the requirement of having a zero response at instants kT for all $(i \neq k)$th pulses. (We shall not need the second and third Nyquist criteria so these will not be given here. Their detailed treatment can be found for example in Refs [19, 21, 22].)

From filters satisfying the first Nyquist criterion, those having bandwidths falling between the Nyquist frequency $1/2T$ and twice this frequency are of special significance. In order to satisfy (2.59), these filters should have a transfer function which is the sum of the ideal Nyquist transfer function (2.56) and another "rolled-off" function defined in the interval $(0, 1/T)$ and antisymmetrical with respect to $1/2T$. The ideal Nyquist response is thus "rolled-off" and a function shown in Fig. 2.29 is obtained.

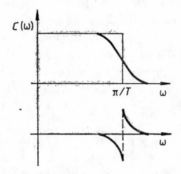

Fig. 2.29. Rolled-off filter characteristic.

Because of mathematical simplification, a cosine roll-off is preferably used for the analysis of these filters:

$$
C(\omega) = \begin{cases} 1; & 0 \leq |\omega| < \dfrac{\pi}{T}(1-\alpha) \\[2ex] \dfrac{1}{2}\left\{1 - \sin\left[\dfrac{T}{2\alpha}\left(\omega - \dfrac{\pi}{T}\right)\right]\right\}; & \dfrac{\pi}{T}(1-\alpha) \leq |\omega| \leq \dfrac{\pi}{T}(1+\alpha) \\[2ex] 0; & |\omega| > \dfrac{\pi}{T}(1+\alpha) \end{cases} \quad (2.60)
$$

and the impulse response is given by

$$
c(t) = \frac{\sin \pi t/T}{\pi t/T} \cdot \frac{\cos \alpha \pi t/T}{1 - 4\alpha^2 t^2/T^2} \quad (2.61)
$$

α is the roll-off parameter; a filter with $\alpha=1$, *i.e.* 100 per cent roll-off, is called a raised cosine filter. The occupied bandwidth is increased with increasing α. It is seen from the above expression of $c(t)$ that the zeros of the ideal Nyquist characteristic have not been displaced but additional zeros can, of course, be generated.

We should add two remarks to the above considerations. First of all, it is seen from (2.57) or (2.61) that the ideal filters or filters with cosine roll-off cannot be realized as their impulse response, $c(t) \neq 0$ for $t < 0$. However, allowing a delay, these filters can be approximated, and this delay can normally be tolerated. Design methods for filters introducing no intersymbol interference are given in Refs [24, 25] for example.

As a second remark, note that the impulse response of an "ideal" Nyquist filter has an amplitude linearly decreasing with increasing t. Assuming the pulse train according to (2.58) as input signal, the response value at $t = kT$ is equal to a_k but will not be bounded at other time instants. An ideal Nyquist response is therefore practically only applicable if the timing is perfect. The impulse response of raised cosine rolled-off filters and of all round-off filters for which the transfer function and its derivative are continuous, goes to zero at infinity as $1/t^3$. The response of such a filter to an input signal given in (2.58) is bounded. Consequently, the intersymbol interference introduced by rolled-off filters is small even with imperfect timing.

Let us now investigate how far the situation changes if the pulse generator in Fig. 2.28 does not generate Dirac impulses but signals with waveforms $b(t)$. If we consider the waveform $b(t)$ as the impulse response of a filter, it can be seen directly that the condition for obtaining a response without intersymbol interference is given by a channel transfer function

$$C'(\omega) = C(\omega)/B(\omega) \tag{2.62}$$

where $C(\omega)$ is a filter characteristic giving no intersymbol interference with Dirac impulses, and $B(\omega)$ is the Fourier transform of $b(t)$. As an example, assume that the pulse generator generates NRZ signals (square pulses of T duration); the corresponding transfer function C' of the channel (without roll-off) then has a bandwidth of $1/2T$ and a response according to the function $x/\sin x$.

The above results for baseband transmission can easily be extended for the amplitude modulated case: the filter characteristic satisfying the first Nyquist criterion is simply shifted from $\omega = 0$ to $\omega = \pm \omega_c$. This kind of filter, similarly to the original low-pass filter, will not introduce intersymbol interference because the spectrum of the AM signal is derived from the modulating baseband spectrum by a similar frequency shift. Moreover, this kind of bandpass filter is evidently symmetrical; consequently, it does not generate components which are in phase quadrature to the input signal. (We will return to this question in the next section.) This means that there will be no crosstalk between the signals which are in phase quadrature. It can thus be concluded that the above defined Nyquist filter will provide an ideal transmission for QAM signals and so also for PSK signals as these are special variants of the former.

We have not covered the single sideband AM and FSK carrier modulation methods. The former is treated, among others, in Ref. [22] while the continuous phase FSK is investigated in [21, 22].

Next, the optimal division of the overall transmission function $C(\omega)$ into transmitter and receiver parts will be determined. For simplicity, baseband transmission will be assumed but the result can be applied directly to QAM.

It has been shown in Section 2.1.3 that for transmission in additive Gaussian noise, the lowest error probability is given by a matched filter with an impulse response of $h(t)=s(T-t)$ and a transfer function of

$$H(\omega) = e^{j\omega T}S^*(\omega) \tag{2.63}$$

where we now have the notation

$$S(\omega) = A(\omega)B(\omega) \tag{2.64}$$

and the asterisk designates the complex conjugate value. Taking into account (2.62), we have

$$C(\omega) = B(\omega)C'(\omega) = B(\omega)A(\omega)H(\omega) \tag{2.64a}$$

or, after substituting (2.63),

$$C(\omega) = |B|^2|A|^2 e^{j\omega T}. \tag{2.65}$$

It is seen from (2.65) that the phase response of C should be linear, and that the following condition should be met for an intersymbol interference free reception which is simultaneously optimum in the sense of Section 2.1.2:

$$|S(\omega)| = \sqrt{|C(\omega)|} \tag{2.66}$$

where (as a reminder) $C(\omega)$ is a transfer function satisfying the first Nyquist criterion.

These considerations prove the importance of the statement in Section 2.1.3 according to which the waveform of signals $s(t)$ has no significance regarding the reception as this is determined by the vector constellation. This can be understood by noting that theoretically there will be no difference between the "optimum" reception of signal sequences and a single signal if the signal vector set corresponds to Section 2.1.5, the receiver corresponds to Section 2.1.2, and finally, the signal waveform set corresponds to this section. However, it should be remembered even in this ideal case that the error probability depends on the signal energy E which is the output signal energy of the transmitter filter $A(\omega)$ and not of the pulse generator.

2.2.2 Simultaneous effect of Gaussian noise and band limiting
on the error probability

In the previous section, we determined the overall transfer function which does not introduce intersymbol interference in spite of a finite transmission bandwidth. The problem has thus been solved in principle and further investigations in this field seem to be unnecessary. However, for various reasons, the actual situation is much less favourable. First of all, we have seen that the ideal transfer function cannot be realized, and should it be approximated by realizable networks, the approximation has often to be carried out in the frequency domain. However, the error in the time domain normally requires further investigations. Moreover, the timing of practical transmission systems is not perfectly accurate. This means that in spite of the fact that an ideal filter provides an intersymbol interference free reception at the ideal sampling time instants, it will evidently not provide this freedom at the real sampling time instants.

The problem of intersymbol interference has been investigated since the beginning of the digital transmission period [22]. The problem was first connected with wire transmission for which reason it is not covered by early books on radio transmission such as [1, 4 and 11]. Since then, the problem of intersymbol interference as a special case of the general interference problem has gained significance, both in satellite and terrestrial transmission systems. (The general interference problem will be covered in Section 2.3.) The problem has been investigated by several publications such as [26, 27, 28, 29]. In these references, primarily PSK systems are dealt with, but their results are usually applicable to other two-dimensional signal sets. The absence of investigations for higher than two-dimensional systems is not a limitation because if bandwidth economy is required, two-dimensional signal sets are only applicable as shown in Section 2.1.6.

Before proceeding with the theoretical investigations, a qualitative picture is first given. The problem is illustrated in Fig. 2.30 in an example of binary NRZ

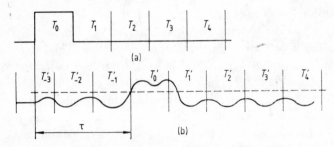

Fig. 2.30. Intersymbol interference: (a) original signal sequence; (b) delayed and distorted received sequence.

signals. Figure 2.30a shows the transmitter signal in the time slot 0 while Fig. 2.30b presents the received signal. τ is the channel delay, and time slots T'_{-1}, T'_0, T'_1 correspond to the time slots T_{-1}, T_0, T_1. We note that the received signal shows transient disturbances both before and after time slot T'_0, thus modifying the noise level resulting in erroneous detection. Note also that the received signal in time slots T'_{-1}, T'_{-2}, etc. is not identically zero, i.e. the bits transmitted in later time slots T_1, T_2, etc. will also interfere with the signal in question. It is seen from the figure that this phenomenon does not contradict the principle of causality according to which the consequence cannot precede the cause.

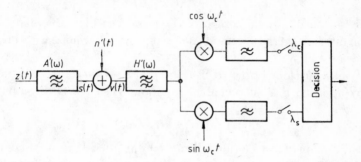

Fig. 2.31. PSK transmission system.

The following analysis is restricted to PSK modulation, and the PSK transmission system model is given in Fig. 2.31. The input signal to the transmit filter is given by

$$z(t) = \sum_{k=-\infty}^{\infty} b(t-knT) \cos(\omega_c t + \Phi_k). \tag{2.67}$$

The information content of the kth bit is comprised in Φ_k which can have the following possible values for the M-ary case:

$$\Phi_i = \frac{\pi}{M}(1+2i), \quad i = 0, 1, ..., M-1. \tag{2.68}$$

The time function $b(t)$ represents the optionally applied pulse shaping and differs from zero in a time interval nT. For a pulse of constant amplitude, $b(t)$ is a square pulse of duration nT. (As earlier, $n = \log_2 M$.)

The pulse response of the transmit filter is $a'(t)$ which can be expressed by a complex envelope form:

$$a'(t) = \text{Re}\,[a(t)e^{j\omega_c t}]; \quad a(t) = a_c(t) + ja_s(t). \tag{2.69}$$

The transmitter output signal thus has the following complex envelope:

$$s(t) = \frac{1}{2} \int_{-\infty}^{\infty} a(\tau) \sum_k b(t-\tau-nkT) e^{j\Phi_k} \, d\tau =$$

$$= \frac{1}{2} \sum_k e^{j\Phi_k} \int_{-\infty}^{\infty} a(\tau) b(t-\tau-nkT) \, d\tau. \tag{2.70}$$

The noise process $n'(t)$ can also be written in a complex envelope form:

$$n'(t) = \text{Re}\,[n(t) e^{j\omega_c t}]$$

$$n(t) = n_c(t) + j n_s(t). \tag{2.71}$$

It can be assumed from the complex noise process $n(t)$ introduced in the above expression that it has zero expectation and represents white noise with a two-sided spectral density of $N_0/2$.

The complex envelope of the receive filter pulse response can be expressed as $h_c + j h_s$ by applying a similar kind of notation. Then, the filter output signal will be

$$y(t) = [s(t) + n(t)] * h(t) \tag{2.72}$$

where the operation of convolution is denoted by an asterisk. Its noise component, denoted by $N(t)$, will be a Gaussian of zero expectation with independent sine and cosine components at a given time instant t. The variance of the noise is given by

$$\sigma_N^2 = \frac{N_0}{2} \frac{1}{2\pi} \int_{-\infty}^{\infty} |H(\omega)|^2 \, d\omega. \tag{2.73}$$

Let us now introduce the following notation:

$$R_k(t) + j I_k(t) \overset{\triangle}{=} a(t) * b(t - knT) * h(t). \tag{2.73a}$$

With this notation, the complex envelope of the signal component of $y(t)$ will be

$$x(t) = \sum_k e^{j\Phi_k} [R_k(t) - j I_k(t)]. \tag{2.74}$$

It will be assumed that the function of the low-pass filters in Fig. 2.31 is merely the elimination of the second harmonic component, without introducing signal distortion. The two filter output signals will thus be given by the real and imaginary parts of $y(t)$. At the sampling time instant t_0, these will be given by

$$\lambda_c = \frac{1}{2} \sum_k (a_k R_k - b_k I_k) + N_c$$

$$\lambda_s = \frac{1}{2} \sum_k (b_k R_k + a_k I_k) + N_s \tag{2.75}$$

where the following notations have been introduced:

$$a_k = \cos \Phi_k; \quad b_k = \sin \Phi_k; \quad R_k = R_k(t_0)$$
$$I_k = I_k(t_0); \quad N_c = N_c(t_0); \quad N_s = N_s(t_0). \tag{2.75a}$$

Let us now calculate the probability of the two-dimensional received signal vector having coordinates λ_c, λ_s falling outside the decision range of the transmitted vector (a_0, b_0). The difference between this case and the case investigated in Section 2.1.1 is clearly shown by (2.74). The received vector is now given by

$$\lambda = s + N + g \tag{2.76}$$

where s is the transmitted signal vector having components

$$\frac{1}{2}(a_0 R_0 - b_0 I_0, \; a_0 I_0 + b_0 R_0) \triangleq (\alpha_0, \beta_0). \tag{2.76a}$$

N is the additive noise and g is the distortion due to intersymbol interference. N has components (N_c, N_s) while the components of g are given by

$$\frac{1}{2}\left[\sum{}' (a_k R_k - b_k I_k), \; \sum{}' (a_k I_k + b_k R_k)\right] \triangleq (\alpha, \beta). \tag{2.76b}$$

The prime following the symbol $\sum$ indicates that the term $k=0$ should be omitted from the addition. The vector g is the additional term compared with the terms given in Section 2.1.1. Its effect on the error probability will now be investigated. The factors R_k stand for the filter responses to the signal $b(t)$ transmitted k time slots earlier; factors I_k represent the crosstalk between the two quadrature components. The terms with negative subscripts represent the responses to the signal transmitted k time slots later.

Without loss of generality, assume that $\varphi_0 = \pi/M$. Also note that the conditional probability distribution of λ is Gaussian (with g as a condition). Thus

$$p(x, y|\alpha, \beta) = \frac{1}{2\pi\sigma_N^2} \exp\left\{\frac{-1}{2\sigma_N^2}[(x-\alpha_0-\alpha)^2 + (y-\beta_0-\beta)^2]\right\}. \tag{2.76c}$$

The conditional probability of the correct decision is therefore given by

$$P_c(g) = \frac{1}{2\pi\sigma_N^2} \iint_D \exp\left\{\frac{-1}{2\sigma_N^2}[(x-\alpha_0-\alpha) + (y-\beta_0-\beta)]\right\} dx \, dy. \tag{2.77}$$

Here the region of correct decision is D; this region is bounded by the straight line sections $\varphi=0$ and $\varphi=2\pi/M$ in the xy plane. The probability of the correct decision is given by

$$P_c = \iint_{-\infty}^{\infty} P_c(g) p(\alpha, \beta) \, d\alpha \, d\beta \tag{2.78}$$

yielding the following error probability:

$$P_E = 1 - P_c. \tag{2.79}$$

The main problem of evaluating (2.78) is the determination of the probability density function $p(\alpha, \beta)$ which has been exactly calculated in Ref. [26] for random signals, *i.e.* for consecutive bits which are statistically independent. The calculation procedure given in the above reference is outlined in the following.

Let the characteristic function of the random variables α, β be denoted by $U(u, v)$. This can be expressed for equi-probable random signals. The characteristic function of the sum of independent random variables is the product of the respective characteristic functions:

$$U(u, v) \triangleq E\{\exp(j\alpha u + j\beta v)\} =$$
$$= \prod_k{}' E\left\{\exp\left\{\frac{j}{2}[(\alpha_k R_k - \beta_k I_k)u + (\alpha_k I_k + \beta_k R_k)v]\right\}\right\}. \tag{2.79a}$$

The prime following the symbol Π again denotes that the term $k=0$ is omitted. Here the expectation value has to be calculated for the possible values of a and b, and φ, respectively. For even M values this expectation is given by

$$\frac{2}{M} \sum_{i=1}^{\frac{M}{2}-1} \cos\left[\frac{1}{2}(R_k u + I_k v)\cos\frac{\pi}{M}(1+2i) + \frac{1}{2}(R_k v - I_k u)\sin\frac{\pi}{M}(1+2i)\right]. \tag{2.80}$$

Knowing the characteristic function and noting that it is the Fourier transform of the probability density, (2.78) can be written as

$$P_c = \frac{1}{(2\pi)^2} \int\int_{-\infty}^{\infty} \int\int_{-\infty}^{\infty} \int\int_D P(x, y|\alpha, \beta)U(u, v)e^{-j\alpha u - j\beta v}\, dx\, dy\, d\alpha\, d\beta\, du\, dv. \tag{2.80a}$$

Performing the integrations with respect to α and β, we have

$$P_c = \frac{1}{(2\pi)^2} \int\int_{-\infty}^{\infty} \int\int_D \exp[-j(x-\alpha_0)u - j(y-\beta_0)v] \cdot$$
$$\cdot \exp\left[\frac{-1}{2\sigma_N^2}(u^2+v^2)\right] U(u, v)\, dx\, dy\, du\, dv. \tag{2.81}$$

This is an exact expression but it cannot be evaluated analytically. However, in Ref. [26], a numerical evaluation method having an arbitrarily small error is given. The method is based on the expansion and term-by-term integration of the characteristic function, and further on the calculation of the coefficients of this series by a recursive formula.

As already stated, the factors I_k in (2.75) represent the crosstalk between the two quadrature components of the carrier. $I_k=0$ for a symmetrical transfer function; this allows an additional interesting conclusion to be drawn for this case.

Rearranging (2.81) and substituting $M=2$ and $M=4$ we have

$$P_E(R|M=2) \approx \frac{1}{2} P_E(2R|M=4) \qquad (2.81a)$$

showing that the error probabilities of the two-state and four-state systems, resulting from intersymbol interference, are related in the same way as the error probability resulting from Gaussian noise only. In the above expressions, R denotes the signal-to-noise ratio within the overall filter band.

For practical evaluation, the bit energy over noise spectral density, denoted by E_b/N_0, is frequently more significant than the signal-to-noise ratio R. Figure 2.32 shows the relation between the bandwidth, E_b/N_0 and the error probability for a specific example.

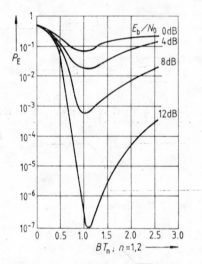

Fig. 2.32. Error probability as a function of a three-element Butterworth filter bandwidth, in a PSK system.

2.2.3 Distortion equalization

It has been shown in the previous section that the performance of the digital transmission can be substantially impaired by the transfer function of the filters, especially for high M values. In many cases, the equalization of the transfer function is more practical than the better approximation of the ideal filter transfer function. The previous section also shows that a time domain equalization is more effective. In the following, the so-called decision feedback nonlinear equalization procedure will be presented, which is extremely effective and related to the investigations of the previous section. This procedure has the effect of compensating the interfering effect of the preceding bits on the bit in question.

Assume that for $k > L$, $R_k = I_k \equiv 0$, i.e. the number of interfering symbols·is L. Neglecting the noise and the terms pertaining to negative k values, the received signal vector is calculated from (2.76) as

$$\lambda = \frac{1}{2}(a_0 R_0 - b_0 I_0, \; a_0 I_0 + b_0 R_0) + \frac{1}{2}\Big(\sum_{k=1}^{L}(a_k R_k - b_k I_k), \; \sum_{k=1}^{L}(a_k I_k + b_k R_k)\Big). \quad (2.81b)$$

Investigating first the binary case, the equalizer shown in Fig. 2.33 will perfectly cancel the effect of intersymbol interference as the equalized λ_c is given by

$$\lambda_c = \frac{1}{2}a_0 R_0 + \frac{1}{2}\sum_{k=1}^{L}a_k R_k - \frac{1}{2}\sum_{k=1}^{L}\hat{a}_k R_k = \frac{1}{2}a_0 R_0. \quad (2.82)$$

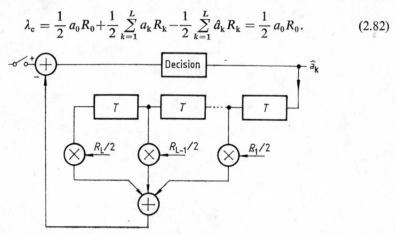

Fig. 2.33. Decision feedback equalizer.

This is explained by the fact that $a_k = \hat{a}_k$ without noise when all decisions are perfect. It is also seen from our investigations that the equalization will be ideal even with noise, as long as the decision is perfect. The same is practically valid for error probabilities less than about 10^{-2}. This is the property by which the decision feedback equalizer is basically distinguished from linear equalizers which result in increased noise due to the modified frequency response.

Extending our investigations for more than two phases, the following procedure has to be applied: a feedback circuit similar to that shown in Fig. 2.33, connected to the input of component λ_s has to be used; further, output a_k has to be fed back to input λ_s, and also output b_k to input λ_c.

A combination of a decision feedback equalizer with a transverse filter can be used to eliminate the interfering effect of the terms with negative subscripts [30].

In the above equalization procedure, exact knowledge of the factors R_k, I_k (*i.e.*, the channel transfer function) is necessary, assuming that this is time invariant. This is valid with good approximation if the distortion introduced by only the filters has to be equalized. However, this is no longer true if the distortion is generated by the transmission medium itself, which may be the case for high bit rates and multipath propagation in terrestrial or satellite microwave links. This situation may require the use of adaptive equalization which will be dealt with in Chapter 4 (see, for example, Refs [9, 30, 31, 32]).

2.2.4 Partial response signal set

In the foregoing, the condition of eliminating intersymbol interference has been investigated, and further the effect of the intersymbol interference and the possibility of equalization have been dealt with. In the following, a transmission system will be outlined which is not free from intersymbol interference but rather introduces a specified amount of interference between a few consecutive bits, thereby allowing a smaller channel bandwidth and thus a better utilization of the frequency range.

In Section 2.2.1, the transfer function resulting in a response a_k at the instant kT to an input signal $a_k(t-kT)$ has been calculated. In the binary case, $a_k=0$ or 1. In a so-called partial response transmission system, the response at the kth time instant is given by

$$c_k = \sum_{i=0}^{m} b_i a_{k-i}. \tag{2.83}$$

The application of such a transmission characteristic results in a multilevel output signal even for a binary input signal. This method has the effect of a reduced band occupancy, similar to the method outlined in Section 2.1.6 according to which symbols comprising n bits of the bit stream have been generated. It has been shown in Refs [9, 20] that in the frequency range 0 to $1/2T$, the filter response can be expressed in the following sampled form:

$$c_s(t) = \sum_{i=0}^{m} b_i \delta(t-mT). \tag{2.84}$$

As an example, Fig. 2.34a shows the response function for the case $b_0=-1$, $b_1=0$, $b_2=1$. The corresponding channel transfer function is given by

$$C(\omega) = \begin{cases} 2\sin \omega T; & 0 \leq \omega \leq \dfrac{\pi}{T} \\[2mm] 0; & \omega > \dfrac{\pi}{T}. \end{cases} \tag{2.85}$$

The above properties of partial response signals, related to baseband transmission, can be utilized for the evolution of microwave transmission systems in several ways. One of these is the quadrature partial response system QPRS. According to this system, the two carrier components in quadrature are amplitude modulated by homochronous binary bit streams similar to the QPSK system but the channel transfer function is chosen according to (2.85). The complete two-dimensional signal space will thus have nine signal states according to the vector configuration shown in Fig. 2.34b. The occupied bandwidth of this system is practically equal to the bandwidth of the 16PSK or 16QAM system but the comparison with Fig. 2.9 shows that the power required is less (because the distance between the points is larger than for either PSK or QAM).

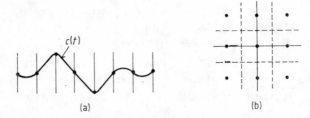

Fig. 2.34. Partial response signals: (a) a typical waveform; (b) 9QPRS signal set.

The decoding process of PRS signals requires the inverse operation of expression (2.83):

$$a_k = \frac{1}{b_0}\left(c_k - \sum_{i=1}^{m} b_i a_{k-i}\right). \tag{2.86}$$

This decoding process involves error propagation which will be dealt with in Chapter 10. In this chapter, the digital realization of the channel transfer function (2.84) will also be presented. In conclusion, we should note that the condition of a specified intersymbol interference and optimum reception for a given additive noise is the realization of the overall channel transfer function according to expressions (2.63) to (2.66).

2.3 Multichannel transmission

In this section, the somewhat generalized model of Fig. 2.1 will be used to investigate the interference effect of additional channels in a multichannel digital transmission system. Figure 2.1 shows that the unwanted carrier frequency may be different from or identical to the useful carrier frequency. The interference caused by disturbing signals is called interchannel interference which may be either co-channel or adjacent channel interference.

In terrestrial and satellite systems, the interchannel interference is a fundamental limiting factor of the digital transmission, similar to the noise. This is explained by the fact that the usable frequency range is finite and has to be utilized economically. An effective method of frequency economy is the use of a single frequency for the transmission of several signals. Excluding special solutions here, two signals at most can be transmitted by a single carrier frequency over identical or geographically nearby paths, by utilizing orthogonal polarizations. A further method to reduce co-channel interference between signals arriving from different directions is the utilization of suitable antenna directional patterns.

For channels with different carrier frequencies, the interference is obviously caused by the fact that part of the disturbing signal spectrum falls into the disturbed channel. Thus, in addition to the application of orthogonal polarizations and suitable antenna characteristics, we have two additional possibilities of reducing the interference: apply a modulation method and signal set which occupies a narrow frequency band, and decrease the residual interference power by the use of sharp cut-off filters. We have investigated some relations between the signal set and the band occupancy in Sections 2.1 and 2.2. As far as filters are concerned, these sections imply a relatively simple method of obtaining interference-free transmission over adjacent channels: it has been shown in Section 2.2.1 that for a roll-off parameter α, intersymbol interference will be eliminated by a filter which has identically zero transmission outside the band $(1+\alpha)/nT$. This system is thus free from any kind of interference (including intersymbol and interchannel interference) provided the separation between adjacent channel frequencies is at least $(1+\alpha)/nT$.

However, these ideal circumstances cannot be realized practically: the real stop-band attenuation is not infinite so interchannel interference will appear. This may be substantial in the case of the interference between transmitters and receivers of a radio relay system utilizing common antennas for reception and transmission (the received power may be even 100 dB less than the transmitted power) and during selective deep fades, when the relative level of the unwanted signal may increase as much as 10 to 30 dB.

In this section, the rate of spectrum level decrease at high detuning will be investigated for a few signal sets. Following this, the results of Section 2.2.2 will be extended for arbitrary interferences, and numerical results will be given showing the simultaneous effect of several interferences. Finally, the spread-spectrum method, allowing the multiple utilization of a given frequency band, will be dealt with. A linear transmission channel will be assumed throughout our investigations.

2.3.1 Spectral properties of some signal sets

The transmission of the digital information over the channel to be investigated can be assumed to be a stochastic process. Moreover, with only slight exaggeration we can consider this process as Markovian. The spectral density of this process follows from the knowledge of the $s(t)$ signal waveform corresponding to all m_i messages $(i=1, 2, ..., M)$, the *a priori* probability of the individual messages, and further the transitional probabilities p_{ik} (these are the probabilities of a message m_k following a message m_i, where now $i, k = 1, 2, ..., M$). The spectral density calculation of Markovian signal sources can be found, for example, in Refs [11, 33, 34]. In the general case, the spectrum has both discrete and continuous parts, the spectral lines of the discrete part being at multiples of the symbol frequency $1/nT$.

In most cases, a signal set generating no discrete spectral lines is chosen in order to reduce the adjacent channel interference. According to Ref. [33], a sufficient condition to eliminate the line components of the spectrum is the following: for each element $s_i(t)$ of the signal set there is also an element $s_{-i}(t) = -s_i(t)$; $p_i = p_{-i}$ for all i values; and for all i and j values, $p_{ij} = p_{-ij}$. This kind of signal set is called a "negative equiprobable" or NEP signal set. Most of the previously investigated signal sets fall into this category: for even M, the MPSK, the MQAM, the bi-orthogonal, the quadrature partial response signal sets, *etc*. The spectral density of the NEP signal sets can be calculated from the expression

$$S(\omega) = \frac{1}{T_s} \sum_{i=1}^{M} p_i |G_i(\omega)|^2 \tag{2.87}$$

where T_s is the time interval of one signal (in our examples, $T_s = nT$), and

$$G_i(\omega) = \int_0^{T_s} s_i(t) e^{-j\omega t} \, dt. \tag{2.88}$$

In Table 2.3, the spectral density expressions for a few signal set complex envelopes are summarized, assuming that the probability of the individual signals is identical. These spectra are all purely continuous. The interfering power in the adjacent channel can be calculated from a knowledge of the spectrum.

The signal set corresponding to sinusoidal frequency shift keying (SFSK) [35] can be regarded as a generalized version of the MSK signal set treated in Section 2.1.4. For both cases, the phase is continuously increased from 0 to $\pi/2$ during interval T, the phase function being linear for MSK and sinusoidal for SFSK.

Comparison of the expressions given in the table allows a few interesting conclusions to be drawn. In agreement with our previous results, the decrease of the spectrum envelope within the main lobe is approximately proportional to $1/n$.

Table 2.3. Complex envelope spectral density of a few signal sets.
Number of states: $M = 2^n$, carrier phase: Φ

Signal set	Description of signal set	Spectral density
MPSK–NRZ	$\Phi = \dfrac{(1+2i)\pi}{M}; \quad t \in (0, nt)$ $i = 0, 1, \ldots, M-1$	$nT\left(\dfrac{\sin \omega nT/2}{\omega nT/2}\right)^2$
MPSK biphase	$\Phi = \begin{cases} (1+2i)\dfrac{\pi}{M}; & t \in \left(0, \dfrac{nT}{2}\right) \\ -(1+2i)\dfrac{\pi}{M}; & t \in (nT/2, nT) \end{cases}$	$nT\left(\dfrac{\sin^2 \omega nT/4}{\omega nT/4}\right)^2$
MSK	$\Phi = \pm\dfrac{\pi t}{2T}; \quad t \in (0, t)$	$4\pi^2 T\left(\dfrac{\cos \omega T/2}{\pi^2 \omega^2 T^2}\right)^2$
SFSK	$\Phi = \pm\dfrac{t}{2T} \mp \dfrac{1}{4}\sin\dfrac{2\pi t}{T}$	$4\pi T\left[\displaystyle\sum_{n=-8}^{\infty} J_n\left(\dfrac{1}{4}\right)\cos\dfrac{\omega T}{2}\times \right.$ $\left. \times\dfrac{4n-1}{\omega^2 T^2 \pi^2 (4n-1)^2}\right]$

The half-power band of a QPSK signal is thus considerably smaller than the same band of an MSK signal. On the other hand, the rate of remote side lobe amplitude decrease is much more pronounced for MSK than for QPSK. The situation is even more favourable for SFSK. The spectrum relations are illustrated in Fig. 2.35 as a function of detuning from the carrier frequency. The spectral density envelope is plotted in Fig. 2.35a while Fig. 2.35b shows the spectrum power percentage falling outside the frequency range given by the abscissa.

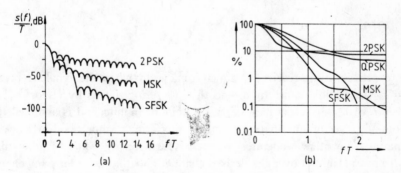

Fig. 2.35. (a) Spectral density functions of a few modulation systems in the case of random modulating signals; (b) power percentage outside the bandwidth given by the abscissa.

The rate of spectral density envelope decrease for $\omega \to \infty$ is proportional to $1/\omega^2$ for QPSK, to $1/\omega^4$ for MSK, and to $1/\omega^6$ for SFSK. In Ref. [34], a general method is presented for specifying the phase time function in order to meet a given asymptotic behaviour of the spectral density.

Let us now draw a few practical conclusions from the previous considerations. For approximately equal signal powers of the wanted and unwanted signals, the frequency difference between adjacent carriers can be reduced if the number of signal states is increased and thus the symbol frequency is decreased, thereby improving frequency band economy. On the other hand, a wider frequency separation should be applied if the interfering signal power is much higher than the desired signal power. In this case, the interference effect can be reduced by suitable choice of modulation system, *i.e.* the phase function $\varphi(t)$. In linear systems, the interfering part of the spectrum components can be arbitrarily reduced; however, this reduction is limited by the intersymbol interference generated. Nonlinear channels comprising limiters need further consideration, and will be presented in Section 2.4.

2.3.2 Simultaneous effect of Gaussian noise, intersymbol interference and interchannel interference on the error probability

In Section 2.2.2, we investigated the effect of noise and intersymbol interference on the error probability. Let us now include the effect of the disturbance originating from the interchannel interference. The frequency of the interfering channel may be either equal to the useful carrier frequency or may differ from it. In principle, our results will be applicable to an arbitrary dimensionality as no explicit restrictions will be given for the dimension of the signal state. One- and two-dimensional cases will be investigated according to the method given in Ref. [36].

Let

$$\mathbf{r} = \mathbf{s}_i + \mathbf{n} + \sum_{k=1}^{L} \mathbf{a}_k \qquad (2.89)$$

be the expression of the received D-dimensional vector where the signal is represented by $\mathbf{s}$, the additive noise by $\mathbf{n}$, and the interfering signals by vectors $\mathbf{a}_k$. The number L of these vectors may be either finite or infinite, and is assumed to be the sum of L_1 vectors originating from the interfering channels and of L_2 vectors originating from intersymbol interference. It is further assumed that random digital signals are transmitted over the desired channel and that the interfering channels are statistically independent. This means that the components in the second and third term of (2.89) are statistically all independent. In most practical cases, this condition is met so our results will not be restricted significantly. On the other hand, it has a fundamental significance for the applied mathematical procedure, similarly to the procedure of Section 2.2.2.

We assume further that the characteristic function of $\mathbf{n}$ and $\mathbf{a}_k$ exists, and the expectations of these random variables and the moments of all their terms of odd subscripts equal zero. The characteristic functions are thus all real functions. The characteristic function of $\Sigma \mathbf{a}_k$ exists even if $L = \infty$.

Assuming that the M individual signals in the wanted channel have equal *a priori* probabilities, we have the following expression for the probability of correct decision:

$$P_c = \frac{1}{M} \sum_{i=1}^{M} P[\mathbf{r} \in R_i | \mathbf{s} = \mathbf{s}_i]. \tag{2.90}$$

Here R_i stands for the region of the ith signal which is part of the D-dimensional space. Again, P is the probability of the bracketed event. In more detail, the conditional probability in (2.90) is given by

$$P\left[\mathbf{r} \in R_i \Big| \sum_{k=1}^{L} \mathbf{a}_k = \mathbf{t}, \mathbf{s}_i\right] = \int_{R_i} p_\mathbf{n}\left(\mathbf{r} \Big| \sum_{1}^{L} \mathbf{a}_k = \mathbf{t}, \mathbf{s}_i\right) dv = \int_{R_i} p_\mathbf{n}(\mathbf{r} - \mathbf{s}_i - \mathbf{t}) \, dv \tag{2.91}$$

and the unconditional probability is given by

$$P[\mathbf{r} \in R_i] = \iint_{R_i} p_\mathbf{n}(\mathbf{r} - \mathbf{s}_i - \mathbf{t}) \, dG_L(\mathbf{t}) \, dv \tag{2.92}$$

where G_L denotes the distribution function of $\Sigma \mathbf{a}_k$. As earlier, (2.92) can be expressed by the characteristic functions. The characteristic function of $\mathbf{n}$ is $\varphi_0(u)$, and that of the random variable Σa_k is given by

$$\Phi_L(u) \triangleq \int e^{\mathrm{j}tu} \, dG_L(t) = \prod_{k=1}^{L} \varphi_k(u). \tag{2.92a}$$

By expressing the probability density functions and distribution functions in (2.92) as the inverse Fourier transforms of the characteristic functions, we obtain

$$P[\mathbf{r} \in R_i] = \frac{1}{(2\pi)^D} \int_{R_i} \int_{-\infty}^{\infty} \varphi_0(\mathbf{u}) \Phi_L(\mathbf{u}) e^{-\mathrm{j}(\mathbf{r} - \mathbf{s}_i)\mathbf{u}} \, d\mathbf{u} \, d\mathbf{r}. \tag{2.93}$$

In principle, (2.93) is the general solution of the problem but practically it is only a formal result. This $2D$-fold integral cannot be evaluated analytically, also taking into account the fact that in the general case, Φ_L is an infinite product and R_i can also be an infinitely large region. Several approximate solutions have been published from which the procedure of Ref. [36] will be outlined in the following.

Φ_L is substituted by Φ_k comprising a finite number of factors where $k < L$. This approximating characteristic function is further approximated by its series expansion, and finally, the integral in (2.93) is extended for a finite D-dimensional cube with a centre at $\mathbf{s}_i$, instead of the whole region R_i. With these approximations,

$$P[\mathbf{r} \in R_i] \approx \sum_{l=0}^{m-1} \frac{b_1}{(2\pi)^D} \int_{R_i'} \int_{-\infty}^{\infty} \mathbf{u}^{2l} \varphi_0(\mathbf{u}) e^{-j(\mathbf{r}-\mathbf{s}_i)\mathbf{u}} \, d\mathbf{u} \, d\mathbf{r} \tag{2.94}$$

where R_i' denotes the common part of the original region of signal vector $\mathbf{s}_i$ and the above defined cube. Φ_k has been expressed by a series expansion written symbolically in the form

$$\Phi_k = \sum_{l=0}^{\infty} b_l \mathbf{u}^{2l}. \tag{2.94a}$$

It is proved in Ref. [36] that the error of this approximate equation can be reduced to an arbitrarily small value.

The significance of this or other methods for the evaluation of (2.90) (see, for instance [37, 38]) depends on how far it can be applied to the numerical solution of practical problems. In the following, a few numerical results will be presented. It should be noted that in radio relay systems, the adjacent channel interference is of primary importance but co-channel interference between orthogonally polarized channels may also be significant; this latter is of special importance in satellite transmission systems.

In order to investigate the efficiency of this method, let us consider the error probability of an MPSK system due to filter distortion, additive noise and co-channel narrow-band interference. The kth co-channel interfering signal is expressed by

$$A_k \cos(\omega_c t + \psi_k) \tag{2.94b}$$

and it is assumed that ψ_k has a uniform distribution between 0 and 2π.

Some results of the calculation are shown in Figs 2.36a and 2.36b. Note that the error probability for a given signal-to-noise ratio is not uniquely determined by the carrier-to-interference ratio C/I but also depends on the number of interferers, more individual interferers giving a higher error probability. It should also be noted that the overall effect of co-channel interference and intersymbol interference cannot be calculated from the superposition of the two effects. It should be mentioned that if a QPSK signal is transmitted over a symmetrical filter, then the error due to thermal noise and also to simultaneous thermal noise and intersymbol interference, can be calculated from the results of the binary transmission; however, this is no longer true for sinusoidal interferers. This is explained by the fact that the characteristic function of these interferences cannot

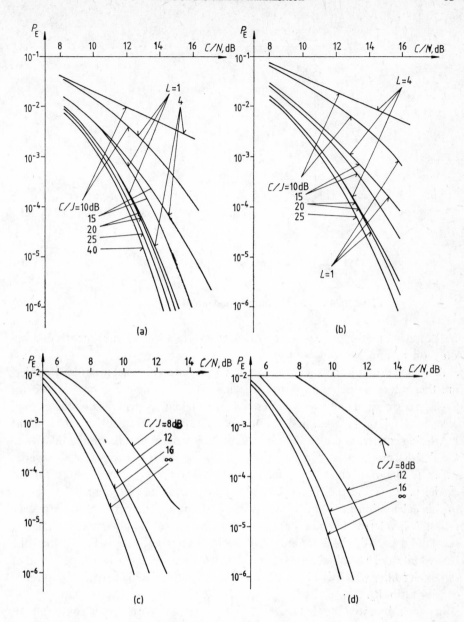

Fig. 2.36. Simultaneous effect of noise, intersymbol interference and interchannel interference: (a) QPSK, three-element Butterworth filter, BT=0.625; (b) as above, but BT=0.5; (c) 2PSK, with a plesiochronous interferer of the same speed; (d) as above, with two interferers.

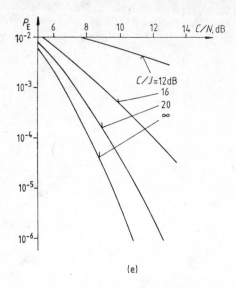

(e)

Fig. 2.36. (e) as for (c), but with a tenth of the interferer speed.

be expressed as the product of two factors so that one of the factors depends only on u_1 and the other only on u_2.

As a next example, consider the case in which the co-channel interfering carriers are themselves modulated by digital signals. This problem has been solved for 2PSK transmission and a single interferer by Ref. [39], by generalizing the method given in Ref. [36]. This solution can be generally applied inasmuch the phase of the interfering carrier, and also the bit rate of the interferer and its timing relative to the desired modulation can be arbitrary. The solution is expressed by a series which is integrated term-by-term, thus averaging the error probability for the carrier phase ψ and the time shift τ of the interfering clock signal. A few results of this calculation are shown in Figs 2.36c, d and e. The carrier-to-interference ratio C/I is referred to the output of the receive filter. The filter may attenuate a substantial part of the interference if the interfering bit rate is higher than the desired bit rate. The abscissa is again the signal-to-noise ratio in the passband of the receive filter which is about 0.5 dB lower than the signal energy/noise spectral density ratio R_0.

In conclusion, we note that recently, highly intense research has been carried out to investigate the behaviour of digital signals in an interfering environment. Reference [40] presents a good summary of the results attained up to 1977. The method outlined above has the advantage over other methods of being more-or-less analytical and thus can be followed relatively easily. However, the analytical nature of all these methods has the shortcoming of not yielding design methods which could be applied to specify the filters responsible for the interference. Filter design will be covered in Section 7.8.1, but we note here that the performance of the filters thus designed should be controlled by a subsequent error probability check (by analytical or simulation methods) before realizing the filter.

2.3.3 The spread spectrum method [41]–[48]

2.3.3.1 General considerations

In this section, a special transmission method applied to reduce the interchannel interference will be surveyed. This method does not follow the well established practice according to which the transmission spectrum is narrowed by the use of suitable signal set; on the contrary, a spectrum much wider than would be required for the transmission of the given bit rate is transmitted. In the receiver, this spread spectrum is de-spread again, thus eliminating a substantial part of the interference superimposed on the signal.

The main features of the spread spectrum method are as follows:

(1) It is a signal processing method for which

$$W \gg B_i$$

where B_i is the frequency band practically occupied by the information signal to be transmitted and W is the spread spectrum signal frequency band which is actually transmitted;

(2) A spectrum spreading signal independent of the information signal is utilized;

(3) This signal is chosen to result in a noise-like overall spectrum.

In principle, the transmitted information may be either analog or digital, the latter being applied more frequently. The spectrum spreading signal is some kind of a pseudo random sequence in order to fulfil the third of the above requirements.

The second property above excludes some transmission systems. From the analog systems, the wideband frequency modulation and the pulse position modulation utilizing high-slope pulses are excluded, and from the digital systems, the transmission utilizing an M-ary orthogonal signal set with $M \gg 1$ is out of question. The condition $W \gg B_i$ applies for these latter systems but without the use of a separate spectrum spreading signal. The third property excludes the so-called chirp-frequency system in which the spectrum is spread by additional frequency modulation according to a time function which is not noise-like, such as a saw-tooth wave or a sine wave.

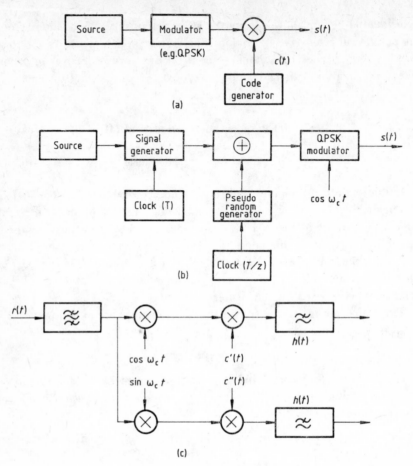

Fig. 2.37. Spread spectrum systems: (a) principle of the transmitter; (b) DS transmitter; (c) receiver.

    The spread spectrum signal to be transmitted is generated by multiplying the information bearing signal by the pseudo random signal applied for spreading the spectrum (Fig. 2.37a). There are two ways of doing this: the information signal is either actually multiplied by a (fast) pseudo random sequence or the carrier frequency is varied in a wide range in a pseudo random manner. The former is the direct sequence (DS) or pseudo noise (PN) system while the latter is called frequency hopping (FH) system.

    The transmitter of the DS system is shown in somewhat more detail in Fig. 2.37b. In practical systems, NRZ signals are generated by the signal source. The clock frequency of the pseudo random signal used for spectrum spreading is z

times the information signal clock frequency; z being an integer, the two clocks are homochronous. Practically, z is in the range of 30 to 1000 which means that the spectrum is spread by a factor of 30 to 1000. The receiver structure is shown in Fig. 2.37c. $c'(t)$ and $c''(t)$ are essentially dibit-components of the transmit side $c(t)$. (In some cases, $c'(t)=c''(t)=c(t)$.) For simplicity, it is assumed that $|c'(t)|=|c''(t)|=1$, *i.e.* the pseudo random sequence $c(t)$ is made up of NRZ pulses of amplitudes ± 1. Let us first assume that the channel is free from noise and interference, and that the spectrum spreading signal $c(t)$ has been perfectly restored at the receive side. In this case, the difference between the shape of the received signal $v(t)$ and that of the transmitted information signal $d(t)$ will only be due to the distortion of the channel filters. For simplicity, assume that no distortion is introduced by the filters. For QPSK modulation, the complex envelopes will then be given by the following expressions:

$$s(t) = c(t)d(t)e^{j\omega_c t} \tag{2.95}$$

for the transmitted signal, and

$$v(t) = \mathrm{Re} \int_{-\infty}^{\infty} s(u)c^*(u)e^{-j\omega_c u}h(t-u)\,du =$$

$$= \int_{-\infty}^{\infty} |c(u)|^2 d(u)h(t-u)\,du = \int_{-\infty}^{\infty} d(u)h(t-u)\,du \tag{2.96}$$

for the received signal.

It is feasible to apply an $h(t)$ matched to $d(t)$. It will be shown that in this case, the performance of reception is barely affected by the spectrum spread if additive Gaussian noise is the only source of distortion. The performance of reception depends only on the energy of an element of the information signal d, *i.e.* a pulse of duration $2T$ and not $2T/z$.

Let us now investigate the effect of interferences which are generated by sources which are stationary in a wide sense and which are independent of the useful signal. Either wideband or narrow-band interference can be considered. Assume that their expectation is zero, and their effective spectrum width is smaller than the bandwidth of the receiver input filter. The received signal can then be expressed as

$$r(t) = s(t) + i(t) \tag{2.97}$$

where the interference signal is given by

$$i(t) = \mathrm{Re}\,[n(t)e^{j\omega_c t}] \tag{2.98}$$

and the power of $n(t)$ is given by

$$P_n = E|n(t)|^2 = \frac{1}{2\pi} \int_{-\infty}^{\infty} S_n(\omega)\, d\omega \qquad (2.98a)$$

where S_n denotes the spectrum density of the interfering process. The useful output signal is of course equal to $v(t)$ as given in (2.96) while the complex envelope of the interference appearing at the output is given by

$$v_n(t) = \int_{-\infty}^{\infty} n(u)c^*(u)h(t-u)\, du. \qquad (2.99)$$

The power of this is equal to the expectation of its squared value, *i.e.*

$$\sigma_{v_n}^2 = E[|v_n(t)|^2] = \iint_{-\infty}^{\infty} h(u)h^*(u)R_n(u-w)R_c(u-w)\, du\, dw \qquad (2.100)$$

where R_n is the correlation function of $n(t)$ and R_c is the correlation function of the spectrum spreading signal.

Equation (2.100) can also be expressed in a form which is more descriptive:

$$\sigma_{v_n}^2 = \frac{1}{2\pi} \int_{-\infty}^{\infty} |H(\omega)|^2[S_n(\omega) * S_c(\omega)]\, d\omega \qquad (2.101)$$

where S_c denotes the spectral density function of $c(t)$.

The interference suppression obtained by the spread spectrum method is explained in Fig. 2.38. Figure 2.38a shows the receiver channel filter output spectrum

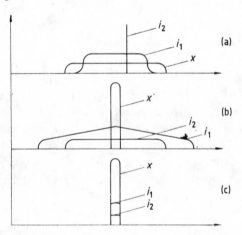

Fig. 2.38. Interference suppression in a spread spectrum system: (a) spectrum of the useful signal (x) and of the two interferers (i_1 and i_2); (b) the same after de-spreading; (c) filtered spectrum.

made up of the useful signal, a wideband interfering signal and a narrow-band interfering signal. Figures 2.38b and 2.38c show the input and output spectrum of the receiver low-pass filter, the latter corresponding to $v+v_n$. The interference suppression can easily be seen by considering that at the receiver output, the useful signal spectrum is concentrated to the actual spectrum width of the information while the power of the wideband interference is reduced according to the bandwidth ratio. The narrow-band interference is similarly reduced by the same ratio, taking into account the spectrum spread shown in Fig. 2.38b. It can actually be shown that almost independently of the spectral density function S_n,

$$\sigma_{v_n}^2 = \frac{B_i}{W} P_n. \tag{2.102}$$

This can also be expressed by the following equation:

$$(\text{signal/noise})_{\text{output}} = \frac{W}{B_i} \ (\text{signal/noise})_{\text{input}}$$

The ratio W/B_i is called processing gain, denoted by PG.

In deriving expression (2.102), it has been implicitly assumed that the spectrum of $c(t)$ is nearly constant in the receiver bandwidth. This condition is approximately met for pseudo random signals in spite of the fact that these are periodic signals having discrete spectrum lines. However, these lines are very close to each other.

FH systems are somewhat different but have similar numerical relations. Because of the processing in the receiver, the disturbing effect of the interferences is limited to the time intervals during which the frequency of an interfering signal falls into the vicinity B_i of $s(t)$. As this occurs only in the time proportion B_i/W, the average interference power is reduced by the same ratio, so again we have

$$PG = W/B_i.$$

In a slowly hopping system in which the frequency remains approximately constant for several bit intervals, we have $W = N\delta f \approx NB_i$ where N is the number of frequencies and δf is their separation. In this case alone, the processing gain is given by

$$PG = N.$$

2.3.3.2 Modulation; error probability due to Gaussian noise and interchannel interference

The modulator shown generally in Fig. 2.37 can, in principle, be implemented by an arbitrary digital modulation method. In the DS system, there is no argument against the use of 2PSK or 4PSK, so in most cases it is feasible to apply these modulation schemes.

Furthermore, as soon as code synchronization is established the spectrum is de-spread; consequently, normal coherent demodulation methods can be applied; the interference noise appearing in the receiver bandwidth B_i can then hardly be distinguished from thermal noise. The apparent signal-to-noise ratio which is related to the error probability is then given by

$$R_a = \frac{C}{N + \dfrac{I}{PG}} \tag{2.102a}$$

where C is the carrier power, N is the noise power and I is the interference power.

Assume now that the processing gain and power increase applied for overcoming the interference are given; the permitted interference-to-carrier ratio can then be calculated as follows:

$$\frac{I}{C} = \frac{PG}{R_s}(1 - 1/L) \tag{2.103}$$

where R_s is the signal-to-noise ratio needed for the given error probability and L is the transmit power increase required by the interference.

The situation can be illustrated by the following numerical example. Assuming voice transmission by using delta modulation, an error ratio of 10^{-3} is acceptable. This corresponds to $R_s = 10\,\text{dB}$; allowing $L = 1\,\text{dB}$ and assuming a processing gain of 30 dB, *i.e.* increasing the 16 kHz bandwidth needed for delta modulation transmission to 16 MHz, $I/C = 13\,\text{dB}$ is also acceptable. On the other hand, higher transmission requirements, *e.g.* an error probability of 10^{-6}, can be met by assuming $R_s = 14\,\text{dB}$. $L = 3\,\text{dB}$ has now to be chosen if $I/C = 13\,\text{dB}$ is again permitted. This means that even in the case of this high quality transmission, a 3 dB increase in transmitter power is sufficient to cope with an interference level which is 16 dB higher than the desired signal level.

It can be seen from (2.103) and the preceding examples that as long as the interference-to-desired-signal ratio is less than PG/R_s, an arbitrary specified error probability can be realized, provided the available transmitter power is sufficiently high. On the other hand, for a given required signal power, there is no difference between the effects of the interference and the thermal noise on the increase of the error probability.

Considering now the FH system, the properties of the generator and the transmission channel do not allow the de-spread signal to be coherent *per se, i.e.* at the end points of the chip-time intervals, the phase of the received signal will show uncontrollable phase jumps. It thus follows that in FH systems, only frequency modulation and non-coherent demodulation, or phase modulation and differentially coherent demodulation, can be applied.

The behaviour of an FH system under interference conditions differs basically from the behaviour of the DS system discussed above. Let us first investigate the slow system. Assuming identical carrier and interfering frequencies, the transmission will be interrupted as soon as the narrow-band interference power reaches say half of the transmitter power. The error probability is thus given by

$$P_E = \frac{1}{2PG} = \frac{1}{2N}; \quad I/C > -3 \text{ dB.} \tag{2.103a}$$

Assuming again $PG = 30$ dB, the error probability cannot be reduced below 5×10^{-4} by a reasonable increase in transmitter power. In slow FH systems, a further reduction in error probability can be obtained only by the application of error correcting coding which is a rather complicated procedure because of the burst structure of the errors encountered. On the other hand, if $P_E = 1/2N$ is acceptable, the transmission performance will not be impaired further by an arbitrary increase in interference power.

The spread spectrum system is especially advantageous in terrestrial mobile transmission systems which are shared by many participants for relatively short periods. In this case, individual pseudo random codes which are mutually quasi-orthogonal (*i.e.,* their cross-correlation is nearly zero) are assigned to each subscriber. All transmitters utilize the same carrier frequency for the transmission of spread spectrum signals having the above mentioned individual codes. It is shown, for example in Ref. [43], that for a given frequency band, the traffic capacity of the above system will be higher than the capacity obtained with individual frequency bands allocated for each subscriber. The conditions of such an FH system are much the same as the traffic conditions of a switched telephone network. However, a unique feature of this system is the behaviour under overload conditions: overload results only in the reduction of transmission performance and not in the need for queueing.

2.4 Transmission of digital signals over nonlinear channels

The digital transmission system described in Chapter 1 and discussed in the preceding is basically nonlinear (as the decision operation is, obviously, non-linear). However, all operations and circuits preceding the decision operation are linear so a corresponding superposition rule could be applied for the discussed distortions and disturbances. In this section, the effect of circuit nonlinearities will be investigated and we believe that the above reasoning will justify this section title.

The nonlinearity of the circuits results in two kinds of effects: it modifies the error probability and the signal spectrum, thus also modifying the interference introduced into adjacent channels. The effect of nonlinearity on the error proba-bility is different for systems with constant amplitude and non-constant amplitude. (The first system comprises the PSK and various FM signal sets while the second system covers the signal sets with additional amplitude modulation; practically, these are the QAM signal sets.) In systems with constant amplitude, the effect of nonlinearities practically encountered is not too high; it will even be shown that in certain cases, the error probability can be improved by the nonlinearity of the receiver. In the case of QAM transmission, the error probability will be con-siderably increased by the nonlinear distortion of the channel. The spectrum modification also depends on the amplitude conditions (which is not so evident).

2.4.1 Nonlinearity of the transmission path

The "small" nonlinearities of the transmission channels have negligible effects only on the transmission performance of the digital transmission systems. There-fore, it is sufficient to apply linear models for several systems (as done earlier), and it is justified to neglect the nonlinear distortion of basically linear amplifiers, mixers and other circuits. However, there are two basically nonlinear circuits, the limiter and the transmitter power amplifier, widely applied in microwave trans-mission systems, which will be analyzed in the following (the latter operating in the vicinity of the saturation point).

The limiter can frequently be regarded as an ideal memoryless nonlinearity having the following output–input characteristic:

$$y = \begin{cases} \dfrac{C}{\lambda}x, & |x| \leqq \lambda \\ \pm C, & |x| > \lambda \end{cases} \tag{2.104}$$

where x stands for the input signal and y for the output signal. In the limiting case,

$\lambda=0$; this corresponds to the so-called hard limiter, whose characteristic can also be expressed in the following more convenient form:

$$y = C \operatorname{sign} x. \tag{2.105}$$

Practical limiters always have a bandpass response, and are made up of a nonlinear circuit defined by (2.104) or (2.105), followed by a so-called zonal bandpass filter for transmitting the spectrum around the fundamental frequency without distortion and attenuating harmonic frequency components. Thus if

$$x(t) = B(t) \cos\left[\omega_c t + \varphi(t)\right] \tag{2.105a}$$

then the output signal of a bandpass hard limiter is given by

$$y(t) = C \cos\left[\omega_c t + \varphi(t)\right]. \tag{2.105b}$$

It is seen that the amplitude fluctuation is eliminated but the phase information is retained without distortion. The bandpass hard limiter is a fairly good model of the injection locked amplifier which is sometimes applied at higher microwave frequencies.

It is feasible to analyze the effect of the transmitter output amplifier in two configurations. In radio relay systems, regenerator type repeater stations are frequently applied, corresponding to the block diagram shown in Fig. 2.39a. Satellite transponders are non-regenerative in most cases, so the complete satellite transmission system is represented by the block diagram shown in Fig. 2.39b. Note that the satellite system comprises two noise sources, the receiver input being made up of the sum of a noisy signal and the noise. In up-to-date systems, the noise source n_1, shown in Fig. 2.39b, is generally negligible as the earth station transmitter power is 30 to 35 dB higher than the satellite traveling wave tube output power. In other cases, n_1 and n_2 may have equal significances. It is possible that future systems will be characterized by the reverse situation and n_2 will be negligible; this might be the case of a high power satellite operating with very small earth stations.

The transmitter output amplifier, e.g. the traveling wave tube (TWT), is more complicated than the limiter since, in addition to the nonlinearity, it also has a substantial AM-to-PM conversion. These parameters can be represented by an output power which is nonlinearly related to the input amplitude, and by a phase shift depending on the input amplitude, as shown by the characteristics in Fig. 2.39c. This type of nonlinearity is best analyzed by the so-called quadrature model [49] shown in Fig. 2.39d. If

$$x(t) = B(t) \cos\left[\omega_c t + \vartheta(t)\right] \tag{2.105c}$$

then

$$y(t) = Z_P(B) \cos\left(\omega_c t + \vartheta\right) - Z_Q(B) \sin\left(\omega_c t + \vartheta\right). \tag{2.106}$$

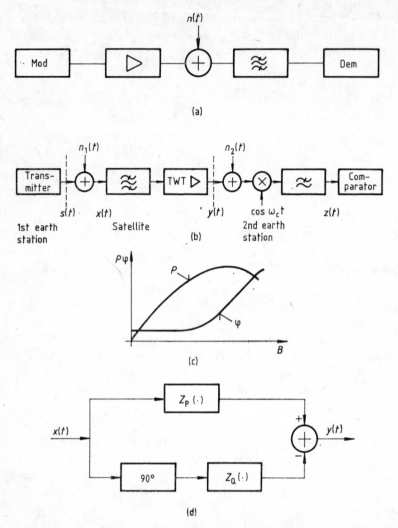

Fig. 2.39. (a) Single-hop system with nonlinear power amplifier; (b) satellite system with on-board TWT; (c) power and phase characteristic of the TWT; (d) model of a nonlinear amplifier.

In Ref. [49], the above Z functions are expressed by the following analytical form:

$$Z_P(B) = C_1 e^{-C_2 B^2} I_0(C_2 B^2)$$

$$Z_Q(B) = S_1 e^{-S_2 B^2} I_1(S_2 B^2)$$

$$(2.107)$$

where I_0 and I_1 are modified Bessel functions and C_1, C_2, S_1 and S_2 are experimental constants which can be determined from the tube characteristics. Transistor output amplifiers can be described by a similar model but have generally more favourable parameters.

There is a third type of nonlinearity, the so-called direct phase regenerator, which is sometimes applied to improve the transmission performance. The structure and operation of this nonlinearity are basically different from the structure and operation of the regenerator discussed in Chapter 1 and will be analyzed further in Chapter 4. The operation of the direct phase regenerator is based on the nonlinearity of the applied circuits and will be treated separately in Section 2.4.7.

It can be seen from Figs 2.39a and 2.39b that nonlinear circuits always have memories in spite of the fact that the nonlinear circuit itself is memoryless, as in most cases its bandwidth is much higher than $1/nT$; however, the memory is provided by the filters preceding and following the nonlinearity. Numerical results show that the effect of the memory is only essential for the system shown in Fig. 2.39b (satellite system without on-board processing). Therefore, the nonlinear amplifier in terrestrial systems will be taken as memoryless. Considering satellite systems, the next section presents an analysis of nonlinearities possessing memories.

2.4.2 Transmission of PSK signals over a channel comprising a bandpass hard limiter

The model shown in Fig. 2.39b will be investigated, utilizing a hard limiter instead of the traveling wave tube. For simplicity, only the BPSK system will be considered [50]; the more general case is investigated by an essentially similar method, *e.g.* in Ref. [51].

The noise process n_1 is Gaussian with standard deviation of σ_1. The complex amplitude of a sample function of n_1 is given by

$$n_1(t) = n_{c1}(t) + jn_{s1}(t) \tag{2.108}$$

while the transmitter signal is given by

$$s(t) = A \cos \Phi_k; \quad \Phi_1 = 0; \quad \Phi_2 = \pi; \quad k = 1, 2. \tag{2.109}$$

Assuming that the input filter of the satellite transponder has a transmission band wide enough to be neglected, the complex envelope of the limiter input signal is thus given by

$$x(t) = B(t)e^{j\vartheta(t)} \tag{2.110a}$$

where

$$B(t)^2 = (A \cos \Phi + n_{c1})^2 + n_{s1}^2 \tag{2.110b}$$

$$\text{tg } \vartheta(t) = \frac{n_{s1}}{A \cos \Phi + n_{c1}}. \tag{2.110c}$$

The complex envelope of the limiter output signal is given by

$$y(t) = Ae^{j\vartheta(t)} \tag{2.111}$$

while the complex expression for the receiver input noise is

$$n_2(t) = n_{c2}(t) + jn_{s2}(t) \tag{2.112}$$

and the comparator input signal is

$$z(t) = A \cos \vartheta(t) + n_{c2}. \tag{2.112a}$$

$n_2(t)$ is white Gaussian noise similarly to $n_1(t)$, with a standard deviation of σ_2. Its magnitude is normalized so as to obtain a receiver input amplitude equal to the transmitter signal amplitude A.

The signal source, as before, generates random signals of equal probability. The error probability is then given by

$$P_E = P[z < 0 | \Phi = 0]. \tag{2.113}$$

The calculation of this error probability requires the knowledge of the probability density of z which is the convolution of $p(\vartheta)$ and the Gaussian distribution of n_{c2}:

$$q(z) = \frac{1}{\sigma_2 \sqrt{2\pi}} \int_0^{2\pi} \exp\left[-\frac{(z + A \cos \vartheta)^2}{2\sigma_2^2}\right] p(\vartheta) \, d\vartheta \tag{2.114}$$

where $p(\vartheta)$ is the density function of the signal + noise phase at the limiter output; its expression can be found in Ref. [52]. Taking this into account, the error probability will be given by

$$P_E = \frac{1}{2} - \frac{1}{2} \int_0^{2\pi} \operatorname{erf}\left[\frac{A}{\sigma_2 \sqrt{2}} \cos \vartheta\right] p(\vartheta) \, d\vartheta \tag{2.115}$$

where $\operatorname{erf}(x) = 1 - \operatorname{erfc}(x)$ is the error function. It is seen from (2.115) without further calculation that if either the noise component n_1 at the limiter input or the noise component n_2 at the limiter output is zero, the limiter does not modify the error probability. This is seen from the following reasoning. If $n_1 = 0$ then $p(\vartheta) = \delta(\vartheta)$ thus

$$P_E = \frac{1}{2} \operatorname{erfc}\left[\sqrt{R_2}\right]; \quad R_2 = \frac{A^2}{2\sigma_2^2}. \tag{2.115a}$$

On the other hand, if $n_2 = 0$ then

$$z(t) = A \cos \vartheta = \frac{A + n_{c1}(t)}{B(t)}. \tag{2.115b}$$

The probability of this being negative is equal to the probability of $n_{c1} < -A$. This similarly produces the relation

$$P_E = \frac{1}{2} \operatorname{erfc}\left[\sqrt{R_1}\right]; \quad R_1 = \frac{A^2}{2\sigma_1^2}. \tag{2.115c}$$

The latter statements have fairly significant practical consequences. It is seen that in a PSK radio relay link, an injection locked oscillator as transmitter output amplifier or a limiter in the receiver can be applied without any harmful effects.

Considering now the general case in which both noise components differ from zero, (2.114) can be evaluated by a rather large number of difficult calculations presented in Ref. [50]. The expression of the error probability is given by

$$P_E = \frac{1}{2}\left[1 - \sqrt{1 - k^2} \, I_c(kx)\right] \tag{2.115d}$$

where

$$k = \frac{R_1 - R_2'}{R_1 + R_2'}; \quad x = \frac{R_1 + R_2'}{2};$$

$$R_1 = \frac{A^2}{2\sigma_1^2}; \quad R_2' = \frac{A^2}{2\sigma_2^2}; \tag{2.115e}$$

$$I_c(kx) \triangleq \int_0^x e^{-t} I_0(kt) \, dt.$$

It has to be emphasized that R_1 is actually a signal-to-noise ratio but R_2' is a (signal+noise) to noise ratio because $A^2/2$ in the expression of R_2' is the resultant power of the signal and noise components. The function I_c defined by Rice being relatively unknown, it is worthwhile presenting simplified expressions for a few special cases. For $R_1 = R_2' = R$, we have

$$P_E = \frac{1}{2} e^{-R};$$

for

$$R_i > R_k \gg 1 \quad \text{and} \quad R_i - R_k \gg 1; \quad (i, k = 0, 1) \tag{2.115f}$$

$$P_E = \frac{1}{2}\sqrt{\frac{R_i}{R_k}} \frac{e^{-R_k}}{\sqrt{\pi(R_i - R_k)}}.$$

Finally, for the case of $R_1 R_2' \ll 1$,

$$P_E \approx \frac{1}{2}\left(1 - \sqrt{R_1 R_2'}\right). \tag{2.115g}$$

In Figure 2.40, the error probability is plotted against R_1, R_2' being the parameter, showing also the error probability of the linear system. For very small

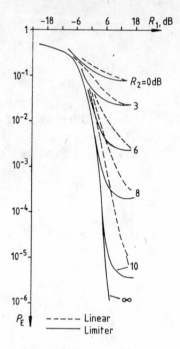

Fig. 2.40. Error probabilities of a repeater with an ideal limiter and of a linear repeater.

R values, the error probability of the linear system is given by

$$P_E \approx \frac{1}{2}\left(1 - \frac{2}{\sqrt{\pi}}\sqrt{R_1 R_2'}\right) \qquad (2.115\text{h})$$

which requires for a given error probability a product $R_1 R_2'$ which is 1 dB less than for the system with a limiter. However, it is seen from Fig. 2.40 that for large R values, the limiting system results in a lower error probability, and this is valid even for several limiting repeater stations and for M-ary systems with $M > 2$. It can thus be generally concluded that for non-regenerative repeater stations, the transmission performance is lower for stations with linear amplifiers as compared with stations with limiting amplifiers. Non-regenerative repeaters will be dealt with in Section 2.4.7.

2.4.3 Effect of a traveling wave tube on binary PSK transmission

In addition to the nonlinear effects treated in the previous section, the effect of AM-to-PM conversion will be investigated in the following. A memoryless AM-to-PM conversion, as defined in connection with Fig. 2.39, will be assumed. The signals and noises appearing at the individual terminals will again be described by their complex envelope. The given signal, noise n_1 and their sum $x(t)$ can again be calculated from expressions (2.108) to (2.110). The amplifier output signal is now

$$y(t) = [Z_P(B) + jZ_Q(B)]e^{j\vartheta(t)}. \tag{2.116}$$

Its "signal component" can be obtained by calculating the conditional expected value, the condition being a given Φ value. $\Phi = 0$ or π, with an *a priori* probability of 0.5:

$$s'(t) = A'e^{j\psi};$$

$$A' = \sqrt{E\{[Z_P(B)\cos\vartheta|\Phi]\}^2 + E\{[Z_Q(B)\cos\vartheta|\Phi]\}^2}$$

$$\operatorname{tg}\psi = \frac{E[Z_Q(B)\cos\vartheta|\Phi]}{E[Z_P(B)\cos\vartheta|\Phi]}. \tag{2.117}$$

Here $E\,[|]$ denotes, as earlier, conditional expectation.
The noise component of the nonlinear amplifier is given by

$$n_1'(t) = (Z_P + jZ_Q)e^{j(\vartheta - \psi)}n_{c1}' + jn_{s1}'. \tag{2.118}$$

At the receiver input, again the Gaussian noise $n_2(t)$ is added to the $y(t)$ signal, and finally, due to multiplication by $\cos\omega_c t$, the real part reaches the comparator input:

$$z(t) = A'\cos\psi + n_{c1}' + n_{c2}. \tag{2.119}$$

The error probability is again defined by expression (2.113). For the case investigated in Section 2.4.1, the probability density function of the overall noise and the error probability could be expressed in closed form (although the functions involved are not tabulated with suitable details). However, in this case, the probability density of both n_1' and $n_1' + n_2$ can be expressed by only an infinite series. In Ref. [49], the former is expressed with the aid of the Gram–Charlier series (see for instance Ref. [53]), and then derives the density function of $n_1' + n_2$. In a similar way, the error probability can also be expressed by an infinite series.

Figure 2.41 presents numerical results. In the example investigated, the input level of the TWT (Traveling Wave Tube) is 2.5 dB less than the saturation level ("back-off" of 2.5 dB); the complete system bandwidth is $3/T$ (a system with such a high bandwidth is practically memoryless). The receiver comprises a majority decision network, the decision being made following three samples (at the com-

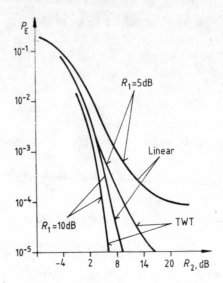

Fig. 2.41. Error probabilities of satellite links with a linear transponder and with a transponder comprising a TWT.

parator output in Fig. 2.39b). Under these circumstances, when the signal time is T, and the bandwidth is three times $1/T$, the samples taken per time interval of $T/3$ are statistically independent. The error probability calculated from nonlinear theory is always less than the error probability of a linear receiver with the same decision network. On the other hand, if the nonlinear system is compared with a linear system with matched filter (the latter having a bandwidth which is the third of the nonlinear system bandwidth), an intersection of the two error probability curves will be found. A lower error probability is obtained for small signal-to-noise ratios with the nonlinear system and for high signal-to-noise ratios with the matched filter system.

Note that in Fig. 2.41 we have R_2' as compared with R_2 in Fig. 2.40. R_2' is the ratio of the signal power defined in formula (2.117) to the receiver input noise. This difference in parameters explains the fact that in Fig. 2.41, the nonlinearity always improves the error probability while in Fig. 2.40, the linear system is better than the nonlinear system for small $R_1 R_2$ values.

The situation is much more complicated if the memory has to be taken into account (*i.e.*, the filter bandwidth hardly exceeds $1/nT$). This problem is treated for example by Ref. [55]. The investigated model is shown in Fig. 2.42. The problem can be solved for example by the method of Volterra series. Without going into detail, Figs 2.43a and 2.43b show the error probability of a linear and a nonlinear QPSK system (based on Ref. [55]). System parameters: $B_1 = 1.8/2T$,

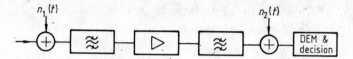

Fig. 2.42. Satellite link with on-board nonlinearity having memory.

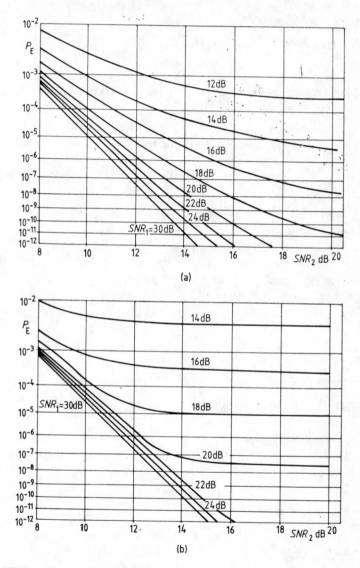

Fig. 2.43. QPSK error probability *versus* down-link signal-to-noise ratio, with up-link signal-to-noise ratio as parameter: (a) linear system; (b) nonlinear system.

$B_2 = 1.1/2T$ (both are Butterworth filters). The traveling wave tube amplifier is characterized by a polynomial of seventh degree. The independent variable is the down-link signal-to-noise ratio, and the parameter in the up-link signal-to-noise ratio. Comparison of the two figures shows that the error probability of the linear channel differs only slightly from that of the nonlinear channel for large (e.g., higher than 24 dB) signal-to-noise ratios, i.e., if the up-link noise is negligible. This is frequently the case for satellite systems, and is always valid for terrestrial radio relay systems corresponding to Fig. 2.39a.

2.4.4 Determination of the optimum nonlinear characteristic

It has been shown in the preceding two sections that in the satellite system shown in Fig. 2.39b, the performance of the nonlinear channel can be better than that of the linear channel. On the other hand, the actual characteristic is practically given, e.g. according to the characteristic of the traveling wave tube given by (2.107). In this section, we will investigate whether it is possible to find an optimum shape of the characteristic, and the transformation of the given characteristic into the optimum characteristic will be detailed [54].

Let us first investigate the case without AM-to-PM conversion; this means $Z_Q = 0$ with the notation of (2.107). The task is thus to find a characteristic of a nonlinear non-regenerative repeater station which will minimize the error probability of a PSK system. Suitably normalizing certain variables and taking into account the expression for $q(z)$ given in (2.114), the error probability will be given by

$$P_E = \frac{1}{2} \frac{e^{-R_1}}{8\pi R_1} \int_0^{2\pi} \int_0^\infty b \exp\left[\frac{-b^2 - 4bR_1 \cos \vartheta}{4R_1}\right] \mathrm{erf}\left[\frac{h(b)}{\sqrt{2}} \cos \vartheta\right] db\, d\vartheta. \quad (2.119a)$$

This expression has to be minimized by a suitable choice of $h(b)$ so as to provide a constant expected value for the satellite transponder output power. Here b is proportional to the amplitude B, $b = AB/\sigma_1^2$. To find the suitable form of $h(b)$ is a so-called isoperimetric problem of variational calculus. However, the functional which is to minimize being a double integral, the solution of the problem yields an integral equation for $h(b)$. It is difficult to find a general solution for this equation but if it is assumed that $R_1 \gg R_2$ then

$$h(b) = \text{constant } I_1(b)/I_0(b). \quad (2.119b)$$

This function, which has a specific form which depends on σ_1, is similar to the transfer function of a soft limiter; with increasing R_1, the function approximates the function of a hard limiter. It can also be shown that the same function will maximize the signal-to-noise ratio at the repeater station output. On the other hand, if $R_2 \gg R_1$, $h(b)$ tends asymptotically to a smaller positive constant. This is fairly similar (in an interesting way) to the output/input characteristic of the traveling wave tube amplifier.

Similar calculations can be performed for nonlinearities having AM-to-PM conversion. In this case, however, no analytical closed expression can be given for the error probability, thus allowing no minimization procedure to be carried out; instead of this, the signal-to-noise ratio can be maximized. In this procedure, the result obtained for the limiter is quasi extrapolated. It has been previously found for the limiter that the function minimizing the error probability will also maximize the signal-to-noise ratio. It will be assumed that this property holds for the present case as well. The result of optimization (see Ref. [54]) is fairly close to the previous one:

$$Z_{P\,opt} = I_1(b)/I_0(b); \tag{2.120a}$$

$$Z_{Q\,opt} = \beta Z_{P\,opt}. \tag{2.120b}$$

The latter result is rather surprising. The quadrature component, responsible for the AM-to-PM conversion, does not have to disappear identically, but it is sufficient that this component is proportional to the optimum Z_p. Of course, β may have the value of zero which corresponds to the condition of no conversion, expected to be optimal.

The relations (2.120) allow compensation of a given traveling wave tube nonlinearity: a predistortion circuit, yielding together with the tube characteristic an optimum overall characteristic, has to be applied at the amplifier input. There are several difficulties in realizing this characteristic, the most important being the fact that it is a function of the signal-to-noise ratio. The actual (noise independent) characteristic can thus only approximate the optimum. A signal-to-noise reduction of about 0.5 dB can be obtained by an actually realized suboptimal compensation. A possible variant of the compensation circuit will be presented in the next section.

2.4.5 Nonlinear power amplifier in QAM systems

The QAM system, or more generally systems also applying amplitude modulation, differ substantially from the systems having constant amplitude previously investigated. In the latter, the memoryless nonlinearity frequently has only second-order, sometimes improving, effects. The effect of nonlinearities with memory is also small if the up-link noise is negligible. In QAM systems, the output amplifier operating near the saturation region has the effect of shifting the position of the

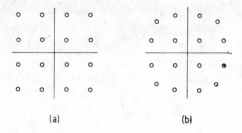

(a)   (b)

Fig. 2.44. 16QAM signal set: (a) without distortion; (b) with distortion.

signal vectors having higher amplitude, thus decreasing the distance between neighbouring signal vectors and substantially reducing the transmission performance.

Figure 2.44 shows an ideal and a distorted QAM signal space diagram. The analytical investigation of this kind of distortion would be very complicated but there are a few simulation procedures (see for instance Ref. [56]) from which the distortion mechanism can be understood. The nonlinear distortion will increase the error probability, primarily because of the situation shown in Fig. 2.44. Specifically, both the AM compression and the AM-to-PM conversion will contribute to the increase of the error probability, not just in an additive way, but in a synergistic manner. Figure 2.45 shows the performance degradation of different amplifiers as a function of the back-off. By "performance degradation" is meant the signal-to-noise ratio increase ΔR which is needed to obtain a given error probability (in our example 5×10^{-4}). By "back-off" is meant the power difference by which the operational output power is less than the saturation power of the amplifier. (The back-off is zero when the power of a signal of maximum amplitude coincides with the saturation power.)

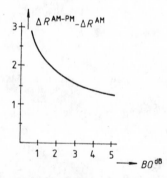

Fig. 2.45. Synergistic effect of AM compression and AM-to-PM conversion in 16QAM performance degradation, for a particular traveling wave tube.

The figure shows that the application of back-off is an effective way of reducing the nonlinear distortion. However, it should be taken into account that while ΔR decreases monotonically with the increasing back-off (the amplifier becomes more and more linear with reduced output power), the back-off has an optimum value regarding transmission performance. This is explained by the fact that a decreasing transmitter power decreases the signal-to-noise ratio R at the receiver input. Near saturation, ΔR is reduced more rapidly while the reduction is negligible at high back-off values. Thus best transmission performance is characterized by a minimum value of the sum $\Delta R + BO$. Such a simulation result is shown in Fig. 2.46, based again on Ref. [56]. It is seen that the optimum back-off is about 3 dB but even here the loss (due to performance degradation) is nearly 6 dB.

The above loss can be reduced further by applying a pre-distortion circuit preceding the amplifier, suitably in the i.f. or baseband part, as mentioned in the previous section for other purposes (see also Refs [57, 58, 59, 60]). Figure 2.47 shows a relatively simple pre-distortion circuit [57] which is suitable for fully compensating either the effect of the AM compression or the AM-to-PM conversion or, to some extent, the simultaneous effect of both phenomena.

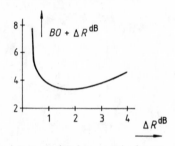

Fig. 2.46. *BO + ΔR versus ΔR* for a particular traveling wave tube.

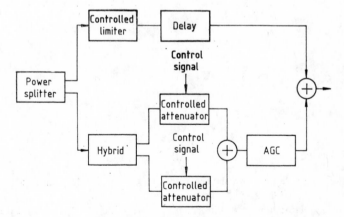

Fig. 2.47. A pre-distortion circuit.

2.4.6 Effect of the channel nonlinearity on the signal sequence spectrum

In some cases, the link error probability is improved by the nonlinearities, but at the same time, the spectrum is widened, thus increasing adjacent channel interference. A few examples were given in Section 2.3.1 to illustrate the selection of the transmit filter characteristic for obtaining reasonable channel spacing, without excessive interchannel or intersymbol interference. Evidently, this filter will limit the spectrum of the radiated signal. In the following, we will investigate how far this limited spectrum will again be widened if the nonlinear circuit (amplifier) lies between the transmit filter and the antenna.

We will show from this viewpoint, that the amplitude change due to the band limiting has an exceptionally high significance. The qualitative relations are illustrated in Fig. 2.48, which shows a binary PSK signal with and without filtering. At the 180° phase jump, the filter has the effect of reducing the amplitude to zero. Driving a hard limiter with this signal, the unfiltered waveform shown in Fig.

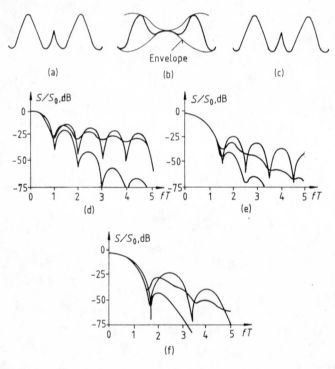

Fig. 2.48. Effect of nonlinearity on the spectrum and on the signal waveform: (a) PSK time function; (b) filtered version of the above function; (c) as for (b), but at the output of a hard limiter; (d) PSK spectra: unfiltered, filtered, filtered with additional effect of TWT; (e) and (f) the same spectra for MSK and SFSK signals.

2.48a and consequently also its spectrum are practically restored. This result has been fully verified by detailed calculation; this is illustrated in Fig. 2.48d showing the spectra of an ideal PSK signal, and also the distorted spectra after the signal has traversed a four-element Butterworth filter, a limiter and a traveling wave tube amplifier.

We could conclude from these figures that from the point-of-view of spectrum spread, the signal sets exhibiting large amplitude changes at phase jumps are especially unfavourable. Let us investigate the most frequently applied QPSK signal set. If the phase of a dibit differs by 180° from the preceding one, the amplitude is decreased to zero. If the phase jump between consecutive dibits is only ±90° then we have an amplitude reduction of only 3 dB (because one of the quadrature components is decreased to zero while the other remains unchanged). It can thus be concluded that the structure of the QPSK signal should be modified so that only one of the quadrature components changes phase at a time. This signal set is called offset PSK: its dibits are halved, and the phase changes of one of the quadrature components occur only at the dibit edges, while the phase changes of the other component occur at the dibit centre. It can be shown that the spectrum of an offset PSK signal sequence is equal to that of the conventional QPSK sequence but the spectrum spread at the filter and limiter output is much less than for a QPSK sequence.

The amplitude of the MSK and SFSK signals discussed in Section 2.3.1 is practically unchanged at the phase changes. Thus the spectrum spread will also be small as shown in Figs 2.48e and 2.48f.

2.4.7 Direct phase regenerators

In this section, a special nonlinear signal processing circuit, improving definitely the transmission performance of PSK systems, will be investigated.

The receiver of the digital transmission system presented in Chapter 1 performs the following operations: it amplifies the received signal, transfers it back to the baseband (*i.e.,* it is demodulated), decides symbol by symbol what the transmitted symbol was, and finally generates a regenerated signal as a result of this decision. As explained in Section 2.1.3, the decision requires the knowledge of the timing. In Chapter 3, methods for clock signal recovery from the received signal, *i.e.* timing recovery, will be presented.

In multi-hop transmission systems, the above operations have to be repeated at each receiver: each repeater station has to apply a demodulator, a clock recovery circuit, a decision device (*i.e.,* a regenerator), and a modulator. Terrestrial radio relay systems perform this procedure almost without exception, in spite of the fact that the above circuits are rather expensive, and have rather large dimensions and rather high power consumption. However, as will be shown in Chapter 4, carrier frequencies above 10 to 15 GHz require that the repeater spacing should be less and less, which means that their number will be higher and higher. The cost of regenerative type repeater stations above 100 to 200 Mbit/s becomes excessive, so that a limit is reached above which the exclusive use of regenerative repeater stations will not be cost effective. Even at lower carrier frequencies and lower bit rates, the use of non-regenerative type repeaters may be justified by the reduction of the dimensions and the power consumption. In satellite transmission systems, a regenerative transponder cannot be accommodated on board since this would bring about excess weight.

In these cases, a trivial solution would be the omission of the demodulator, the baseband regenerator and modulator, and the application of linear repeater stations, thereby tolerating the reduced transmission performance. However, the performance of non-regenerative phase modulated systems can be substantially improved without excessive additional cost, by supplementing the linear amplifier by a so-called direct phase regenerator [61—65]. This is a nonlinear circuit limiting the amplitude and having a phase which has two possible states in the BPSK system and four possible states in the QPSK system. Figure 2.49 shows the ampli-

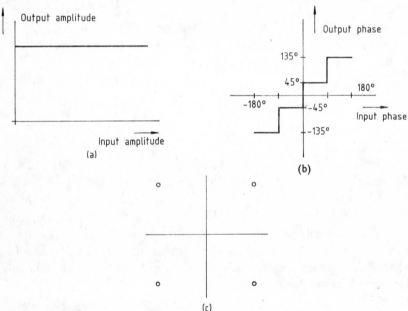

Fig. 2.49. (a) Ideal amplitude characteristic of direct phase regenerator; (b) ideal phase characteristic of direct phase regenerator; (c) signal space diagram of QPSK for which the above characteristics apply.

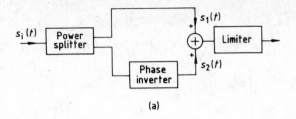

(a)

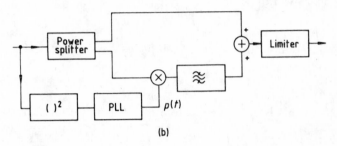

(b)

Fig. 2.50. BPSK direct phase regenerator: (a) principle; (b) block diagram.

tude and phase characteristics of an ideal QPSK direct phase regenerator, together with the signal space diagram corresponding to the four states. The binary variant has the same amplitude characteristic and a two-state phase characteristic (*e.g.*, 0 and π). The figure shows the advantage and disadvantage of the direct regenerator: the amplitude and phase errors due to noise, intersymbol interference and other causes are equalized but the correct symbol durations will not be restored.

Binary and quaternary phase regenerators have been described, in Refs [61 and 62] among others. Figure 2.50a shows the equivalent circuit while in Fig. 2.50b, the phase inverter is also shown to some detail.

In the following, the direct phase regenerator in the ideal case will be analyzed.

Let the input signal be expressed by

$$s_1(t) = a(t) \cos [\omega_c t + \varphi(t)] \tag{2.121}$$

where $a(t)$ is the amplitude fluctuating because of the filter distortion and the additive noise, and $\varphi(t)$ is the phase comprising the useful modulation and also the phase noise. If the amplitudes of $s_1(t)$ and $s_2(t)$, defined in Fig. 2.50a, are equal then the limiter input signal will be given by

$$s_1(t) + s_2(t) = b(t) \cos [\omega_c t + \varphi(t)] + b(t) \cos [\omega_c t - \varphi(t)] =$$
$$= b(t) \cdot \cos \omega_c t \cdot \cos \varphi(t). \tag{2.122}$$

The latter can be written in the following form:

$$s_1 + s_2 = b(t)|\cos \varphi(t)| \cos (\omega_c t + \varPhi) \tag{2.123}$$

where

$$\varPhi = \begin{cases} 0; & \cos \varphi(t) \geqq 0; & |\varphi(t)| \leqq \dfrac{\pi}{2} \\[2mm] \pi; & \cos \varphi(t) < 0; & \dfrac{\pi}{2} < |\varphi(t)| \leqq \pi. \end{cases} \tag{2.124}$$

If this signal is used to drive a hard limiter then the following regenerated limiter output signal will appear:

$$s_0(t) = \cos (\omega_c t + \varPhi). \tag{2.125}$$

The phase inverter has the effect of multiplying the signal proportional to $s_i(t)$ by the reference signal

$$p(t) = 2 \cos 2\omega_c t. \tag{2.126}$$

The product signal is then directed to a zonal filter (see Fig. 2.50b). Actually,

$$s_2(t) = \{2b(t) \cos 2\omega_c t \cos [\omega_c t + \varphi(t)]\}_{\text{f.c.}} = b(t) \cos [\omega_c t - \varphi(t)] \tag{2.127}$$

where f.c. denotes the fundamental component. The generation of the reference signal $p(t)$ is similar to the generation of the "phase information" in the coherent system. This will be treated in Chapter 3.

The practical operation is not fully covered by the above idealized analysis: the amplitudes of s_1 and s_2 may differ, and the limiter may have finite AM-compression and some AM-to-PM conversion. The AM compression allows the change of amplitude, while the AM-to-PM conversion has the effect of distorting the phase positions.

A QPSK direct regenerator may be derived from the parallel connection of two BPSK regenerators according to Fig. 2.51.

In the simplest case, the direct phase regenerator may be analyzed according to the arrangement shown in Fig. 2.52. In the two-hop transmission system shown, the first receiver comprises a direct regenerator while the second receiver comprises a baseband regenerator, the latter also having the function of timing recovery.

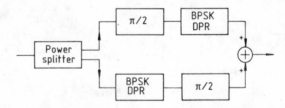

Fig. 2.51. Principle of QPSK direct phase regenerator.

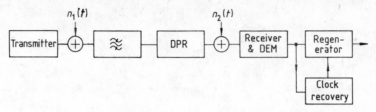

Fig. 2.52. Two-hop link comprising direct phase regenerator.

In order to correspond to the actual circuit operation, the analysis should take into account the effect of the filters, thus the memory of the nonlinearity. This investigation could probably be carried out by a Volterra series procedure which was briefly mentioned in Section 2.4.2. To the knowledge of the authors, such an analysis has not yet been published in the literature. For instance, Ref. [62] investigates the simultaneous effect of the up-link and down-link Gaussian noise while in Ref. [63], the effect of the phase noise in the reference signal $p(t)$ is investigated. The method of investigation is, in principle, similar to the method presented in Section 2.2.3 and 2.3.2: the conditional error probability is calculated from the integral of the density with conditional Gaussian distribution, taken for the appropriate domain of the signal space. This is then averaged from the knowledge of the conditional density function. In both references, the regenerator is assumed to be memoryless, and the (non-Gaussian) interchannel interference is not covered. This kind of investigation is only a rough approximation of the actual conditions and therefore will not be presented in detail.

It will be shown in Chapter 4 that in terrestrial systems, the simultaneous appearance of substantial thermal noise in more than one receiver has a negligible probability. Therefore, in a two-hop link, only the interchannel and intersymbol interferences give rise to phase fluctuation which has to be suppressed by the regenerator in one of the hops, while in the other hop the phase fluctuation is generated by the effect of these and of the thermal noise. The substitution of this complex procedure by a Gaussian process is an oversimplification.

Numerical results show that in the case of Gaussian up-link and down-link noise, the performance of the ideal direct phase regenerator is only slightly worse than that of the baseband regenerator. To obtain a given error probability, the signal-to-noise ratio has to be increased by only a few tenths of a dB.

2.5 References

[1] Wozencraft, J., Jacobs, J. M.: *Principles of Communication Engineering.* Wiley, New York 1965.

[2] Hellstrom, C.: *Statistical Theory of Signal Detection.* Pergamon Press, New York 1960.

[3] Middleton, D.: *Introduction to Statistical Communication Theory.* McGraw-Hill, New York 1966.

[4] Schwartz, M., Benett, W., Stein, S.: *Communication Systems and Techniques.* McGraw-Hill, New York 1966.

[5] Weber, C. L.: *Elements of Detection and Signal Design.* McGraw-Hill, New York 1968.

[6] Viterbi, A., Omura, J.: *Principles of Digital Communication and Coding.* McGraw-Hill, New York 1979.

[7] Arthurs, E., Dym, H.: On the optimum detection of digital signals in the presence of white Gaussian noise. *IRE Trans. Commun. Syst.,* Vol. CS-10, pp. 336–372, Dec. 1962.

[8] De Buda, W.: The Fast FSK Modulation System, *ICC 72,* pp. 41.25–41.27, 1972.

[9] Spilker, J.: *Digital Communication by Satellite.* Prentice-Hall, Englewood-Cliffs 1977.

[10] Osborne W., Lutz, M.: Coherent and noncoherent detection of CPFSK. *IEEE Trans. Commun.,* Vol. Com-22, No. 8, pp. 1023–1036, Aug. 1974.

[11] Lindsey, W., Simon, M.: *Telecommunication Systems Engineering.* Prentice-Hall, Englewood-Cliffs 1973.

[12] Fleck, A., Trabka, K.: *Error Probabilities of Multiple-State Differentially Coherent PSK Systems.* ASTIA Document No. AD 256 584, 1961.

[13] Miller, L., Lee, J.: The probability density function of the output of an analog cross correlator. *IEEE Trans. Inf. Theory,* Vol. IT-20, pp. 433–490, July 1974.

[14] Bussgang, J., Leiter, M.: Error rate approximations for differential phase shift keying. *IEEE Trans. Commun. Syst.,* Vol. CS-12, pp. 18–27, March 1964.

[15] Frigyes, I., Szabó, Z.: A simple method for the transmission of binary information. In: *Proc. 4th Colloq. Microwave Commun., Budapest,* Vol. 1, pp. ST-11/1–ST-11/9, 1970.

[16] Balakrishnan, A.: A contribution to the sphere packing problem of communication theory. *J. Math. Anal. Appl.,* Vol. 3, pp. 485–506, Dec. 1961.

[17] Salz, J., Sheehan, J., Paris, D.: Data transmission by combined AM and PM. *Bell Syst. Tech. J.,* Vol. 50, pp. 2399–2419, Sept. 1971.

[18] Landau, H. J., Pollak, H. O.: Prolate spheroidal wave functions, Fourier analysis and uncertainty, III. *Bell. Syst. Tech. J.,* Vol. 41, pp. 1295–1336, July 1962.

[19] Lucky, R., Salz, J., Weldon, E.: *Principles o Data Communication.* McGraw-Hill, New York 1967.

[20] Nyquist, H.: Certain topics in telegraph transmission theory. *Trans. AIEE,* Vol. 47, pp. 617–644, Apr. 1928.

[21] Sunde, E. D.: Theoretical fundamentals of pulse transmission. *Bell Syst. Tech. J.,* No. 4, pp. 987–1010, July 1954.

[22] Bennett, W., Davey, J.: *Data Transmission.* McGraw-Hill, New York 1965.

[23] Gobby, R., Smith, W.: Some extensions of Nyquist's telegraph theory. *Bell Syst. Tech. J.,* Vol. 44, pp. 1487–1510, Sept. 1965.

[24] Lind, L., Nader, S.: Design tables for a class of data transmission filters. *Electron Lett.,* Vol. 13, No. 19, pp. 564–566, Sept. 1977.

[25] Nader, S., Lind, L.: Optimum data transmission filters. *IEEE Trans. Circ. Syst.,* Vol. CAS-26, pp. 36–45, Jan. 1979.

[26] Shimbo, O., Fang, R., Celebiler, M.: Performance of M-ary PSK systems in Gaussian noise and intersymbol interference. *IEEE Trans. Inf. Theory*, Vol. IT-19, pp. 44–58, Jan. 1973.

[27] Benedetto, S., Vincentiis, C., Luvison, A. A.: Error probability of intersymbol interference and additive noise for multilevel digital signals. *IEEE Trans. Commun.*, Vol. Com-21, pp. 181–190, March 1973.

[28] Saltzberg, B.: Intersymbol interference error bounds with application to ideal bandlimited signalling. *IEEE Trans. Inf. Theory*, Vol. IT-14, pp. 563–568, July 1968.

[29] Prabhu, V.: Error probability performance of M-ary CPSK systems with intersymbol interference. *IEEE Trans. Commun.*, Vol. Com-21, pp. 97–109, Febr. 1973.

[30] George, D., Bowen, R., Storey, J.: An adaptive decision feedback equalizer. *IEEE Trans. Commun.*, Vol. Com-19, pp. 281–293, June 1971.

[31] Gitling, R., Ho, E., Mazo, J.: Pass-band equalizer of differentially phase modulated data signals. *Bell Syst. Tech. J.*, Vol. 52, pp. 219–238, Febr. 1973.

[32] Hartmann, P., Allen, E.: An adaptive equalizer of multipath distortion in a 90 Mbit/s 8PSK system. *ICC 79 Boston*, Vol. 1, pp. 5.6.1–5.6.5, June 1979.

[33] Titsworth, J., Welch, R.: *Power spectra of signals modulated by random and pseudorandom sequences*. Jet Propulsion Lab. Tech. Report, No. 32–140, Oct. 1961.

[34] Marsan, A., Biglieri, E.: Power spectra of complex PSK for satellite communication. *Alta Freq.*, Vol. XLVI, No. 6, pp. 123E–131E, Giugno, 1977.

[35] Amoroso, F.: Pulse and spectrum manipulation in the MSK format. *IEEE Trans. Commun.*, Vol. Com-24, pp. 381–384, March 1976.

[36] Fang, R., Shimbo, O.: Unified analysis of a class of digital systems in additive noise and interference. *IEEE Trans. Commun.*, Vol. Com-21, pp. 1075–1091, Oct. 1973.

[37] Benedetto, S., Biglieri, E., Castellani, V.: Combined effects of intersymbol, interchannel and co-channel interferences in M-ary CPSK systems. *IEEE Trans. Commun.*, Vol. Com-21, pp. 997–1008, Sept. 1973.

[38] Prabhu, V.: Bandwidth occupancy in PSK systems. *IEEE Trans. Commun.*, Vol. Com-24, pp. 456–462, Apr. 1976.

[39] Celebiler, M., Coupé, G.: Effects of thermal noise, filtering and co-channel interference on the probability of error in binary coherent PSK systems. *IEEE Trans. Commun.*, Vol. Com-26, pp. 257–267, Febr. 1978.

[40] Jeruchim, D.: A survey on interference problems and applications to geostationary satellite networks. *Proc. IEEE*, Vol. 65, pp. 317–331, March 1977.

[41] Special issue on spread spectrum communications. *IEEE Trans. Commun.*, Vol. Com-25, Aug. 1977.

[42] Dixon, R.: *Spread Spectrum Systems*. Wiley, New York 1976.

[43] Utlaut, J.: Spread spectrum. *IEEE Com. Soc. Magazine*, pp. 21–31, Sept. 1978.

[44] Eckert, A., Kelly, J.: Implementing spread spectrum technology in land mobile radio services. *IEEE Trans. Commun.*, Vol. Com-25, pp. 867–869, Aug. 1974.

[45] Holmes, J.: *Coherent Spread Spectrum Systems*. Wiley, New York 1982.

[46] Dixon, R.: Spread spectrum techniques. *IEEE Press*, 1978.

[47] Special issue on spread spectrum systems. *IEEE Trans. Commun.*, Vol. Com-30, May, 1982.

[48] *IEEE MILCOM-82 Conf. Record., Boston, Mass.*, Oct. 1982.

[49] Hetrakul, P., Taylor, D.: The effects of transponder nonlinearity on binary CPSK signal transmission. *IEEE Trans. Commun.*, Vol. Com-24, pp. 546–553, May 1976.

[50] Jain, P., Blachman, N.: Detection of a PSK signal transmitted through a hard-limited channel. *IEEE Trans. Inf. Theory*, Vol. IT-19, pp. 623–630, Sept. 1973.

[51] Mizuno, T., Morinaga, N., Namekawa, T.: Transmission characteristics of an M-ary

coherent PSK signal via a cascade of N bandpass hard limiters. *IEEE Trans. Commun.*, Vol. Com-24, pp. 540–545, May 1976.

[52] Schwartz, M.: *Information-Transmission, Modulation and Noise.* McGraw-Hill, New York 1970.

[53] Marcum, J.: A statistical theory of target detection by pulsed radar. *IRE Trans. Inf. Theory,* Vol. IT-6, pp. 59–267, Apr. 1960.

[54] Hetrakul, P., Taylor, D.: Compensation for band-pass nonlinearities in satellite communication. *IEEE Trans. Aerosp. Electron. Syst.*, Vol. AES-12, pp. 509–514, July 1976.

[55] Benedetto, A.: Optimization and performance evaluation of digital satellite transmission systems. In: *Proc. 6th Summer Symp. Circuit Theory, Prague,* pp. 187–205, 1982.

[56] Amadesi, J.: Including a nonlinear amplifier and a predistorter in a bandlimited 16QAM system. *ICC 83, Boston,* paper No. 25.7, 1983.

[57] Schwarz, W., Slade, R., Kenny, J.: Radio repeater design for 16QAM, *ICC 81, Denver,* paper No. 13.5, 1981.

[58] Kenney, J.: Digital radio for 90 Mbit/s 16QAM. *Microwave J.,* pp. 71–80, Aug. 1982.

[59] Gerard, D., Feher, I.: A new baseband linearizer, *GLOBECOM 82, Miami, Florida,* paper No. 15.2, 1982.

[60] Takenaka, S., Takeda, J., Sakane, T., Nakamura, M., Toyonaga, N.: A new 4 GHz 90 Mbps radio system using 64QAM. *ICC 84, Amsterdam,* paper No. 22.3, 1984.

[61] Komaki, S., Kurita, O., Mamita, O.: GaAs MESFET regenerator for PSK signals at the carrier frequency. *IEEE Trans.,* Vol. MTT-24, pp. 367–372, June, 1976.

[62] Komaki, S., Kurita, O., Mamita, O.: 400 Mbit/s QPSK MIC regenerator at the carrier frequency using GaAs MESFET. *IEEE Int. Symp.,* pp. 326–328, 1976.

[63] Komaki, S., Akeyama, K., Kurita, O.: Direct phase regenerator of a 400 Mbit/s QPSK signal at 1.7 GHz. *IEEE Trans.,* Vol. Com-27, pp. 1829–1836, Dec. 1979.

[64] Mattheis, N., Riris, A.: Performance analysis of QPSK transmission through a direct phase regenerator. *IEE Proc.,* Vol. 128, Part F, No. 2, pp. 91–95, Apr. 1981.

[65] Mattheis, N., Riris, A.: Performance degradation of two-link binary CPSK system incorporating a direct phase regenerator. *Electron. Lett.,* Vol. 16, No. 17, pp. 650–651, 14th Aug. 1980.

CHAPTER 3

CARRIER AND CLOCK RECOVERY

3.1 Introduction

Figure 2.2 of Chapter 2 shows the principle of digital transmission corrupted by Gaussian noise. This figure shows a separate connection between transmitter and receiver, in parallel with the main path, intended for the transmission of the bit (or symbol) timing information. In the discussion of the optimal receiver, the utilization of this information has also been given (*e.g.,* in a matched filter realization, the sampling gate has to be closed at the time instant $t=T$). Figure 2.13 shows the principle of coherent transmission, comprising a further connection for the transmission of the carrier phase information. In this chapter, the realization of these further connections will be discussed.

In the practical realization, the receiver should have suitable signal sources exactly repeating the corresponding transmit side signals; not only the frequency but also the phase of the transmit and receive side signals should be exactly identical in the ideal case. Usually, a square wave is used as a clock source and a sine wave is used as a carrier source. This requires the synchronization of the receive side signal sources (in most cases, oscillators) to the reference signal which should be made available for this purpose. In practice, the exact coherence of the signals can only be approximated, resulting in a degradation of the transmission performance.

In accordance with the above, the main parts of this chapter are the following: the characteristics of the widely used synchronization system, the phase-locked loop will be surveyed, together with the special PLL variant, the digital phase-locked loop. Following this, the main problems of carrier and clock synchronization will be dealt with, and finally, the effect of synchronization shortcomings on the transmission performance will be treated.

3.2 Phase-locked loops

The phase-locked loop, having the well known abbreviation PLL, is widely used in telecommunications, control, measurement and other fields of electronics, and has a significant role in digital transmission too. The theory of phase-locked loops has been thoroughly investigated during the recent 10 to 15 years, manifested by many papers and other publications. Most detailed treatments are given in References [1, 2, 3 and 8]. In the following, a few results will be presented without detailed theoretical investigation.

The principle of an analog phase-locked loop is shown in Fig. 3.1. The input signal is expressed by

$$s(t) = A\sqrt{2} \sin{[(\omega_c t + \vartheta(t)]}$$ (3.1)

where the amplitude of the signal is considered to be constant, and it is assumed that the rate of change of the phase function $\vartheta(t)$ is slow with respect to the carrier frequency. The noise process $n(t)$ is assumed to be a narrow-band process with a Gaussian distribution. The low-pass filter $F(p)$ is called a loop filter, p denoting the operator d/dt. It is also assumed that the voltage controlled oscillator, denoted by VCO, has a linear control characteristic, *i.e.* its frequency change is proportional to the control signal. The multiplying circuit at the loop input is also regarded as ideal.

The input narrow-band noise process can be expressed as follows:

$$n(t) = n_c \cos \omega_c t - n_s \sin \omega_c t = N_c \cos{[\omega_c t + \vartheta(t)]} - N_s \sin{[\omega_c t + \vartheta(t)]},$$ (3.2)

where

$$N_c(t) = n_c \cos \vartheta(t) - n_s \sin \vartheta(t)$$
$$N_s(t) = n_c \sin \vartheta(t) + n_s \cos \vartheta(t).$$ (3.3)

The oscillator output signal is given by

$$r(t) = \sqrt{2} K_1 \cos{[\omega_c t + \hat{\vartheta}(t)]}; \quad \hat{\vartheta}(t) = K_v \int^t z(u)\, du$$ (3.4)

where K_v denotes the tuning slope of the oscillator.

The error signal is expressed by

$$\varepsilon(t) = K_m r(t) s(t)$$ (3.5)

and comprises DC and second harmonic frequency terms. It is assumed that the loop filter has a high attenuation at frequency $2\omega_c$ so the essential part of $\varepsilon(t)$ will be proportional to the sine of $\varphi(t)$ which represents the phase difference between $s(t)$ and $r(t)$. It follows from (3.4) and (3.5) that

$$\varphi(t) \triangleq \vartheta(t) - \hat{\vartheta}(t) = \vartheta(t) - \frac{KF(p)}{p}\{A \sin \varphi(t) + N[t, \varphi(t)]\}$$ (3.6)

where $K=K_1K_vK_m$; and $N\overset{\triangle}{=}N_c\cos\varphi-N_s\sin\varphi$; N is called phase noise process.

Equation (3.6) is a nonlinear stochastic integro-differential equation for $\varphi(t)$. Preceding its interpretation, the baseband equivalent circuit of the PLL is shown in Fig. 3.1b which is more meaningful than Fig. 3.1a. The $1/p$ block, representing integration, is justified because the frequency of $r(t)$ is proportional to $z(t)$ while in the multiplying circuit, the phases of s and r are compared.

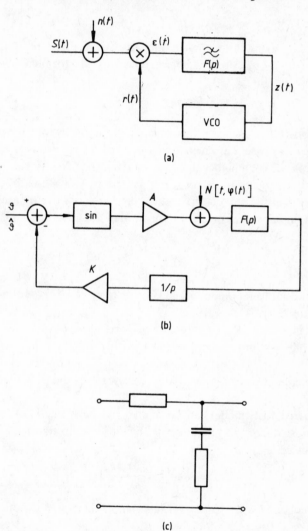

Fig. 3.1. (a) Phase-locked loop; (b) baseband equivalent circuit; (c) loop filter with non-ideal integration.

The phase-locked loops are normally classified according to the type of the loop filter $F(p)$. If $F(p)=1$, *i.e.* there is no loop filter, a first-order differential equation is obtained for $\varphi(t)$ by differentiating (3.6). This kind of loop is called a first-order PLL. For various reasons, the second-order loop with non-ideal integration is used most widely, the loop filter of which is characterized by the expression

$$F(p) = \frac{1+\tau_2 p}{1+\tau_1 p} \ . \tag{3.7}$$

An RC circuit having this characteristic is shown in Fig. 3.1c. Finally, the filter of the ideal second-order loop is characterized by the formula

$$F(p) = \frac{1+\tau_2 p}{\tau_1 p} \ . \tag{3.7a}$$

The noise process $N(t, \varphi)$ is rather complicated, and cannot be handled practically in the general case. However, it can be shown that if the rate of change of processes N_c, N_s is much higher than that of $\varphi(t)$, then $N(t, \varphi)$ can be substituted by a process independent of φ. Also, $N(t, \varphi)$ can be substituted by a white noise process of identical spectral density if the spectral density of $n(t)$ is constant. This means that in this case, the correlation function of the noise in Eq. (3.6) and shown in Fig. 3.1b is given by

$$R_N(\tau) \cong \frac{N_0}{2} \delta(\tau). \tag{3.8}$$

Practically all information published on the noisy PLL is based on the approximate equation (3.8).

Let us first investigate Eq. (3.6) in the noise-free case, and let us assume that the loop signal follows "fairly well" the phase of the input signal. $\varphi(t)$ will then be small allowing $\sin \varphi \sim \varphi$. Substituting this into (3.6) we obtain a linear equation. In the equivalent circuit of Fig. 3.1b, this simply means the omission of the "sine" block.

This linearized PLL model can be treated by the methods of linear network theory. Its most important parameter is the closed loop transfer function which can be expressed as

$$H(s) \triangleq \frac{\hat{\vartheta}_L(s)}{\vartheta_L(s)} = \frac{AkF(s)}{s+AkF(s)} \tag{3.9}$$

where s is the complex frequency variable, and subscript L denotes the Laplace transform. For a first-order loop we have

$$H(s) = \frac{Ak}{s+Ak} \tag{3.9a}$$

and for a non-ideal second-order loop we have

$$H(s) = \frac{1+\tau_2 s}{1+\left(\tau_2+\dfrac{1}{Ak}\right)s+\left(\dfrac{\tau_1}{Ak}\right)s^2}. \tag{3.9b}$$

In the linearized model, it is relatively easy to take into account the additive noise. In this case, $\varphi(t)$ can be calculated from Eq. (3.9), and is a Gaussian noise added to $(\vartheta-\hat{\vartheta})$. The variance of $\varphi(t)$ is given by

$$\sigma_\varphi^2 = \frac{N_0}{2A^2}\frac{1}{2\pi}\int_{-j\infty}^{j\infty}|H(s)|^2\,ds \triangleq \frac{N_0 W_L}{2A^2}. \tag{3.10}$$

This formula is also the definition of the two-sided noise bandwidth of the loop which is one of the most important loop parameters both in linear and nonlinear theory. Instead of the two-sided noise bandwidth W_L, the single-sided noise bandwidth B_L is also used: $W_L = 2B_L$.

The noise bandwidth of a first-order loop is given by

$$W_L = \frac{AK}{2} \tag{3.11}$$

and that of a second-order loop with non-ideal integration is given by

$$W_L = \frac{r+1}{2\tau_2(1+\tau_2/r\tau_1)} \tag{3.12}$$

where $r \triangleq \tau_2 AKF_0$, and $F_0 \triangleq \tau_2/\tau_1$.

It should be noted that the loop parameters depend not only on the circuit parameters $F(s)$ and K but also on the signal voltage A. The quantity KA which has a dimension of frequency is called loop gain. We have seen that the linear approximation is justified if $\varphi(t)$ is small. It is thus important to ascertain in principle the types of $\vartheta_L(s)$ functions which can be tracked by the VCO of the loop. This is even more important from the practical point of view as the PLL is intended for tracking a signal, or in other words, the synchronization of the VCO with the incoming signal.

Let us first investigate the synchronization by unmodulated signals. Let ω_N be the free running oscillator frequency (*i.e.*, the frequency when $z=0$), and ω_c the frequency of the $s(t)$ input signal. The tracking range is that range of $\omega_c-\omega_N$ within which the detuning of the oscillator results in a small phase error. The pull-in range is that range within which the steady-state phase error is zero or small.

According to the linear theory, the steady-state phase error in a first-order loop (without noise) is given by

$$\varphi(t=\infty) = \frac{\omega_c - \omega_N}{2W_L}. \tag{3.12a}$$

The practical pull-in range is that range of $\omega_c - \omega_N$ within which $|\varphi| < 1$ radian. For an imperfect second-order loop,

$$\varphi(t=\infty) = \frac{\omega_c - \omega_N}{KA} \approx \frac{\omega_c - \omega_N}{W_L} \frac{r+1}{2r} F_0. \tag{3.12b}$$

In a second-order loop with ideal integration, the phase error will be zero for an input signal with an arbitrary frequency ω_c.

Without presenting equations, we note that the phase error of the loop for an input signal with linearly changing frequency will also change linearly (*i.e.*, it will go to infinity), for a non-ideal second-order loop, it will go to a finite value for ideal second-order loop, and it will go to zero only for a third-order loop.

The PLL is a feedback circuit requiring the investigation of the stability conditions. According to the linear theory, the loop will be stable if the poles of the closed loop transfer function are located on the left half plane of the s plane. Equation (3.9) shows that the location of the poles is dependent on factor KA. It can be shown that the poles of the first- and second-order loops are always located on the left half plane (in the point $s=0$ in the limit case), *i.e.* these loops are stable (according to linear theory). However, the poles of a third-order loop may be located on the right-half plane for small KA values. Even the second-order loop may not be unconditionally stable if the loop filter transfer function is not a rational function but also has a delay which cannot be excluded in practical cases.

After this summary of conclusions following from the linear model, let us now consider the nonlinear theory giving a better description of the actual circuit, first neglecting the effects of noise. Assuming an unmodulated input carrier, the phase error differential equation of a first-order loop is given by

$$\dot{\varphi} = -KA\sin\varphi + \omega_c - \omega_N. \tag{3.13}$$

This equation can be represented on the so-called phase plane; this has an abscissa φ and an ordinate $\dot{\varphi}$. At a point where $\dot{\varphi}=0$, φ is constant and stays at this value in a first-order loop because the acceleration $\ddot{\varphi}=0$ if $\dot{\varphi}=0$. This means that at points $\dot{\varphi}=0$, the loop will be in equilibrium state (see Fig. 3.2 — the arrows show the direction of the shift of φ and point to the right for a positive velocity). φ will be in the equilibrium state, *i.e.* its derivative will be zero if

$$\varphi = n\pi + \sin^{-1}\frac{\omega_c - \omega_N}{KA}. \tag{3.13a}$$

These equilibrium states will be stable for even values of n, and unstable for odd values of n. It is seen that the phase error in the equilibrium state depends on the initial detuning $\omega_c - \omega_N$. It is also seen that for $\omega_c - \omega_N > KA$, the loop will not reach the equilibrium condition, $i.e.$ the PLL cannot be synchronized to the input signal frequency. In other words, the pull-in range of the first-order loop is given by $(2/\pi)W_L$. The pull-in time, $i.e.$ the time needed to attain practically the equilibrium state, depends both on the detuning and on the initial phase difference. It is longest if the loop is near an unstable equilibrium state at the time of switch-on.

It should be noted that if there exists a stable equilibrium condition this will be attained at all conditions. During the pull-in period, the phase error φ will change by less than 2π, $i.e.$ less than one cycle, independently of the initial phase.

A significant and rather general property of nonlinear oscillators is the infinite number of stable equilibrium states. Assume that for some reason, the system is transferred from equilibrium state A to B. In this case, the system returns to state A. However, if the system is transferred to the unstable state C, it then returns, with equal probabilities, either to state A or to a different equilibrium state D. This latter phenomenon is called cycle slip which is caused for example by noise. The effect of cycle slip in the recovered carrier on the digital transmission will be investigated in Section 3.4.

The nonlinear behaviour of the first-order loop was treated in the foregoing in more detail than justified by its practical significance, because the loop properties can be relatively easily interpreted on the phase plane. The second-order loop has a much wider application field but its analysis is more complicated and will therefore not be dealt with. The best method would again be the characterization on the phase plane. A few major results, based on the investigations published in Refs [1, 2, 4], are given in the following.

Provided the loop filter is described by Eq. (3.7), the expression of the system differential equation is given by

$$\omega_c - \omega_N = \tau_1 \ddot{\varphi} + (1 + AK\tau_2 \cos \varphi)\dot{\varphi} + AK \sin \varphi. \tag{3.14}$$

Similarly to the first-order loop, the solutions of the differential equation yield stable and unstable equilibrium points, or no equilibrium points at large detuning. However, certain initial conditions may have the effect that the system will not reach the equilibrium points but will tend to a limit cycle. Under other initial conditions, the equilibrium state is only reached by cycle slips over several periods.

The pull-in range of the second-order loop is inversely proportional to F_0, $i.e.$ a higher pull-in range is obtained if the loop operates close to ideal integration. The pull-in range may be several times the loop bandwidth at small F_0 values but at large detuning, the pull-in time may be excessive. The pull-in time can be expressed as

$$T = T_f + T_p$$

where T_f is the frequency pull-in time and T_p is the phase pull-in time. During the frequency pull-in time, the phase change may be several times 2π. Following the frequency pull-in time interval, the steady-state phase is approximated within 2π, and is reached during the phase pull-in time. With good approximation,

$$T_f = \frac{\pi^2(r+1)^3(\Delta f)^2}{2r^2 W_L^3}; \quad T_p < \frac{5(r+1)}{W_L r} \tag{3.15}$$

where $\Delta f = (\omega_c - \omega_N)/2\pi$. A lowest T_f is obtained with $r=2$, so that in practice this value is chosen in most cases.

The following upper limit can be given for the pull-in range:

$$\frac{|\Delta f|}{W_L} < \frac{2}{\pi} \frac{r}{r+1} \frac{\sqrt{2F_0 - F_0^2}}{F_0}. \tag{3.16}$$

It can be seen from (3.15) and (3.16) that a usable pull-in time can only be obtained with a detuning which is not much higher than the half of the loop bandwidth, though the pull-in range can be arbitrarily large.

The behaviour of phase-locked loops in the presence of noise will now be investigated by analyzing the stochastic differential equation (3.6). As already mentioned, $N(\varphi, t)$ will be assumed to be white Gaussian noise. Evidently $\varphi(t)$ itself will also be a stochastic process, but will not have Gaussian distribution because of the nonlinear differential equation describing its properties. In the analysis of the linear model, the expected value and variance had to be only determined while in the present nonlinear analysis, the distribution or probability density function of φ has to be calculated. For simplicity, the first-order loop will again be discussed in detail.

The rather complicated nature of the problem is shown by the following qualitative considerations. Assume that the first-order loop has a detuning of $\Delta f = 0$, and is in the equilibrium state at the time instant of switch-on. Because of the noise, φ will move around the equilibrium position, will arrive sooner or later at point C in Fig. 3.2, and finally, by slipping one cycle, at point D. From now on, φ will move around point D as long as the next cycle slip occurs, bringing it either back to point A or to the next stable equilibrium state to the right of point D. During an infinite time interval, φ may have an arbitrary abscissa, i.e. a probability density of identically zero and a variance of infinity. As φ cannot be handled statistically owing to its periodical nature, a more efficient model has to be introduced.

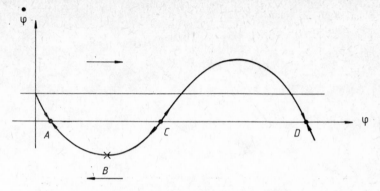

Fig. 3.2. First-order PLL trajectory on the phase plane.

This model can be found by the following considerations. Let us define an interval of width $\pm\pi$ around point φ_0 which corresponds to a stable equilibrium state (*e.g.*, point A). $\varphi(t)$ can then be expressed in the following form:

$$\varphi(t) = \xi(t) + 2\pi J(t); \quad \xi\in[\varphi_0-\pi, \varphi_0+\pi] \tag{3.17}$$

where $\xi(t)$ is the so-called phase error reduced to 2π, and $J(t)$ is a stochastic process with permitted values of $0, \pm1, \pm2, \ldots$, changing its value at any time instant. The expression according to (3.17) has the advantage that the variance of the $\xi(t)$ reduced phase error is finite, and its probability density function can be calculated. The process $J(t)$ comprises the cycle slips, though a cycle slip does not necessarily take place at a transition of ±1. The statistical description of $J(t)$ has to be carried out separately. However, it will be shown that in several applications, these statistical properties are not needed. In other cases, ξ and J are both of importance.

The relation between the total phase error φ and the reduced phase error ξ is illustrated in Fig. 3.3. It is obvious that the stochastic differential equation (3.13) is suitable also for defining ξ: $\sin\varphi = \sin\xi$ and $\dot\varphi = \dot\xi$, with the exception of the phase jumps where the latter equation is valid only for limiting values. Finally, $N(t, \varphi) = N(t, \xi)$ because in the equation defining N we have only $\sin\varphi$ and $\cos\varphi$ terms.

It is seen from Eq. (3.13) that $\xi(t)$ is a Markovian process. The differential equation is a first-order equation, and $N(t)$ has been substituted by a white noise process (see expression (3.8)). This means that the value of $\xi(t+h)$ depends only on the value of $\xi(t)$. However, as we have seen, ξ is not a Gaussian process so its probability density function has to be calculated for its statistical description. $\xi(t)$ being a first-order Markovian process, the transitional probability density $p(\xi, t|\xi_0, t_0)$ yields the complete statistical description of ξ.

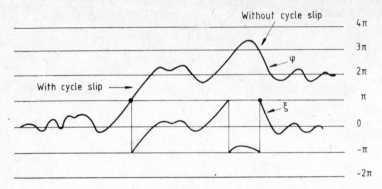

Fig. 3.3. Phase angle φ and reduced phase angle ξ of a noisy PLL as time functions.

Utilizing the stochastic differential equation (3.13), the Fokker–Planck differential equation for $p(\xi, t)$ can be given as

$$\frac{\partial(p \sin \xi)}{\partial \xi} + \frac{\partial^2 p}{\partial \xi^2} = 4 \frac{1}{K^2 N_0} \frac{\partial p}{\partial t}. \tag{3.18}$$

In the above equation, it has been assumed for simplicity that $\omega_c = \omega_N$. The time-dependent solution of this equation has no great significance but the behaviour in the steady-state condition, *i.e.* the probability density function $\lim_{t \to \infty} p(\xi, t)$, is of interest.

In the steady-state condition, ξ is a stationary process for which the solution of (3.18) is the following:

$$p(\xi) = \frac{e^{\alpha \cos \xi}}{2\pi I_0(\alpha)}; \quad |\xi| \leq \pi \tag{3.19}$$

where α is the loop signal-to-noise ratio:

$$\alpha = R_L = \frac{2A^2}{N_0 W_L}; \tag{3.19a}$$

$p(\xi)$ is usually called the Tikhonov distribution.

For a second-order loop, ξ can be defined similarly to the preceding first-order loop definition. ξ can now be represented by a vector Markovian process. It can be shown that for sufficiently high loop signal-to-noise ratio R_L, ξ will again follow a Tikhonov distribution, but α will not be exactly equal to R_L but

$$\alpha = \frac{r+1}{r} R_L - \frac{1 - F_0}{\sigma_{\sin \xi}^2}; \quad \sigma_{\sin \xi}^2 = \frac{1}{2} \left[1 - \frac{J_2(\alpha)}{J_0(\alpha)} \right]. \tag{3.19b}$$

"Sufficiently high" here means only $R_L > 1.5$ to 2 dB. It is seen that α can only be calculated by solving a rather complicated nonlinear equation. However, a small numerical error is obtained by applying the approximate equation of

$$\alpha \approx R_L - 0.2 \approx R_L. \tag{3.19c}$$

Finally, the statistical properties of the cycle slipping events will be investigated. The description of the complete process is given in Ref. [5] for the first-order loop and in Ref. [6] for the general case. The theoretical background of these investigations will not be presented here, but the following result of these investigations is of importance: the distribution of ξ, even according to this more general theory, is a Tikhonov distribution in the case of $\Delta f = 0$, or it is the generalization of a Tikhonov distribution, not expressible in closed form for $\Delta f \neq 0$.

It will be shown that from the statistical properties, the error probability is influenced by the expected value of the time between consecutive cycle slips. For a first-order loop, this is given by

$$T_c = \frac{R_L [I_0(R_L)]^2 \pi^2}{W_L}. \tag{3.20}$$

For a second-order loop, simulation results are known. Figure 3.4 shows $\dfrac{1}{T_c W_L}$ as a function of R_L for different values of r. The curves can be approximately described by the empirical formula

$$T_c = \frac{a}{W_L} e^{bR_L}. \tag{3.21}$$

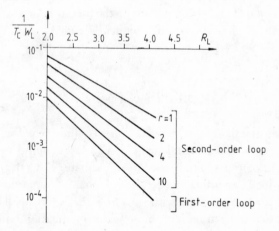

Fig. 3.4. Average cycle slip frequency of a noisy PLL as a function of the loop signal-to-noise ratio, for different values of r.

Table 3.1. Values of constants in Eq. (3.21) as functions of loop parameter r
$(F=0.02)$

r	1	2	4	10
a	1.18	0.65	0.93	1.10
b	1.28	1.71	1.79	1.93

The values of a and b in Eq. (3.21) are given in Table 3.1 for several parameter values of a second-order loop. According to Fig. 3.3, the number of cycle slips is not equal to the number of phase jumps. For a first-order loop, the number of phase jumps is approximately twice the number of cycle slips. For a second-order loop, there is a high probability of cycle slips occurring in bursts. In this case, it may be assumed that a cycle slip occurs at each phase jump.

3.3 Implementations of digital phase-locked loops

In digital equipment, phase-locked loops are applied for a variety of purposes. During the last decade, the structure and application of the digital phase-locked loop (DPLL) in frequency synthesizers and in various measuring instruments have been dealt with in an increasing number of publications [26, 31, 34, 35, 38, 45, 48]. In the following, the application of DPLL's in clock recovery circuits will be surveyed. It should be noted that by DPLL we mean a phase-locked loop of all-digital operation. This is in contrast with the literature in which phase-locked loops with partially digital circuits are frequently called "digital" phase-locked loops.

3.3.1 Circuit elements

The structure of the digital phase-locked loop is similar to the analog phase-locked loop inasmuch as the digital phase-locked loop comprises a phase comparator, a loop filter and a digitally controlled oscillator (DCO). In Ref. [46], the DPLL's are classified according to the applied phase comparator type:
— FF (flip–flop) type comparator with sawtooth characteristic. The comparator output signal is a pulse train with pulse width modulation, and the time difference between the local reference signal and the synchronizing incoming signal is represented by the pulse width deviation.

— Nyquist rate type comparator. The sampling of the input signal by the reference clock signal is performed at just the Nyquist rate.
— Zero crossing type comparator. The loop control signal is derived from the zero crossings of the incoming signal.

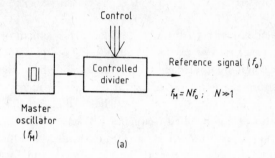

(a)

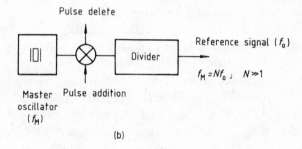

(b)

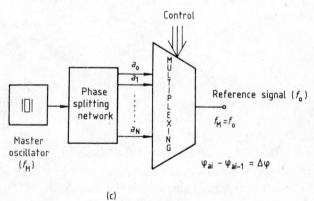

(c)

Fig. 3.5. Variants of a digitally controlled oscillator: (a) oscillator with a controlled frequency divider; (b) oscillator with a fixed frequency divider; (c) oscillator with a phase splitting network.

— LL (lead–lag) type comparator. The comparator output signal depends on the relative time difference between incoming digital signal transitions and local reference signal transitions (whether the incoming signal leads or lags the reference signal). This type of comparator is also called "early–late" detector.

The loop filter is made up either from a digital filter arrangement or a so-called sequential filter. The latter will be treated in Section 3.3.6.

The variants of the digitally controlled oscillator are shown in Fig. 3.5. In all variants, the DCO comprises a master oscillator of high frequency stability. In the arrangement of Fig. 3.5a, the control signal has the effect of changing the frequency division ratio of a controlled frequency divider. In the variant of Fig. 3.5b, a frequency divider of constant division ratio is applied, and the pulse train of the master oscillator is varied by either adding pulses to the pulse train or deleting pulses from the pulse train, according to the control information. Both above arrangements are characterized by a master oscillator frequency which is a multiple of the free running loop frequency. In Fig. 3.5c, the master oscillator frequency is equal to the free running frequency, and a phase splitting network is applied to generate several pulse trains $a_0, a_1, ..., a_N$ having different relative phases. These pulse trains are connected to a multiplex which is driven by the control signal, thus selecting the suitable pulse train for clock reference.

Several combinations of the presented loop elements are possible to form different digital PLL's. According to our experience, DPLL's with a lead–lag (or early–late) type comparator, a sequential filter with a binary detector and a controlled frequency divider used as a DCO are used most frequently for clock recovery purposes. This variant will therefore be discussed in the following.

3.3.2 Operation

The DPLL is a phase-locked loop comprising a phase comparator and a controlled frequency divider, acting as a voltage controlled oscillator. For improving the performance, the loop may also comprise a filter arrangement for effectively decreasing the disturbing effect of noise.

The structure of the DPLL is shown in Fig. 3.6. The controlled divider is applied to divide the frequency of the clock signal generated by the master oscillator which operates at a frequency $F = q f_0$, where f_0 is the recovered clock frequency and q is the division ratio of the controlled divider. The operation of the DPLL is based on the fact that q can be varied in a small range in response to the phase difference between the received and recovered clock signals. This control

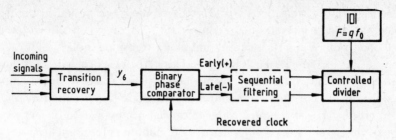

Fig. 3.6. Structure of a DPLL.

of the division ratio results in a phase error which has a discrete value of $\pm a \dfrac{T}{q}$
where a is a constant characterizing the type of DPLL and T is the symbol time of
the received digital signal.

The operation of the DPLL is illustrated by the comparator characteristics
shown in Fig. 3.7. The phase comparator carries out a symbol-to-symbol com-

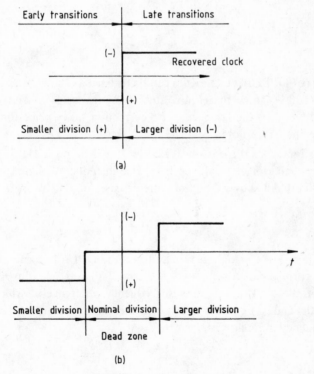

Fig. 3.7. Phase comparator characteristics: (a) without dead zone; (b) with dead zone.

parison of specified transitions in the recovered clock signal and in the received clock signal, and as a result of this comparison, a suitable correction of the controlled divider division ratio is carried out at the transition instants of the received signal. If the incoming signal transition leads the clock signal transition (*i.e.* it is an early transition) then a $(+)$ control takes place: the next division ratio will be $q-1$ instead of q, thus shortening the next clock signal period by T/q. The control process can be regarded as being carried out by a fixed divider receiving an input pulse train comprising one or more additionally inserted pulses, hence the $(+)$ sign. If the incoming signal transition lags the clock signal transition (*i.e.*, it is a late transition), then a $(-)$ control takes place: the next division ratio will be $q+1$ instead of q, thus lengthening the next clock signal period by T/q. The control process can then be regarded as being carried out by the same fixed divider receiving an input pulse train from which one or more pulses are deleted, hence the $(-)$ sign.

No correction will take place if there is no transition in the received signal in the vicinity of the specified clock signal transition. In this case, the recovered clock frequency, generated by the DPLL, will correspond to the free running frequency of

$$f_0 = \frac{F}{q}. \tag{3.22}$$

Two kinds of DPLL characteristics can be realized as shown in Fig. 3.7. Figures 3.7a and Fig. 3.7b show the comparator characteristic with and without dead zone, respectively. In the latter case, no correction takes place for phase errors below a specified threshold.

The operation of a DPLL is illustrated by Figs 3.8 and 3.9. It is assumed that correction can take place at all specified time instants, *i.e.* signal transitions are available. Figure 3.8 shows the ideal extreme cases of DPLL operation, with the notation

$$\Delta f = f - f_0 \tag{3.23}$$

where f is the received signal bit rate and f_0 is the DPLL free running frequency. Figure 3.8a corresponds to $\Delta f = 0$, while Figs 3.8b and 3.8c illustrate the pull-in capability of the DPLL. The highest frequency difference at which pull-in still takes place depends on whether Δf is positive or negative. For positive values, the highest frequency difference is given by

$$\Delta f = \frac{f_0}{q-1} - \varepsilon \tag{3.24}$$

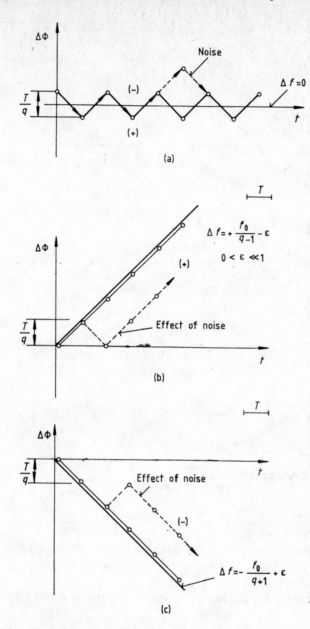

Fig. 3.8. Operation of a DPLL: (a) $\Delta f = 0$; (b) all possible (+) controls are realized; (c) all possible (−) controls are realized.

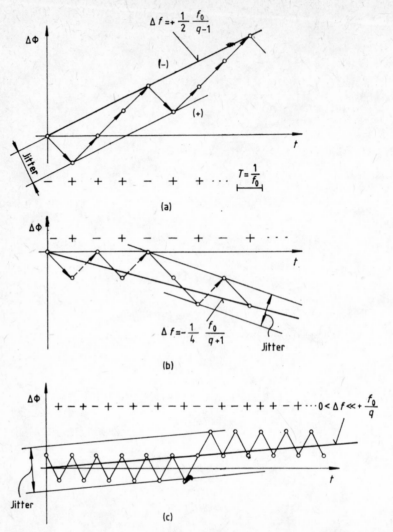

Fig. 3.9. Operation of a DPLL: (a) frequency difference high, $f>f_0$; (b) frequency difference high, $f<f_0$; (c) frequency difference low, $f>f_0$.

and for negative values, the highest frequency difference is given by

$$\Delta f = -\frac{f_0}{q} + \varepsilon \tag{3.25}$$

where ε is an arbitrarily small number. In Fig. 3.8, the disturbing effect of noise is denoted by dashed lines. Figure 3.9 shows similar characteristics for smaller frequency differences which normally occur during operational conditions.

3.3.3 Main characteristics

The most important characteristic of a DPLL is the pull-in range. This is difficult to calculate as it depends on the symbol pattern of the received digital signal stream. Let us investigate the time instants at which transitions resulting in positive or negative corrections may appear in the received signal. Let α be the probability of no transition, β be the probability of a transition giving a positive correction, and γ be the probability of a transition giving a negative correction. As one of these three events will certainly take place, we have

$$\alpha + \beta + \gamma = 1. \tag{3.26}$$

From the above, the expected value of the pull-in range can be expressed as

$$[\Delta f]_{\max} = 2f_0 \left[\alpha + \beta \frac{q-2}{q-1} + \gamma \frac{q+2}{q+1} \right]. \tag{3.27}$$

If there is no transition in the received signal for a considerable time then

$$\alpha = 1, \quad \beta = \gamma = 0, \tag{3.27a}$$

and it follows from (3.27) that $\Delta f_{\max} = 0$, *i.e.* the DPLL will not operate.

In the following, the highest number of missing signal transitions at which the DPLL still operates will be determined. For the case in which only transitions giving $(+)$ correction appear in the received signal,

$$\alpha = 0, \quad \beta = 1 \quad \text{and} \quad \gamma = 0 \tag{3.27b}$$

resulting in

$$[\Delta f]_{\max} = \frac{f_0}{q-1}. \tag{3.28}$$

For the case in which only transitions giving $(-)$ corrections appear,

$$\alpha = \beta = 0 \quad \text{and} \quad \gamma = 1 \tag{3.28a}$$

resulting in

$$[\Delta f]_{\max} = \frac{f_0}{q+1}. \tag{3.29}$$

It thus follows that the value of the pull-in range, as given by Eq. (3.27), will change continuously between the extreme values given by (3.28) and (3.29). If the probability variables α, β and γ are expected values, then $\Delta f_{\max}$ will also be an expected value.

A merit of the DPLL is the short pull-in time which is an important consideration. It can be shown that the highest value of the pull-in time is given by

$$\tau_{be_{\max}} = \frac{q}{2} T \tag{3.30}$$

if the effect of noise is neglected and assuming that in all possible time instants there is a transition giving a correction.

Another important parameter of the DPLL is the permitted highest number of symbols without transitions which has significance when transmission without any code restriction is required. Let this number be denoted by n_{max}. The phase difference accumulated during a symbol time which can still be followed by the DPLL is given by

$$\Delta T \cong \frac{1}{f_0} \pm \frac{1}{f_0 \left(1 \pm \dfrac{\Delta f}{f_0}\right)} \tag{3.31}$$

where Δf is the allowable frequency deviation. The following condition has thus to be fulfilled:

$$n_{max} \Delta T = \frac{T}{q}. \tag{3.32}$$

Using expressions (3.31) and (3.32), we have

$$n_{max} \cong \frac{f_0}{\Delta f q}. \tag{3.33}$$

As an example, assume $\Delta f/f_0 = 10^{-4}$ and $q = 10$. In this case, the transmission of a symbol sequence comprising up to 1000 bits without symbol transitions is possible.

3.3.4 Loop types

The type of digital phase-locked loop is characterized by two parameters: one of the phase characteristic types shown in Fig. 3.7, and the ratio of the master oscillator frequency F to the free running frequency f_0.

For the pn-type loop, $qf_0 = F$, and the phase comparison without dead zone is applied. In this case, $(+)$ and $(-)$ corrections will alternate even for $\Delta f = 0$. For a pon- (positive-zero-negative) type loop, phase comparison with dead zone is applied. A suitable choice of dead zone then results in extremely infrequent corrections, thus reducing substantially the r.m.s. value of the inherent jitter of the DPLL.

For the p-type loop, $F = (q + 1/2)f_0$, and for the n-type loop, $F = (q - 1/2)f_0$. In contrast to the pn- and pon-types having three alternating frequency division ratios, the p- and n-types have only two alternating division ratios during operation of the controlled frequency divider. For instance, in a p-type DPLL, the consecutive division ratios will be $q, q+1, q, q+1, ...$, if no transitions appear for an extended time span, the division ratios corresponding to the individual symbol times. Consequently, the equation

$$f_0 = \frac{F}{q + \dfrac{1}{2}} \tag{3.34}$$

will apply to a p-type DPLL and the equation

$$f_0 = \frac{F}{q - \dfrac{1}{2}} \tag{3.35}$$

will apply to an n-type DPLL, instead of Eq. (3.22). The main characteristics of DPLL's are summarized in Table 3.2.

Table 3.2. Main characteristics of DPLL's

Type	Pull-in range		Inherent jitter	F
	$2\Delta f$	$\sim 2\Delta f$	pp	
pn pon	$2f_0 \dfrac{q}{q^2-1}$	$2\dfrac{f_0}{q}$	$\dfrac{2}{f_0 q}$	qf_0
p	$\dfrac{f_0}{2} \dfrac{2+\dfrac{1}{q}}{q+1}$	$\dfrac{f_0}{q}$	$\dfrac{1}{f_0 q}$	$\left(q+\dfrac{1}{2}\right)f_0$
n	$\dfrac{f_0}{2} \dfrac{2-\dfrac{1}{q}}{q-1}$	$\dfrac{f_0}{q}$	$\dfrac{1}{f_0 q}$	$\left(q-\dfrac{1}{2}\right)f_0$

3.3.5 Comparison with analog phase-locked loops

The advantages of DPLL's when applied in digital transmission systems, as compared to analog PLL's, are summarized in the following.

(1) Applicable for transmission without any code restriction. Assuming clock generators of equal frequency accuracy, the allowed number of missing transitions with APLL would be q times the number with DPLL. However, the accuracy of the free running voltage controlled oscillator frequency is about

10^{-2} while the accuracy of the quartz crystal oscillator in a DPLL is about 10^{-5}, q being approximately 10, meaning that the application of a DPLL results in an improvement of about two orders of magnitude in view of transmission without code restriction.

It should be noted that in principle, a crystal controlled VCO (VCXO) could also be applied in an APLL. However, this is a complicated solution having several disadvantages.

(2) The free running DPLL frequency is crystal controlled owing to the crystal controlled clock generator applied.

(3) Short pull-in time is realizable, without the transient phenomena which are normally present with analog PLL's.

(4) A wide pull-in range can be realized.

(5) Possibility of adaptive realization (programmability) by varying the value of q by simple means. Here q can be chosen to equal q_1 during the pull-in time and to equal q_2 after the pull-in time, by the recognition of a unique word or the frame synchronizing word, with $q_2 > q_1$. This may be useful because the pull-in time, which is proportional to q, can be reduced by choosing a small q_1 value, and the inherent jitter can be reduced by choosing a large q_2 value.

(6) Inexpensive digital integrated circuits having favourable temperature properties can be applied.

(7) For applications in which variable clock rates have to be accepted, the master oscillator of the DPLL can be easily changed while the free running frequency of an analog PLL needs complicated circuit modifications and VCO re-tuning.

(8) The performance of the DPLL can be improved easily and substantially by the application of sequential filtering. However, this will generally result in a degradation when receiving transmissions without code restriction. An adaptive control mentioned under item (5) may be applied in this case.

The price to be paid for the above advantages is the increased jitter of the DPLL. Principally, the jitter can be arbitrarily decreased by increasing q but this will result in a decreased pull-in range and an increased pull-in time. Here p-type and n-type DPLL's are favourable for obtaining small peak-to-peak jitter values while pon-type DPLL's are applied for obtaining small r.m.s. jitter values (see Table 3.2). However, pon-type DPLL's are difficult to realize at high bit rates; in this case, $pn \rightarrow pon$ transformation is required with the application of sequential filtering.

3.3.6 Performance improvement by sequential filtering

In Fig. 3.6 which shows the structure of the DPLL, the sequential filter is shown as an option. Essentially, the sequential filtering process corresponds to a process of majority decision inasmuch as the correction signal driving the controlled frequency divider of the DPLL is validated only if it is deemed beneficial. Owing to the noise generated in the radio channel, the time instants of the recovered signal transitions are shifted, and the harmful effect of this shift on the synchronized clock signal can be diminished by sequential filtering, as shown in the following.

Re-writing Eq. (3.27) with the condition $\beta \approx \gamma$, we have

$$[\Delta f]_{\max} = f_0 \left(\alpha + 2\beta \frac{q^2 - 2}{q^2 - 1} - 1 \right). \tag{3.36}$$

Assume now that in an extreme case, the transmitted symbol pattern and the noise result in the following inequality:

$$\alpha \gg \beta \approx \gamma. \tag{3.37}$$

It is then easily shown that

$$[\Delta f]_{\max} \rightarrow 0 \tag{3.37a}$$

i.e., the pull-in range tends to zero and thus the DPLL does not operate. However, the sequential filtering may have the effect that the relation

$$\beta \approx \gamma \tag{3.37b}$$

should not hold. In this case, the trend of the clock frequency and clock phase variations can be extracted even from an extremely noisy received signal.

3.3.6.1 The "N before M" filter

This filter, shown in Fig. 3.10 and used most widely, comprises three counters. $(+)$ correction signals are counted by counter N^+, $(-)$ correction signals by counter N^-, and both type of signals by counter M. Also, N and M stand for the counter capacity. The overflow of any of the counters results in resetting all three counters thus generating an output signal. Overflow of counter N^+ generates a $(+)$ control signal while overflow of counter N^- generates a $(-)$ control signal.

The general condition of "N before M" filters is given by

$$N \leq M \tag{3.38}$$

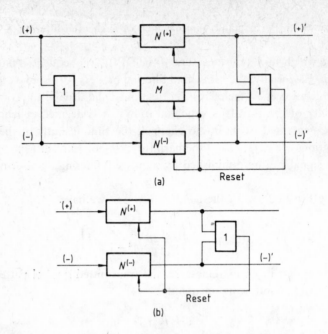

Fig. 3.10. The "N before M" filter: (a) normal case; (b) degenerated (racing type) case.

as no output signal would ever be generated in the opposite case. Different filter types can be obtained by the choice of parameters N and M, as discussed in the following.

Filter based on unanimous decision. For this filter,

$$N = M. \tag{3.39}$$

A (+) or (−) correction signal will appear at the filter output if and only if N consecutive input correction signals have the same sense. It can be shown that for $N \geq 2$, the pn-type DPLL will become a pon-type. The pn-type DPLL generates alternately (+) and (−) correction signals which will not appear at the output of the $N=M$ filter. It can also be shown that the width of the dead zone shown in Fig. 3.7b is proportional to N.

Filter based on specified majority decision. For this filter,

$$\frac{M}{2}+1 < N \leq M-1. \tag{3.40}$$

For instance, $N=5$ and $M=7$ can be chosen, and the simple majority is not sufficient.

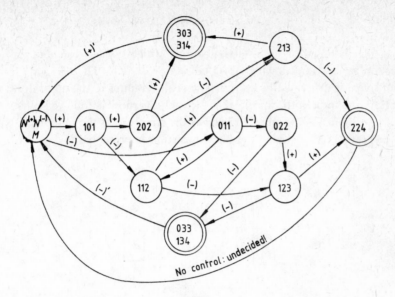

Fig. 3.11. Graph of the filter with $N=3$, $M=4$.

Filter based on simple majority decision. For this filter,

$$\frac{M}{2} < N \leqq M-1. \tag{3.41}$$

This condition is met for example by $N=3$, $M=4$. The graph of this filter is shown in Fig. 3.11. For this filter, at least three of four consecutive input correction signals have to be of the same sense for a decision of generating an output signal. An undecided condition can also be established.

Filter based on competition. For this filter,

$$N \leqq \frac{M}{2}. \tag{3.42}$$

This is the degenerated case of the $N=M$ filter in which overflow of counter M can never occur so this counter can be deleted, resulting in the filter arrangement shown in Fig. 3.10b.

3.3.6.2 The "random walk" filter

This filter is based on an up/down counter which is made to count up by $(+)$ correction signals and down by $(-)$ correction signals. The $(+)'$ or $(-)'$ filter output signal is generated by an overflow of the counter in the up or down direction. For a counter with capacity $2N+1$, each overflow is followed by a counter reset to position $N+1$. Figure 3.12 shows the structure of the random walk filter, and the graph characterizing the filter operation is presented in Fig. 3.13.

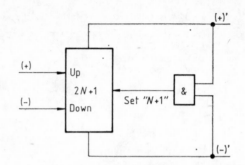

Fig. 3.12. The random walk filter.

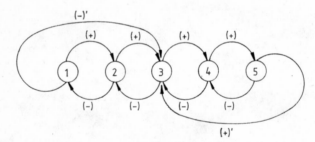

Fig. 3.13. Graph of a random walk filter with $N=2$.

3.3.6.3 Conclusions

The application of sequential filtering has a beneficial effect on the efficiency of clock recovery under noisy conditions; however, it is unfavourable from the point of view of transmission without any code restriction. For this reason, filters with relatively small N and M parameters are generally used.

It is extremely difficult to perform a quantitative analysis of sequential filter performance. Approximate calculations are presented in Refs [22, 34, 35, 36]. The difficulty of calculation is sometimes avoided by a graphical analysis [40].

Another method is the relatively easy realization of the DPLL circuit, together with the sequential filter, and to measure the parameters under specified conditions. There is also the possibility of the operational simulation of the circuits made up from digital circuit elements.

Practical results show that with a *pn*-type DPLL and a sequential filter having parameters $N=3$, $M=4$, a given clock recovery performance could be obtained at a signal-to-noise ratio 8 to 10 dB lower than without the sequential filter.

3.4 Transmission of the phase information: carrier recovery

The phase information for synchronizing the carrier at the receive side can be obtained, in principle, either from a carrier transmitted over a separate channel, as shown in Fig. 2.13, or from a carrier recovered from the received digital signal. In the latter case, which is exclusively used in practice, there are again two possibilities: the transmitted signal may have either an unmodulated carrier component or it may be a suppressed carrier signal. In the following, both possibilities will be discussed.

3.4.1 Systems with a transmitted reference

In this section, we will assume that the transmitted digital signal comprises in some form an unmodulated carrier serving as reference. In this case, the available power is divided, one part transmitting the information, and the other part serving for synchronization. In the spectrum of the transmitted signal, there is a line component at the carrier frequency. A transmitted reference is typically applied in satellite telemetry and relatively low speed data transmission systems, and is not normally used in terrestrial or satellite telecommunication systems.

At the receive side, the frequency and phase of the transmitted reference carrier has to be estimated. Because of the noisy transmission channel, the estimated values will differ from the actual values. One method of optimizing the estimated value is the so-called maximum likelihood estimation according to which the estimated phase angle of the carrier is equal to the maximum value of the conditional probability density $p[y(t)|\vartheta]$ where $y(t)$ is the received signal and ϑ is its phase.

The theory of optimal parameter estimation is dealt with in Refs [3, 9] among others. It is shown in Ref. [3] that the carrier synchronization circuit which is intended to perform an optimal estimation of the carrier phase is best realized by a feedback loop which is similar to the phase-locked loop. The advantage of this closed loop realization is the ability to follow the slow phase changes due to the

instability of the transmitter oscillator. The optimization of the PLL is carried
out by suitable choice of loop filter and phase discriminator characteristics. For
practical reasons, there are not too many design variants as the voltage-phase
characteristic of microwave or i.f. phase discriminators is sinusoidal with good
approximation, leaving practically no alternatives (this kind of phase detector
alone was treated in Section 3.2). Higher than second-order loop filters cannot be
used for stability reasons, and the ideal integration can be realized only approximately. It can be shown that the ideal second-order loop is best suited to lock onto
a signal with a constant frequency differing from the free running loop frequency.

The fact that the carrier frequency to which the loop has to be synchronized
is not unmodulated presents no difficulties because, as assumed, the transmitted
spectrum comprises a substantial carrier frequency component. At the transmit
side, the modulated signal has to be multiplexed with the unmodulated reference
signal. This multiplexing can be performed in time, in frequency, and "in quadrature".

According to time division multiplexing, an unmodulated carrier is transmitted
in a fraction of the time, allowing the loop to acquire synchronization during
this time. During the time slots of information transmission, the input signal is
disconnected from the loop which is thus left free running. This kind of burst
synchronization of a loop is discussed in Ref. [10]. Its results will not be presented
here but it is noted that for successful synchronization, the synchronizing time
slots should have a frequency which is at least twice the frequency difference
$\Delta f = (\omega_c - \omega_N)/2\pi$. The phase of the reference signal thus recovered will exhibit
a certain variance around its steady-state expected value even in the noise-free
case.

According to the practical realization of the frequency division multiplexing,
two subcarriers are applied, one of them being modulated by the information bearing signal and the other being transmitted unmodulated. The two subcarriers
should be coherent in order to generate a coherent reference signal. The spectrum
envelope of this system is illustrated in Fig. 3.14a.

According to the process which may be called "quadrature multiplication", two
subcarriers which are in phase quadrature are applied, one of them being 2PSK
modulated by the information signal, and the other transmitted unmodulated.
The binary signal set then has the following expression:

$$s(t) = \sqrt{2}\, A \cos\left[\omega_c t + \Delta\Phi m(t)\right] \tag{3.43}$$

where $m(t) = \pm 1$. Rearranging (3.43) yields

$$s(t) = \sqrt{2}\left[A \cos \Delta\Phi \cos \omega_c t - Am(t) \sin \Delta\Phi \sin \omega_c t\right]. \tag{3.44}$$

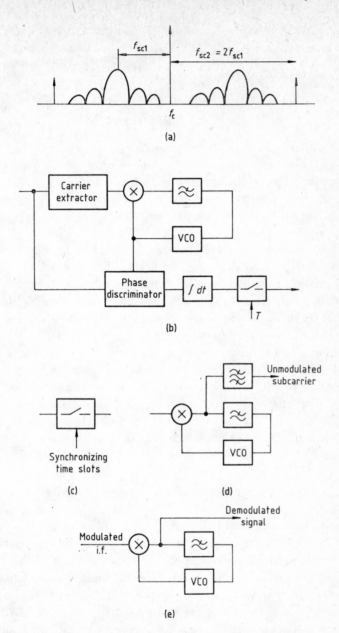

Fig. 3.14. (a) PSK transmission system with frequency multiplexed reference, spectrum envelope; (b) same system, PSK demodulator; (c) carrier recovery circuit, reference signal transmitted by time division multiplexing; (d) carrier recovery circuit, reference signal transmitted by frequency division multiplexing; (e) carrier recovery circuit, reference signal transmitted by quadrature multiplexing.

In this expression, the first term is the unmodulated carrier, and the second term is the information bearing signal. The power of the unmodulated carrier is given by

$$P_c = A^2 \cos^2 \Delta\Phi \tag{3.44a}$$

and the power of the modulated information carrier is given by

$$P_m = A^2 \sin^2 \Delta\Phi. \tag{3.44b}$$

The sum of these powers is the total carrier power A^2.

In order to achieve a lock on the unmodulated quadrature component, a sufficiently small loop bandwidth has to be chosen. A drawback of such a system is the loss of the phase information after a relatively long train of binary ones or binary zeros. The PLL will then lock onto phase $+\Delta\Phi$ or $-\Delta\Phi$ instead of phase zero. Such a system is treated in Refs [11, 8].

The above insufficient phase lock can be eliminated by the application of bi-phase coding instead of NRZ coding (see Section 2.1). In this case, the longest time of any phase position is T, and the bandwidth of the PLL has to be chosen so as to obtain a loop time constant which is much larger than T. Figure 3.14b shows a 2PSK demodulator and carrier recovery circuit which operates on a transmitted signal containing a reference carrier.

Figures 3.14c, d and e show examples of the carrier recovery circuits according to the above three principles. Figure 3.14c shows the circuit of the time division system which is a switch, closed only during the synchronizing time slots. Figure 3.14b shows the circuit of the frequency division system which is a demodulator recovering the unmodulated subcarrier. Finally, Fig. 3.14e shows the circuit of the quadrature multiplicated system which is a PLL, also supplying the demodulated signal.

3.4.2 Carrier recovery from a suppressed carrier

It was shown in Section 3.4.1 that in addition to the information signal, a reference carrier supplying the phase information can be additionally transmitted. However, in most cases, it is more feasible to utilize the total available power for the transmission of the information bearing signal. The transmitted spectrum will then not contain a carrier spectrum line component (at least when the information signal has no DC component), and the PLL according to Section 3.4.1 (regarded as a parameter estimating circuit) cannot be applied directly. The addition of a filter would be of no use since there is no linear operation generating a spectral line from a continuous spectrum.

All signal sets discussed in Section 2.2 result in suppressed carrier modulation systems (with the exception of ASK which, however, is normally used in non-coherent transmission). In these suppressed carrier tracking loops, a nonlinear operation, generating the required spectrum line component, has to be performed on the received modulated carrier, either preceding or within the PLL. A PLL taking the fourth power of the received signal, which is one of the many possible solutions, will be discussed in the following (applicable to 4PSK transmission). Other methods, including those applicable to other modulations will be discussed in Chapter 9.

Taking the fourth power of a 4PSK signal results in an unmodulated carrier if the following conditions are met: phase deviation of $k \times 90°$ ($k = 0, 1, 2, 3$), application of NRZ (or possibly biphase) signals, and a sufficiently large bandwidth resulting in little distortion. The input signal can then be expressed as

$$s(t) = A[M(t)\cos \omega_c t + Q(t)\sin \omega_c t]$$
$$M, Q = \pm 1.$$

(3.44c)

The fourth power of this expression contains a DC component and components of frequency $2f_c$ and $4f_c$. The expression of the $4f_c$ component is given by

$$s_4(t) = \frac{A^4}{2}\left[-\frac{M^4 + Q^4}{4} - \frac{3}{2}MQ\right]\cos 4\omega_c t + \frac{A^4}{2}MQ(M^2 - Q^2)\sin \omega_c t.$$

(3.45)

This is simplified by noting that $M^2 = Q^2 = 1$, independent of the modulation:

$$s_4(t) = -\frac{A^4}{4}\cos 4\omega_c t.$$

(3.46)

This expression is no longer valid if the signals are distorted by band limiting; in this case, $M(t)$ and $Q(t)$ represent rolled-off signals instead of NRZ pulses. In the following, Eq. (3.46) will be regarded as valid, and the intermodulation noise due to band limiting will be discussed later.

It is seen from Eq. (3.46) that the component of frequency $4f_c$ is unmodulated which can be used to synchronize a phase-locked loop. The reference signal needed for the coherent demodulation of a suppressed carrier 4PSK signal can be generated by the circuit shown in Fig. 3.15. (Note that for M-PSK signals, the Mth power has to be taken instead of the fourth power.) The operation of a fourth power loop shown in Fig. 3.15 will first be discussed in the noise-free case. It can be seen from the figure and from Eq. (3.45) that the loop will be stable in four different phase positions of the VCO operating at $4f_c$. This can be understood by considering that, taking the input signal phase as zero, the loop will be in the equilibrium state if the phase $\hat{\vartheta}$, shown in the figure, is around zero, but this will take place when the phase difference between the VCO signal and the input signal is 0, 90, 180 or 270 degrees.

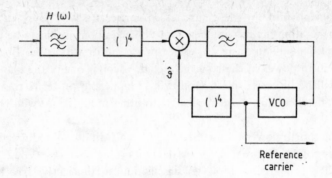

Fig. 3.15. Fourth power carrier recovery circuit for QPSK transmission.

On the other hand, if $\hat{\vartheta}/4 \neq 0$, the quadrants of the two-dimensional signal space will be mixed up in the demodulator; this means that in the demodulated signal, each dibit may comprise one or two errored bits (see Fig. 2.7). By representing the dibits by numbers of a quaternary numerical system, a reference signal phase shift of $k \times 90°$ will have the effect that instead of the correct number D, the number $D + k \pmod 4$ will be demodulated. It is therefore necessary to resolve in some way this four-fold ambiguity in the received digital signal.

There are two methods for resolving the phase ambiguity. In the first method, it is assumed that part of the transmitted information is known at the receive side. This part may be for example the periodically recurring framing word in the PCM bit stream, and this can be recognized, and its errored reception ascertained from a $k \times 90°$ phase shift of the reference signal. The VCO phase has then to be shifted into the correct position or, an equivalent operation, a suitable code transformation has to be carried out on the incorrectly demodulated signal.

According to the second, more widely used method, operations are carried out on the received signals, whose result is invariant with respect to the $k \times 90°$ phase shift of the reference signal. Such an operation is a (mod 4) subtraction: the subtraction of the ith dibit from the $(i-1)$th dibit has a result which is unchanged if an integer in (mod 4) sense is added to both dibit values. This operation is usually

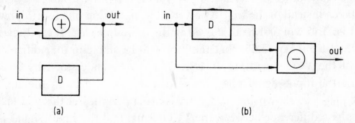

Fig. 3.16. (a) Differential encoding circuit; (b) differential decoding circuit.

called differential decoding. For proper transmission, the inverse operation, *i.e.* differential encoding has to be performed at the transmit side. Figure 3.16 shows a differential encoding and decoding circuit whose operation can be understood as follows. Assume that the transmitted signal is a, the coded signal is b, and the received signal is c; then

$$b_i = a_i \oplus b_{i-1} \quad (\text{mod } 4)$$

$$c_i = b_i \ominus b_{i-1} = a_i \quad (\text{mod } 4). \tag{3.46a}$$

On the other hand, if b_i is changed to $b_i' = b_i + k$ for all i values, then

$$c_i' = b_i' \ominus b_{i-1}' = b_i \oplus k - (b_{i-1} \oplus k) = c_i = a_i. \tag{3.46b}$$

(It is noted that the differentially coherent phase demodulator treated in Section 2.2 also performs differential decoding, thus requiring the differential encoding of the transmitted signal.)

Let us now discuss the fourth power loop in the presence of noise. First of all, the properties of the white noise having originally a spectral density of $N_0/2$ will be changed by the fourth power operation preceding the loop (see Fig. 3.15). These properties can be found by the following method. Let us express the signal + noise at the frequency $4f_c$ as the sum of the signal and noise complex envelopes. The fourth power of this sum, and also the correlation function of the component at frequency $4f_c$ is then calculated, for example by the method given in Ref. [12]. Because of the filtering and the nonlinear transformation, this noise will not be white nor have a Gaussian distribution, invalidating the theorems applying to the noisy PLL. In practice, however, the bandwidth B of the receiver filter having a transfer function of $H(\omega)$ will be much larger than the loop bandwidth W_L. This means that the spectral density of the noise can be taken as constant within the loop frequency range, and will be given by

$$\tilde{N}_4 = \int_{-\infty}^{\infty} R(\tau)\, d\tau. \tag{3.46c}$$

Here $R(\tau)$ denotes the correlation function of the fourth power of the noise which can be calculated by the method presented earlier. The phase error φ can then be regarded as a Markovian process, and the results of Section 3.2 can be applied. According to these results, the loop at frequency $4f_c$ will behave like a "linear" PLL, but the signal-to-noise ratio will be given by

$$\tilde{R}_{4L} = \frac{R_L}{16} \frac{1}{1 + \dfrac{9}{2} \dfrac{B}{R_L W_L}}; \quad R_L = \frac{2A^2}{N_0 W_L}. \tag{3.47}$$

In this and the following expressions, the properties of the noise at frequency $4f_c$ are denoted by the sign $\sim$. The signal-to-noise ratio at frequency f_c is given by

$$\tilde{R}_L \approx R_L \frac{1}{1 + \dfrac{9}{2} \dfrac{B}{R_L W_L}}. \qquad (3.48)$$

The coefficients in these expressions will change if the filter distorts substantially the originally NRZ impulses.

Assuming negligible detuning $\Delta\omega$, the probability density of the reduced phase error can be calculated from Eq. (3.19), but 4ξ has to be written instead of ξ because of the frequency quadrupling. The probability density of the reduced phase error at frequency f_c is thus given by

$$p_4(\xi) = 4p(4\xi) = \frac{2e^{\alpha \cos 4\xi}}{\pi I_0(\alpha)}; \quad |\xi| \leq \frac{\pi}{4}. \qquad (3.49)$$

For M-ary phase modulation and an Mth power loop, similar considerations result in the following expression:

$$p_M(\xi) = \frac{M e^{\alpha \cos M\xi}}{2\pi I_0(\alpha)}. \qquad (3.50)$$

The loop signal-to-noise ratio for M-ary modulation can be expressed as

$$\tilde{R}_{M,L} \approx \frac{R_L}{M^2} \frac{1}{1 + C \dfrac{B}{R_L W_L}}. \qquad (3.51)$$

The coefficient $C = 1/2$ pertains to $M = 2$.

The expected value of the time between consecutive cycle slips can again be calculated from (3.20) or (3.21) but the quantity R_{4L} of Eq. (3.47) has to be substituted instead of R_L. During a cycle slipping process, the value of 4ξ will jump by $360°$, so that ξ will change by $90°$.

3.4.3 Effect of a noisy reference

The transmission performance is further degraded if not only the received signal but also the reference signal recovered in the receiver is noisy. In spite of the fact that both signals are affected by the same physical noise source the two noise processes can be regarded as being uncorrelated. This is explained by the fact that the receiver bandwidth is much higher than the loop bandwidth resulting in a loop delay time which is much longer than the correlation time of the noise at the output of the receiver filter. It has been shown in the preceding sections that the loop noise has two effects: the reduced phase error will show time fluctuations, and the phase will be in the vicinity of different stable states while changing by $2k\pi$ or $2k\pi/M$.

Let us first investigate the effect of the ξ fluctuations. According to our assumption, $B \gg W_L$, and further $B \approx 1/nT$ according to the results of Sections 2.1 and 2.2. It can thus be assumed that the reduced phase error ξ is constant during a symbol time nT. The error probability has thus to be calculated as a function of the signal-to-noise ratio R with ξ as a condition, and this has to be averaged with respect to ξ:

$$\bar{P}_E = \int\limits_{(\xi)} P_E(R|\xi)p(\xi)\,d\xi. \tag{3.52}$$

$P_E(R|\xi)$ can be relatively easily calculated for 2PSK and 4PSK systems. For instance, applying a reference signal of the form $\cos(\omega_c t + \xi)$ in the demodulator shown in Fig. 2.14a, the received signal will seemingly be reduced by the factor $\cos \xi$. The error probability for 2PSK will then be given, by appropriate application of (2.23), by

$$P_{E2}(R|\xi) = \frac{1}{2}\operatorname{erfc}\left[\sqrt{R}\cos\xi\right]. \tag{3.53}$$

For the case of 4PSK modulation, the argument of the reference signals in the demodulator shown in Fig. 2.14b has to be similarly increased by ξ. The error probability is then given by simple rearrangement:

$$P_{E4} = 1 - \left\{1 - \frac{1}{2}\operatorname{erfc}\left[\sqrt{R}\sin(45° - \xi)\right]\right\} \cdot \left\{1 - \frac{1}{2}\operatorname{erfc}\left[\sqrt{R}\cos(45° - \xi)\right]\right\} \approx$$

$$\approx \frac{1}{2}\operatorname{erfc}\left[\sqrt{\frac{R}{2}}(\cos\xi - \sin\xi)\right] + \frac{1}{2}\operatorname{erfc}\left[\sqrt{\frac{R}{2}}(\cos\xi + \sin\xi)\right]. \tag{3.54}$$

If the conditional error probability is known, the error probability can be determined by the numerical evaluation of Eq. (3.52). The probability density of the reduced phase error can be calculated by Eq. (3.19) for the case of a transmitted reference signal, and by Eq. (3.49) or (3.50) for the case of a suppressed carrier signal. The factor α in these expressions is equal to the loop signal-to-noise ratio for first-order loops, and is only slightly different from this for second-order loops. The loop signal-to-noise ratio for a transmitted reference signal is given by $R_L = = 2A^2/N_0W_L$ while for the Mth power loop, the value of R_L given in Eq. (3.48) or (3.49) has to be substituted. For instance, for 4PSK transmission with transmitted reference signal,

$$\bar{P}_E = \frac{1}{2\pi I_0(\alpha)} \int\limits_{-\pi}^{\pi} e^{\alpha\cos\xi}\operatorname{erfc}\left[\sqrt{\frac{R}{2}}(\cos\xi + \sin\xi)\right]d\xi \tag{3.54a}$$

and for 4PSK transmission with suppressed carrier,

$$\bar{P}_E = \frac{2}{\pi I_0(\alpha)} \int_{-\pi/4}^{\pi/4} e^{\alpha \cos \xi} \operatorname{erfc}\left[\sqrt{\frac{R}{2}}(\cos \xi + \sin \xi)\right] d\xi. \tag{3.54b}$$

These two error probabilities are substantially different, the difference being most conspicuous if R goes to infinity for a finite α or R_L. In the case of a transmitted reference, P_E will then be equal to the probability of $|\xi| \geq \pi/M$ or

$$\bar{P}_E(R \to \infty) = 2 \int_{\pi/M}^{\pi} p(\xi) d\xi. \tag{3.54c}$$

On the other hand, for a loop taking the Mth power, the probability of $|\xi| > \pi/M$ is zero, so $\bar{P}_E = 0$ even for finite values of R_L.

The loop noise or part of it may be independent of the data channel signal-to-noise ratio, and an example for this case will be presented in Chapter 11. In most cases, however, the loop noise is proportional to the signal-to-noise ratio. In the transmitted reference case, let the signal power be given by A^2; $m^2 A^2$ is used for information transmission and $(1-m^2)A^2$ is used for reference signal transmission. Then

$$R_L = (1-m^2)\frac{2A^2}{N_0 W_L} = \frac{1-m^2}{m^2} R \frac{B}{W_L}. \tag{3.54d}$$

In a suppressed carrier system,

$$\tilde{R}_{M,L} = \frac{R}{M^2} \frac{B/W_L}{1+C_M/R}; \quad C_2 = \frac{1}{2}, \quad C_4 = \frac{9}{2}. \tag{3.54e}$$

It has been shown that in a suppressed carrier system, the phase ambiguity can be eliminated by differential encoding. If a symbol reception is in error because of the additive noise, then this faulty signal will appear twice in the differential decoder. Such a line error will thus result in two faulty symbols. It has been shown in Section 2.1 that Gray encoding has the effect of producing a bit error from each symbol error meaning that each decision error results in two bit errors in the differentially and Gray encoded output signal. Examination of this situation shows that if a symbol comprising the rth bit of the transmission is received erroneously then two faulty bits will be comprised in the M-bit group starting with the rth bit.

Figure 3.17 shows the error probability P_E of a 4PSK system as a function of the signal-to-noise ratio E/N_0 for different loop bandwidth values. For each curve, the receiver bandwidth B is ten times the loop bandwidth W_L, and E is the energy of one bit.

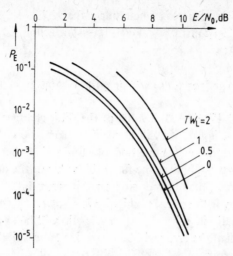

Fig. 3.17. Error probability of QPSK transmission as a function of the signal-to-noise ratio for a fourth power carrier recovery circuit. Parameter is the loop bandwidth; $B=10W_{\mathrm{L}}$.

Another consequence of PLL noise was discussed in Section 3.2: the total phase error φ can differ from ξ by $2k\pi$ or $2k\pi/M$, or, which is equivalent the value of φ exhibits jumps of $\pm2\pi$ or $\pm2\pi/M$ at random time instants. This has different consequences in transmitted reference systems and suppressed carrier systems [13, 14].

In suppressed carrier systems, a single bit error is introduced by every phase jump when the phase error $M\varphi$ is shifted from the original stable position by the value $\pi\leqq\varDelta\varphi\leqq3\pi$. In the following, this will be proved for 4PSK transmission, but it is also true for any value of M. The digits x and y of the dibit B can be expressed, without noise, as follows:

$$x = \operatorname{sign} \int_0^{2T} s(t)\,r_1(t)\,dt,$$

$$y = \operatorname{sign} \int_0^{2T} s(t)\,r_2(t)\,dt$$

(3.54f)

where the transmitted signal can be expressed as

$$s(t) = a(t)\cos(\omega_c t + \varPhi); \quad \varPhi = k\frac{\pi}{2}, \quad k = 0, 1, 2, 3$$

(3.54g)

and the two reference signals can be expressed as

$$r_{1,2}(t) = \cos(\omega_c t + \pi/4 + \varphi).$$

(3.54h)

The correlation time of $\varphi(t)$ is much longer than $2T$ as W_L is much less than B. As in the preceding, $\varphi(t)$ can be taken as constant. Simple rearrangement yields the following result:

$$\begin{Bmatrix} x \\ y \end{Bmatrix} = \text{sign} \left[\cos \Phi (\cos \varphi \mp \sin \varphi) + \sin \Phi (\mp \cos \varphi + \sin \varphi) \right]. \tag{3.55}$$

If φ crosses the $\pm 45°$ lines, the value of the dibit changes by ± 1 (in a mod 4 sense), and all following dibits will also differ from the correct values by ± 1. The final effect depends on the method of eliminating the phase ambiguity. If differential enconding is applied, the first dibit value of the differentially decoded signal will be incorrect, thus introducing one bit error, but all following bits will be correct. However, if the phase ambiguity is resolved by the recognition of the framing word then all bits following the cycle slip will be in error up to the next framing word in which the error will be recognized and suitable corrective action will be taken. The expected value of the bit errors following the cycle slip will be at least $K/2$ where K is the length of the frame. The actual number of bit errors is determined by the correction strategy applied.

It follows from the preceding considerations that in a suppressed carrier system with differential encoding, the error probability, including the effect of the reference signal noise, will be given by

$$P_E = 2\bar{P}_E + (1 - \bar{P}_E) T/T_c \tag{3.56}$$

while the error probability without differential encoding is given by

$$P_E = \bar{P}_E + L(1 - \bar{P}_E) T/T_c; \quad L \geqq K/2 \tag{3.57}$$

where T_c can be calculated from Eq. (3.20) or (3.21), and K is the frame length. The factor 2 in the first term of expression (3.56) shows the effect of the differential encoding while the factor $1 - \bar{P}_E$ in the second term shows that the actual number of bit errors will be reduced by one if the cycle slip and bit error take place simultaneously. In practice, $1 - \bar{P}_E \approx 1$.

We have mentioned that taking the Mth power is only one of the possibilities of eliminating the digital modulation of a carrier. However, the effect of cycle slip on the error probability is similar for other carrier recovery methods too.

Similar considerations can be applied to ascertain the effect of cycle slips in the case of a transmitted reference signal. It is then found that a cycle slip does not introduce bit errors.

Let us now compare the suppressed carrier 4PSK system with the transmitted reference 4PSK system by taking into account the effect of cycle slips. It is found that if only the loop is noisy and there is no additive noise, the ratio of error probabilities for the transmitted reference system to the suppressed carrier system is given by

$$\frac{P_{ET}}{P_{ES}} \approx \text{constant} \frac{\exp(-0.29\alpha)}{\exp(-0.089\alpha)} \tag{3.57a}$$

for large α values. This means that cycle slips will take place at a signal-to-noise ratio which is 5 dB higher than the signal-to-noise ratio at which the PLL reduced phase error $|\xi| \geq \pi/4$. The numerical value of 5 dB refers to a loop parameter value of $r=2$.

Figure 3.18 shows the error probability as a function of the signal-to-noise ratio R both by taking into account and by neglecting cycle slips, for a carrier recovery circuit applying a fourth power loop.

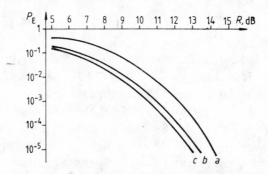

Fig. 3.18. Error probability of differentially encoded QPSK transmission as a function of the signal-to-noise ratio in the receiver band. a — Including the effect of cycle slips; b — neglecting cycle slips; c — with perfectly recovered reference signal.
Parameters: $B=1$; $T=1$; $W_L T=0.1$; $B/W_L=10$; $r=2$.

3.4.4 Comparison of different methods

First, the application field of transmitted reference systems and suppressed carrier systems will be discussed. We have already seen some qualitative comparisons. Quantitative comparisons are based on the error probability or the more easily calculated variance of the phase error φ or ξ, for a given total transmitter power for both systems.

In the transmitted reference system, the information is carried by m^2 times the transmitter power $(m^2 < 1)$. The error probability depends naturally on the value of m; the error probability will be lowest for a given value of m. (This can be understood by noting that the symbol error probability will be unity for both $m=1$ (no transmission of the reference signal) and $m=0$ (no transmission of the in-

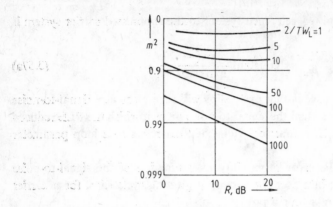

Fig. 3.19. Optimal value of m^2 as a function of the signal-to-noise ratio. The parameter is the relative loop bandwidth.

formation signal); in between these values, the error probability has to be minimal.) This question is discussed in detail in Ref. [3]. The optimal value of m depends on TW_L and R; a few functions are plotted in Fig. 3.19.

On the other hand, also according to Ref. [3], the error probability for a given signal-to-noise ratio will be smaller for a carrier recovery with an Mth power loop than for the case of a transmitted reference signal (at optimal level). The transmitted reference system is superior to the suppressed carrier system only for small signal-to-noise ratios and for relatively large loop bandwidth values. This fact determines the application fields of the two systems: the suppressed carrier system is definitely superior for relatively high bit rates and small error probabilities, as encountered in radio relay systems and communication satellite systems. In low bit rate systems, in which the bit frequency $1/T$ is smaller than the instability of the applied oscillators, the separate transmission of the reference signal may be of advantage. The telemetry and telecommand channels of spacecrafts fall into this category.

For suppressed carrier systems, two methods have been presented for the elimination of the recovered carrier phase ambiguity. The first method, $i.e.$, the application of differential encoding, has the advantage that a part of the transmitted information does not have to be known at the receive side, $i.e.$ this system may be called "transparent", the utilized circuits are simple, and the system is not very sensitive to cycle slips. A disadvantage is the duplication of errors by the appearance of two-bit error bursts. This is mainly disturbing when error correcting codes are applied.

The second method, $i.e.$ the method based on the recognition of the framing word, is rather complicated, and cycle slips are followed by a very long burst of errors. However, if the probability of cycle slipping is negligible then single errors will appear. In order to eliminate the duplication of errors, PLL variants have been published with extremely low probability of cycle slipping [15]; this could eliminate the application of differential encoding.

3.4.5 Effect of oscillator instability

It has been assumed during the above qualitative investigations that the frequencies of the transmitter oscillator and the receiver VCO are constant and their frequency difference Δf is even zero. In practical systems, this is only approximately true. The actual frequency of oscillators differs from the nominal value and shows time fluctuations. Also, phase noise is generated which can be regarded as being fairly stationary and is made up of the following components: frequency flicker noise with a spectral density proportional to $1/f^3$; white frequency noise with a spectral density proportional to $1/f^2$; phase flicker noise with a spectral density of $1/f$, and white phase noise (constant spectral density). A detailed discussion of oscillator noise is presented in Ref. [16].

Transmission performance is affected by the instability and noise of the oscillators in the transmission equipment. We have seen that the pull-in range of the PLL is approximately equal to its bandwidth which can thus not be reduced below a lower limit; this means that the loop noise cannot be decreased arbitrarily. It should also be noted that Eqs (3.19), (3.49) and (3.50) apply only for $\Delta f = 0$. For $\Delta f \neq 0$, $p(\xi)$ can be expressed in an integral form [1]; increasing Δf will also increase the variance of ξ. A further degradation is introduced by the fact that the phase of the transmitter oscillator can be tracked by the receiver PLL only with a certain phase error.

Because of all these effects, the performance parameters described in Section 3.4.3 are valid for the ideal case only. A detailed investigation, though feasible according to Refs [1, 16], will not be carried out. However, it can be briefly stated that the ideal case does not differ much from the actual case, provided the bit rate is sufficiently high, say $1/nT\Delta f > 15$, and the noise of the oscillators stays within reasonable limits. For low bit-rate systems, this is no longer true, and owing to the poor carrier recovery performance, a non-coherent system would yield better results.

The pull-in range can also be increased by some acquisition aid, even without increasing the bandwidth. According to a method which can be regarded as an active acquisition aid, unlocking the PLL enables a sweep oscillator which slowly detunes the VCO. As soon as the incoming signal f_c is sufficiently approximated by the VCO signal frequency f_N so that f_c lies within the pull-in range, locking will

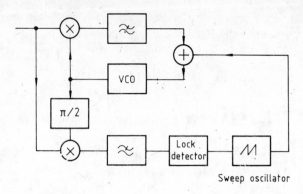

Fig. 3.20. PLL with lock detector and sweep oscillator.

take place, and the sweep oscillator will be disabled. The essential parts of this system are thus the lock detector and the sweep oscillator. The system is shown in Fig. 3.20, and its analysis is presented in Ref. [8].

In Refs [18a, 18b], a passive acquisition aid is described. According to this system, the incoming signal frequency is estimated by a so-called spectrum discriminator, and the result of this estimate is used to tune the VCO so as to bring the input signal frequency within the pull-in range. Figure 3.21 shows a variant of this system which is applicable to the squaring loop utilized for 2PSK transmission. The variant given in Ref. [18b] allows the application of coherent demodulators even for oscillators which have a frequency error higher than $1/T$.

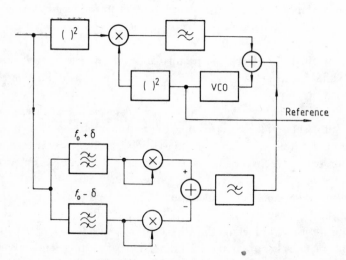

Fig. 3.21. Squaring loop comprising a spectrum discriminator.

3.5 Transmission of the timing information: clock recovery

One of the main advantages of digital over analog transmission is the possibility of regenerating the received digital signal. Regeneration is essentially a sampling procedure, carried out at a suitable time instant (see Sections 2.1 and 2.2). At this "suitable" time instant, the signal-to-noise ratio is highest and the possibility of interference is lowest. To make this time instant known to the receiver, the appropriate information has to be somehow transmitted from the transmitter to the receiver. In this section, the methods of obtaining this timing or clock information at the receive side will be outlined, and the effects of timing errors will be evaluated.

3.5.1 Methods of clock recovery

Theoretically, the clock information can be transmitted over a separate channel but this method is seldom applied. In the following, the widely used methods of recovering the clock information from the digital information signal will be discussed.

The clock information related to the bit time T is comprised in the transitions of the transmitted digital signal which should therefore comprise a sufficiently high number of transitions. On the other hand, it has been shown in Section 2.3.1 that the spectrum of the transmitted signal does not contain a line component of clock frequency $1/T$ if random modulation is applied; this component therefore has to be generated by some nonlinear operation. Variants of clock recovery circuits with nonlinear elements are shown in Fig. 3.22. Taking only a random bit stream

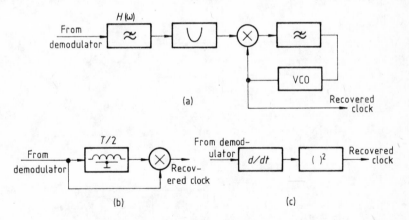

Fig. 3.22. Clock recovery loops: (a) utilizing an even-order nonlinearity; (b) and (c) utilizing other types of nonlinearities.

into account in which the transition probability is 0.5, only the latter problem (the generation of a clock frequency spectrum line) has to be dealt with. In this case, an even-order nonlinearity is applied as shown in Fig. 3.22a. The low-pass filter designated by $H(\omega)$ can represent the receiver filter, e.g., it can be a matched filter. The narrow-band filter following the nonlinear network may be realized by a PLL, and the optimal input–output characteristic of the nonlinear element, as given in Ref. [11], is expressed by $y=\ln \cosh x$. The nonlinearity can be realized not only by a network having a nonlinear input–output characteristic; examples for other variants are shown in Figs 3.22b and 3.22c.

Owing to the noise of the received signal, all circuits shown in Fig. 3.22 will have a timing error ΔT which has, according to Ref. [19], a variance of

$$\sigma_{\Delta T}^2 \approx T^3 B_L/R; \quad R = E/N_0. \tag{3.58}$$

A better solution is obtained by placing the nonlinearity within the closed loop [11, 9]. Figure 3.23 shows such a circuit called "early–late gate circuit". With optimal design, the timing error variance of this circuit is about 5 dB less compared to the circuits given in Fig. 3.22:

$$\sigma_{\Delta T}^2 = \frac{1}{3}\frac{T^3 B_L}{R}. \tag{3.59}$$

In deriving expressions (3.58) and (3.59), it has been assumed that each bit period contains a level transition. (Practically, this could be realized by applying antipodal RZ (Return-to-Zero) signals, i.e. pulses of lengths less than T.)

This is not the case for NRZ signals: level transitions appear only at 0–1 or 1–0 bit transitions. The clock recovery circuit will then receive input information only at these transitions. This situation is similar to the transmitted reference,

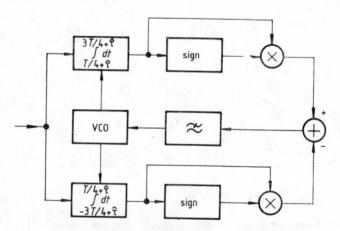

Fig. 3.23. "Early–late gate" clock recovery circuit.

time division multiplexed carrier recovery circuit, as discussed in Section 3.4. This phenomenon introduces an additional timing error, even without noise.

According to the preceding statement, the accuracy of clock recovery depends basically on the transmitted information. We have seen that the stochastic nature of the incoming signal results in an increase in the timing error variance, and the clock synchronization can even be completely lost by a long run of zeros or ones. Some kind of countermeasure is therefore required if the probability of these long runs without transitions is substantial: either the clock synchronizing circuit should be designed to tolerate these long runs without transitions with acceptable σ^2, or else the number of transitions should be increased artificially.

The latter solution can be realized by several methods. For instance, a suitable modulation, comprising transitions at all bit intervals, is the pseudo 8PSK modulation applied frequently in data transmission. In this case, the phase at the time instant of the lth symbol will be

$$\Phi_K = \frac{k\pi}{2}; \quad k = 0, 1, 2, 3 \tag{3.59a}$$

and at the time instant of the next $(l+1)$th symbol will be

$$\Phi_K = \frac{(k+1/2)\pi}{2}; \quad k = 0, 1, 2, 3. \tag{3.59b}$$

It is seen that there will be a phase change of $(2k+1)\pi/4$ at the end of each symbol.

Another possibility is to apply, instead of NRZ signals, a modulating signal set which comprises a transition during each bit interval. The biphase signal set meets this requirement:

$$s_1 = -s_2 = \begin{cases} 1; & 0 \le t \le T/2 \\ 0; & T/2 < t \le T. \end{cases} \tag{3.59c}$$

A further possibility is the application of the so-called scrambler: the signal to be transmitted is suitably combined with a pseudo random signal (of equal bit rate), thus generating an approximately random bit stream.

3.5.2 Effect of the recovered clock signal timing error

The performance of the clock recovery can be expressed by the reduced phase error and the number of cycle slips, similarly to the carrier recovery.

The effect of cycle slips in the recovered clock signal is much more pronounced than for carrier recovery. We have seen that the cycle slips of the recovered carrier

results in one or not even one bit error. However, cycle slips of the recovered
clock have the effect of introducing one additional bit or deleting one bit. In most
cases, this will have the effect of a substantial number of bit errors: frame synchronization may be lost and has to be re-established; in error correcting systems, the
block structure may fail, etc. Numerical values can only be quoted for particular
cases, but it is evident that only a very small frequency of cycle slips can be permitted.

The effect of the clock signal timing error will now be calculated for NRZ
signals. If a binary signal is sampled, owing to the clock signal timing error, at
time instants differing from $t=T$, then the error probability will depend on the
relative deviation $\lambda = \Delta T/T$. If there is no transition, the error probability will
be independent of λ. If there is transition, the error probability will be higher
because of the finite signal rise time. As the transition probability is 0.5, we have

$$P_{\mathrm{E}}(\lambda) = \frac{1}{4}\,\mathrm{erfc}\,(\sqrt{R_{\mathrm{s}}}) + \frac{1}{4}\,\mathrm{erfc}\,[\sqrt{R_{\mathrm{s}}}\,(1 - 2|\lambda|)]; \quad |\lambda| \leq 1/2. \tag{3.60}$$

The expected value of the error ratio is given by

$$\bar{P}_{\mathrm{E}} = \int\limits_{-1/2}^{1/2} P_{\mathrm{E}}(\lambda)p(\lambda)\,d\lambda \tag{3.61}$$

where $p(\lambda)$ is the probability density function of λ. Figure 3.24 shows the dependence of the error probability on the clock signal-to-noise ratio for different σ_λ^2
variance values, by assuming that λ has a Tikhonov distribution discussed in
Section 3.2.

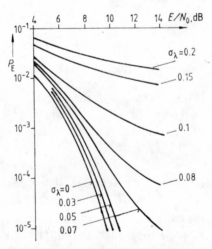

Fig. 3.24. Error probability as a function of signal-to-noise ratio in the case of an imperfectly
transmitted clock signal. NRZ signal, 2PSK modulation.

3.6 References

[1] Lindsey, W.: *Synchronization Systems in Communication and Control.* Prentice-Hall, Englewood-Cliffs 1972.

[2] Viterbi, A.: *Principles of Coherent Communications.* McGraw-Hill, New York 1966.

[3] Stiffler, J.: *Theory of Synchronous Communication.* Prentice-Hall, Englewood-Cliffs 1971.

[4] Van Trees, H.: *Detection, Estimation and Modulation Theory.* Part II, Wiley, New York 1971.

[5] Meyr, H.: Nonlinear analysis of correlative tracking systems using renewal process theory. *IEEE Trans. Commun.,* Vol. Com-23, pp. 192–203, Febr. 1975.

[6] Ryter-Meyr, H.: Theory of phase tracking system of arbitrary order. *IEEE Trans. Inf. Theory,* Vol. IT-24, pp. 1–7, Jan. 1978.

[7] Ascheid, G., Meyr, H.: Cycle slips in phase-locked loops. A tutorial survey, *IEEE Trans. Commun.,* Vol. Com-30, pp. 2228–2241, Oct. 1982.

[8] Holmes, J.: *Coherent Spread Spectrum Systems.* Wiley, New York 1982.

[9] Van Tress, H.: *Detection, Estimation and Modulation Theory.* Part I, Wiley, New York 1971.

[10] Schiff, L.: Burst synchronization of phase locked loops. *IEEE Trans. Commun.,* Vol. Com-21, pp. 1091–1100, Oct. 1973.

[11] Zimmerman, M., Kirsch, A.: The AN/GSV-10 variable data rate modem. *IEEE Trans. Commun. Tech.,* Vol. Com-15, pp. 197–204, Apr. 1967.

[12] Frigyes, I.: Behaviour of fourth power carrier recovery networks in noisy and band limited environment (in Hungarian). *Híradástechnika,* Vol. 30, pp. 23–28, Jan. 1979.

[13] Frigyes, I.: Voice Under Data: A method for the transmission of hybrid information over microwave links. In: *Proc. 5th EMC, Paris,* pp. 615–619, 1978.

[14] Frigyes, I.: Comments on *M*-ary CPSK detection. *IEEE Trans. Aerosp. Electron. Syst.,* Vol. AES-18, pp. 395–397, May 1980.

[15] Kurihara, O.: Carrier recovery with low cycle slipping rates for CPSK/TDMA system. *GLOBECOM 83, San Diego,* 1983.

[16] Lindsey, W.: *Modelling, characterization and measurement of oscillator frequency instability.* NTIS Report AD 785 862, 1980.

[17] Proakis, W., Drouilhet, P., Price, J.: Performance of coherent decision directed channel measurement. *IEEE Trans. Commun. Tech.,* Vol. Com-17, pp. 484–497, July 1969.

[18] Huzii, A., Kondo, S.: On the timing information disappearance of digital transmission systems. *IEEE Trans. Commun.,* Vol. Com-21, pp. 1072–1074, Sept. 1973.

[18a] Frigyes, I., Berceli, T., Szabó, Z.: A new method for carrier acquisition in suppressed carrier digital microwave radio. In: *Proc. 11th Europ. Microwave Conf., Amsterdam,* pp. 341–346, 1981.

[18b] Frigyes, I., Berceli, T., Szabó, Z.: New concepts in narrow-band coherent digita. radio systems, *GLOBECOM 82, Miami, Florida,* paper No. F. 3.2., 1982.

[19] Bennett, W. R., Davey, J. R.: *Data Transmission.* McGraw-Hill, London, pp. 260–267, 1965.

[20] Bylanski, P., Ingram, D. G. W.: *Digital Transmission Systems.* Peter Peregrinus, London, pp. 170–215, 1976.

[21] Spilker, J. J.: *Digital Communications by Satellite.* Prentice-Hall, Englewood-Cliffs, pp. 429–454, 1977.

[22] Lajtha, Gy., Lajkó, S.: *PCM in Telecommunication* (in Hungarian). Műszaki Könyvkiadó, Budapest Sect. 1.7 and 2.7, 1978.

[23] Hobdell, J. L.: High-speed regeneration of phase-shift keying signals phase-script logic. *IEEE,* Vol. 124, No. 12, pp. 1148–1154, 1977.

[24] Komaki, S.: GaAs MESFET regenerator for phase-shift keying signals at the carrier frequency. *IEEE Trans. Microwave Theory Techn.*, Vol. MTT-24, No. 6, pp. 367–372, 1976.

[25] Koga, K.: On-board regenerative repeaters applied to digital satellite communications. *Proc. IEEE,* March, pp. 401–410, 1977.

[26] Gupta, S. C.: Phase-locked loops. *Proc. IEEE,* Vol. 63, No. 2, pp. 219–306, 1975.

[27] Schiff, L.: Burst synchronization of phase-locked loops. *IEEE Trans. Commun.,* Vol. Com-21, No. 10, pp. 1091–1099, 1973.

[28] Takasaki, Y.: Analysis of nonlinear timing extraction in pulse transmission. *Electron. Commun. Jpn.,* Vol. 55-A, No. 12, pp. 1–9, 1972.

[29] Mueller, K. H.: Timing recovery in digital synchronous data receivers. *IEEE Trans. Commun.,* Vol. Com-24, No. 5, pp. 515–530, 1976.

[30] Roza, K.: Analysis of phase-locked timing extraction circuit for pulse code transmission. *IEEE Trans. Commun.,* Vol. Com-22, No. 9, pp. 1236–1249, 1974.

[31] Feher, K.: A new symbol timing recovery technique for burst modem applications. *IEEE Trans. Commun.,* Vol. Com-26, No. 1, pp. 100–108, 1978.

[32] Qureshi, S. U. H.: Timing recovery for equalized partial response systems. *IEEE Trans. Commun.,* Vol. Com-24, No. 12, pp. 1326–1331, 1976.

[33] Hogge, L.: Carrier and clock recovery for 8PSK synchronous demodulation. *IEEE Trans. Commun.,* Vol. Com-26, No. 5, pp. 528–533, 1978.

[34] Cessna, J. R.: Phase noise and transient times for a binary quantized digital phase-locked loop in white Gaussian noise. *IEEE Trans. Commun.,* Vol. Com-20, No. 4, pp. 94–104, 1972.

[35] Holmes, J. K.: Performance of a first-order transition sampling digital phase-locked loop using random walk models. *IEEE Trans. Commun.,* Vol. Com-20, No. 4, pp. 119–132, 1972.

[36] Hurst, G. T.: Quantizing and sampling considerations in digital phase-locked loops. *IEEE Trans. Commun.,* Vol. Com-22, No. 1, pp. 68–72, 1974.

[37] Holmes, J. K.: A second-order all digital phase-locked loop. *IEEE Trans. Commun.,* Vol. Com-22, No. 1, pp. 62–67, 1974.

[38] Kirlin, R. L.: An all digital second-order PLL. *IEEE Trans.,* AES-10, No. 5, Sept. pp. 710–712, 1974.

[39] Yamanoto, S.: Performance of a binary quantized all digital phase-locked loop with a new class of sequential filter. *IEEE Trans. Commun.,* Vol. Com-26, No. 1, pp. 35–45, 1978.

[40] D'Andrea, P.: A binary quantized digital phase-locked loop: a graphical analysis. *IEEE Trans. Commun.,* Vol. Com-26, No. 9, pp. 1355–1364, 1978.

[41] Martin, J. D.: Bit timing for telecontrol communication systems. In: *Proc. Digital Processing of Signals in Communications. Loughborough,* pp. 321–330, 1977.

[42] Ványai, P.: Timing extraction from noisy biphase signals. In: *Proc. Digital Processing of Signals in Communications. Loughborough,* 1977.

[43] Frigyes, I.: Scramblers, bit sequence independence and the design of bit-synchronizing circuits in digital microwave transmission. In: *Proc. 6th Colloq. Microwave Commun., Budapest,* Vol. I. 1978.

[44] Nosaka, K.: PSK demodulator with delay line for the PCM–TDMA system. *IEEE Trans. Commun. Technol.,* Vol. Com-18, No. 4, pp. 427–434, 1970.

[45] Ványai, P.: Bit-synchronization of digital transmission system using digital phase-locked loop. *Proc. 6th Summer Symp. Circuit Theory. Prague,* pp. 615–619, 1982.

[46] Lindsey, W. C., Chak, S.: A survey of digital phase-locked loops. *Proc. IEEE,* Vol. 69, No. 4, pp. 410–431, 1981.

[47] Proakis, J. G.: *Digital Communication.* McGraw-Hill, New York 1983.

[48] Feher, K.: *Digital Communication.* Prentice-Hall, Englewood-Cliffs, 1983.

CHAPTER 4

PROPERTIES OF THE MICROWAVE TRANSMISSION PATH

4.1 Introduction

In Chapters 2 and 3, the theory of digital signal transmission, together with the synchronization tasks involved, has been surveyed, and the more important disturbing effects enumerated. Subsequently, the methods for providing optimal immunity against these disturbing effects and the signal sets minimizing the disturbances themselves have been discussed. In connection with some disturbing effects, the optimization problem has not been dealt with, but instead the disturbances in heuristically assumed structures have been discussed. It has been shown that the error probability, characterizing the performance of the digital transmission, is a monotonically decreasing function of the carrier power appearing at the receiver input; this statement applies for cases in which the properties of the transmission system can be determined by the designer, and is valid for the effects of additive noise, interferences and synchronization inaccuracies. However, there are at least two reasons for carrying out further investigations regarding the properties of the microwave transmission path. The design parameter which can be suitably chosen is the transmitter output power P_t rather than the receiver input power C, and thus a knowledge of the free space attenuation between the transmit and receive antennas is required. On the other hand, the transmission path may introduce additional distortion. In the following, this transmission path will be assumed to cover that part of the microwave link which extends between the input port of the transmit antenna and the output port of the receive antenna. It will be shown that the model of this path is a linear two-port network which has, in general, a frequency dependent attenuation. In low speed systems, this frequency dependence can often be neglected but in high speed systems, the dependence may have a decisive effect on the system performance. The magnitude and the frequency dependence are also time dependent but this change in time is nearly always negligible with respect to the signal speed in question.

The time dependence of the transmission properties of the channel, primarily the fluctuation of the attenuation, is called fading. In this chapter, we will discuss primarily fading phenomena, for the purpose of the design of digital microwave transmission systems. Furthermore, procedures will be presented which can be applied to counteract the performance degradation due to fading. This material will be used to specify the transmission systems and to elaborate a systematic method for the determination of the main system parameters. The physical background of the fading phenomena will not be discussed but relevant references (*e.g.*, [1] and [2]) are given.

If the transmit and receive antennas are placed in a medium which is homogeneous, lossless, and has time invariant material properties, then the path attenuation will be constant and will have a value of

$$a_{s0} \stackrel{\triangle}{=} \frac{P_t}{C_0} = \frac{(D\lambda)^2}{A_1 A_2} = \left(\frac{4\pi D}{\lambda}\right)^2 \frac{1}{G_1 G_2} \tag{4.1}$$

where P_t is the transmitter power, C_0 is the received power in the case of free space propagation, D is the path length, A_1 and A_2 are the effective areas of the transmit and receive antennas, respectively, λ is the wavelength, and G_1 and G_2 are the gains of the transmit and receive antennas, respectively.

A rare example for approximately free-space propagation is the propagation between two satellites which are quite remote from the earth. However, in operating microwave transmission systems, generally no free-space propagation takes place. For instance, in a terrestrial radio relay section, microwaves propagate at most within the lower few hundred meters of the troposphere in which the electrical parameters of the air (permittivity, refractive index and possible attenuation) are inhomogeneous and varying in both space and time; the microwaves are partly reflected, and partly absorbed by the terrain. In an earth-satellite link too, the refractive index of the air is not constant in the vicinity of the earth.

Taking into account the inhomogeneities and their time dependence, the path attenuation will not be constant as given in Eq. (4.1), but will be time dependent and possibly also frequency dependent as given in the following equation:

$$a_s(t, f) = a_{s0} A(t, f) \tag{4.2}$$

where $A(t, f)$ is the so-called fading attenuation. Generally, this cannot be described by deterministic means but has to be regarded as a stochastic process (normally characterized by two independent parameters). Strictly speaking, this process is not stationary but shows seasonal and diurnal statistical changes. However, it has been shown in Ref. [3] that for all practical cases, the fading attenuation can be assumed to be stationary so we shall not complicate our discussions with the treatment of a non-stationary model which would not give additional information.

Equation (4.2) shows explicitly that the fading attenuation is a function of frequency and time. However, in many cases, the frequency dependence is slow which has the effect that the attenuation is nearly constant within the frequency band of one channel. This kind of fading is called wideband or non-selective fading. For low and medium capacity systems, the fading can nearly always be assumed to have wideband or non-selective properties but for high capacity systems, the selectivity of the fading has to be taken into account. On the other hand, the time dependence of the fading is nearly always slow with respect to the bit time or symbol time. Should this (exceptionally) not be the case, then the fading is called fast fading. The fast or slow, and the selective or wideband character of the fading are independent properties which can appear in any of the four possible combinations. In the following, the effects of both wideband and selective fading will be investigated. For our purposes, the distortion effects of the fast fading are always negligible. However, the rate of change of the fading may be significant for the design of some circuits which will be taken into account. Considering the design of transmission systems, the most important effect of the fading phenomenon is to render the transmission medium itself not absolutely reliable. Within a small but finite time percentage, the high path attenuation or significant distortion due to fading will result in a high error probability, *i.e.* interruption of the transmission. One of the most important tasks of system design is to reduce the fraction of this interruption to a tolerable small value. The purpose of this chapter is the presentation of suitable methods to obtain this result.

4.2 Terrestrial radio relay — low and medium capacity systems

It has already been mentioned in the previous section that for these systems, fading in the microwave channel can be considered to be wideband and slow, resulting in a slowly varying channel attenuation. In the following, a suitable statistical description of the fading attenuation will be given, and through the knowledge of this attenuation, the main parameters of the transmission equipment will be determined. Throughout our discussions, line-of-sight transmission will implicitly be assumed.

4.2.1 Sources of fading, statistical properties of the path attenuation

In the microwave frequency range, there are two main sources of fading attenuation: interference generated by the multipath propagation of the waves, and absorption due to heavy rainfall. The first of these sources is present in all microwave frequency bands while the second has significance only in the frequency bands above 10 GHz. Further losses are present at frequencies above 20 GHz due to gas and water vapour absorption, but these will not be discussed because of their smaller significance.

During the last four or five decades, considerable research work has been carried out investigating attenuation due to interference between electromagnetic waves, and detailed results are available for a large number of propagation paths. According to the central limit theorem, the received field strength in this case can be considered to be Gaussian, and accordingly, its amplitude has a Rayleigh distribution. If one of the wave amplitudes is significantly higher than those of the others, the Rice or Nakagami distribution will give the best approximation. These distributions are suitable for characterizing for example an HF channel or a mobile VHF or UHF channel in an urban surrounding. However, the number of propagation paths is not too large in microwave links, resulting in a less active transmission channel, compared with the channel corresponding to the Rayleigh, Rice or Nakagami distribution.

One of the best presentations of the multipath attenuation phenomena in a microwave channel is given in Ref. [4], while the theoretical background is discussed in Ref. [5]. According to these references, the probability of the fading attenuation exceeding a specified value of A (expressed as a power ratio) is given by

$$F(A) = \frac{6 \times 10^{-7} abfD^\delta}{A} \tag{4.3}$$

where a and b are constants characterizing the soil and the climatic zone, f is the frequency in GHz, and D is the length of the radio section in km. The constants are strongly dependent on the geographical location, their approximate values being the following: $a=4$, 1 or 1/4, for a link above a water surface, above an average soil or above an extremely dry soil, respectively; $b=1/2$, 1/4 or 1/8, in a tropical, temperate or extremely dry climatic zone, respectively.

The accuracy of the equation is reasonable if the value of A is higher than 10 dB, *i.e.* the power ratio is higher then 10. (It should be noted that for high attenuations, the distribution function $F(A)$ is inversely proportional to A in the Rayleigh case too, but in this case, it is not proportional to fD^3.) Relation (4.3) is a suitable statistical description of the non-selective multipath microwave channel.

Attenuation due to rainfall has also been extensively investigated, and the first significant results come from as early as the beginning of the century [6]. The effect of rainfall on the wave propagation in the vicinity of the earth has been investigated for nearly all Western European countries in a project reported in Ref. [7], and the results are presented in [8]. The rainfall attenuation properties which are important for our investigations will be summarized in the following.

The absorption of electromagnetic waves by raindrops results in attenuation of the waves. In principle, the dB value of the attenuation is proportional to the section length, *i.e.* $A^{dB} = \alpha D$. The attenuation factor α depends, among others, on the rain rate I which has dimensions of mm/h, the polarization of the waves, and the frequency. In the ideal case when the rain drops are spherical and the rain rate is constant along the whole section, the attenuation factor is given by

$$\alpha = kI^c \tag{4.4}$$

where k and c are functions of the frequency; empirical values for these constants are found in Ref. [8]. In the frequency range of 10 to 20 GHz, the frequency dependence can be expressed by the following simple relation:

$$\alpha = 0.0266I^{1.137}\frac{f-6}{7} \text{ dB/km} \tag{4.5}$$

where I has the dimension of mm/h, and f is written in GHz.

In the practical case, the rain drops are not spherical, and the attenuation depends on the wave polarization; the attenuation of horizontally polarized waves is higher, but the deviation in the given frequency range is within 20 per cent. The attenuation factor depends in practice also on the length of the section because the rain rate is not constant along the section: usually, the extension of heavy rain cells does not cover the complete length, resulting in an attenuation which changes along the section. On the other hand, we can define an equivalent rain rate so that the attenuation can be expressed in the following form:

$$A = D\alpha(I_{eq}) \tag{4.5a}$$

where α can be calculated from expression (4.4) or (4.5) if I_{eq} is known.

It follows from the preceding investigations that the estimation of the rainfall attenuation requires the knowledge of statistical rain rate data. The difficulties of obtaining these data will not be detailed but we indicate two important aspects. First, the number of heavy rainfalls can by no means be concluded from the annual distribution of precipitation. For example, in the western coastal districts of Mediterranean countries, the probability of heavy rainfall is much higher than in the much wetter Western European countries. The second aspect is that during short intervals of a few minutes, the rainfall may be much heavier than given by the data of the Meteorological Institutes, based on measurements integrated over one hour or even longer intervals.

Let us now consider the parameters which are needed for taking into account the rainfall attenuation in the design of digital microwave transmission systems. The most important parameter is the probability distribution function of the rain rate I. In Ref. [8], fairly detailed and accurate data are presented for the previously mentioned European countries. Instead of these, the worldwide data given in Ref. [14] will be presented; these data are less accurate and less detailed, and this latter property makes them more suitable for preliminary system design. Figure 4.1 presents the five distributions which are applicable in individual areas. Further,

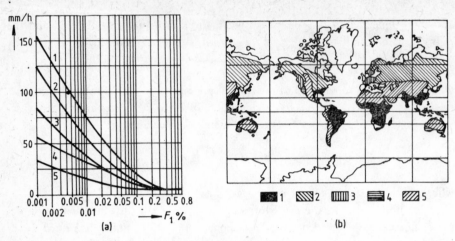

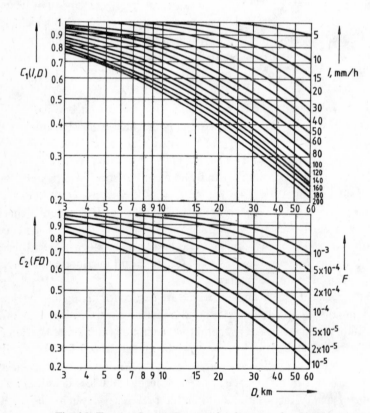

Fig. 4.1. (a) Probability distributions of rain rate; (b) areas corresponding to the distributions given.

Fig. 4.2. Factors C_1 and C_2 appearing in expression (4.5b).

the equivalent rain rate I_{eq} has also to be determined for a given time percentage, section length and rain rate. This can be calculated from a formula given in Ref. [15]:

$$I_{eq} = \sqrt{C_1 C_2}\, I \qquad (4.5b)$$

where the values of C_1 and C_2 are given in Fig. 4.2.

Summarizing, it can be stated that the probability of a rain attenuation of at least A dB on a hop length D is F_1. A is given by the expression

$$A = 0.0266 I_{eq}^{1.137} D\, \frac{f-6}{7}. \qquad (4.5c)$$

I_{eq} is obtained from Eq. (4.5a), C_1 and C_2 are obtained from Fig. 4.2, and I and F_1 are obtained from the appropriate distribution curve of Fig. 4.1. D is substituted in km, f in GHz, and A is obtained in dB.

The distribution of I is well approximated by the log–normal distribution. It then follows from Eq. (4.4) that α will also have a log–normal distribution as α is proportional to some power of I.

4.2.2 Design for reliability

In most cases, microwave transmission systems are specified by the performance of a hypothetical path; for instance, the error probability at the output of a 2500 km radio relay link is specified by the CCIR [9]. In order to take into account the inherent unreliability of the microwave channel, the specification normally has two parts: a relatively small error probability is specified which may be exceeded for small percentage of the time, and a higher error probability is also specified which may be exceeded for even smaller time percentage. As a practical example, in 99 per cent of the time, the error probability may not exceed 10^{-7}, and in 99.95 per cent, the error probability may not exceed 10^{-3}. In the remaining 0.05 per cent of the time, the channel may have an unacceptable quality. For nearly all low and medium capacity systems, the fulfilment of the higher error probability requirement will automatically satisfy the lower error probability requirement as well.

In long distance radio relay links, several repeater stations have to be used because of the earth's curvature and land obstructions. Therefore, the main parameters of the system can only be correctly determined if the specifications for the hypothetical path are broken down for a single radio section. To this end, the following procedure, given in Ref. [10], will be applied.

By considering that the error ratio of a single radio section is a function of the signal-to-noise ratio, the following average error ratio can be defined:

$$\bar{P}_E = \int_0^\infty P_E(R)p(R)\,dR \qquad (4.6)$$

where R is the carrier-to-noise ratio, $p(R)$ is the probability density of R, and $P_E(R)$ is the conditional probability of error, the condition being that the carrier-to-noise ratio is just equal to R.

Expression (4.6) shows that the average error ratio thus defined will be very high because $p(R) \neq 0$ even in the vicinity of $R=0$. On the other hand, the link performance in this vicinity will be unacceptable anyway. It is therefore justifiable to interrupt the connection at small R values. In this case, the corrected expression of the average error ratio will be given by

$$\bar{P}_E = \int_{R_s}^\infty P_E(R)p(R)\,dR \qquad (4.7)$$

where R_s is the carrier-to-noise ratio on interruption of the transmission. This leads us to another performance parameter, the probability of link interruption:

$$P_m = \int_0^{R_s} p(R)\,dR. \qquad (4.8)$$

We shall come back to the determination of $p(R)$ which is possible through the knowledge of the fading source as given in Section 4.2.1. $P_E(R)$ depends on the modulation system, the equipment quality, the interference surrounding and other factors. For a two-dimensional modulation system and for sufficiently high carrier-to-noise ratios which are due to Gaussian noise only, a tight upper bound is given by the formula

$$P_E = e^{-(d/2)^2/N_0} \qquad (4.9)$$

where d is the smallest Euclidean distance between the signal space vectors, and N_0 is the spectral density of the noise (see Section 2.2). This formula can also be expressed with practically more useful parameters:

$$P_E = \exp\left[\frac{\ln P_s}{R_s}R\right] \qquad (4.10)$$

where P_s is the highest error probability at which transmission performance is acceptable (e.g., 10^{-3} in practical systems), and R_s is the carrier-to-noise ratio needed for this error probability. Expressions (4.9) and (4.10) are accurate at small error probabilities, but P_E generally is sufficiently small.

If the further sources of impairment given above have also to be taken into account, then Eq. (4.9) is still valid but R_s is then not an exactly calculated parameter but an empirical constant which is higher than the theoretical R_s and depends on the equipment in question and its surrounding.

It should be noted that if R goes to infinity, then the error probability does not in general go to zero but to a finite small value. This is due to imperfect circuit operation, mains disturbances and other impairments. The universal empirical formula of the error probability is thus given by

$$P_E = \exp\left[\frac{\ln P_s}{R_s} R\right] + P_{E0} \tag{4.11}$$

where P_{E0} can be called the residual error probability.

Let us now assume that the radio relay connection is made up of N radio sections, each having the same length D. The total length of the hypothetical connection is thus ND. (In a real link, the individual radio sections have different lengths, but for design purposes, this assumption is justified.) Let us introduce the allowable interruption ratio per km and denote it by S. The interruption probability of the total path will then be given by $NP_m = NSD$. The permitted value of this parameter will be equal to that time percentage in which the error probability may be higher than P_s (in our above example, higher than 10^{-3}, in 0.05 per cent of the time). The first relation for system design is thus given by

$$P_m = SD. \tag{4.12}$$

There are two kinds of events which may cause interruption, the equipment failures (including those due to mains outage or other infrastructural causes), and deep fades giving rise to a signal-to-noise ratio which is less than R_s, the fading being caused by either rainfall or multipath propagation. Deep fades due to multipath propagation and rainfall are in practice two mutually exclusive events because heavy rain during thunderstorms is accompanied by strong winds while multipath fading occurs primarily at dawn in summer, in quiet weather. Equipment failure is independent of these two events but the simultaneous occurrence of equipment failure with any of the two fadings has a negligible probability. It thus follows that the total probability of interruption can be regarded as being equal to the sum of probabilities due to any of the above reasons.

Let us now introduce the following notation: S_1 — interruption ratio due to rain fading; S_2 — interruption ratio due to multipath fading; S_3 — interruption due to equipment failure. The following relation then holds:

$$S = S_1 + S_2 + S_3. \tag{4.13}$$

One of these interruptions can be expressed directly:

$$S_3 = \frac{MTTR}{MTBF + MTTR} \cdot \frac{1}{D} \qquad (4.14)$$

where the well known abbreviations of MeanTime To Repair and Mean Time Between Failures have been used. The design methods for achieving these reliability parameters are beyond the scope of this book, and S_3 will therefore be assumed to be a given quantity. Its knowledge allows the calculation of the interruption ratio due to fading by the equation $S_1 + S_2 = S - S_3$.

The next design step is the determination of the fade margin. This is defined as a fading attenuation which, in a section of length D, has a probability of $(S_1 + S_2)D$. The two fading sources result in substantially different fading attenuations with equal probabilities, and so for given values of S_1 and S_2, the higher of the two fade margins should be taken into account. The separate values of S_1 and S_2 will be optimal if both require identical fade margins which can be calculated from the formulas in Section 4.2.1. From these quantities, the smallest sufficient fade margin can be determined, and this allows calculation of the equipment parameters satisfying the outage requirement: the transmitter power, the receiver noise figure and the antenna gains. Two of these should be assumed, and the third is calculated.

The second performance parameter is the error probability which should also be determined on a per kilometer basis, as with the interruption ratio; if the error probability/km is denoted by P then the total reference path has an error probability of PND, so the error probability permissible for one section is given by

$$PD = P_{E0} + \int_{R_s}^{\infty} \exp\left[\frac{\ln P_s}{R_s} R\right] p(R)\, dR. \qquad (4.15)$$

Substituting the integral by an asymptotic expression this can be expressed as

$$PD = P_{E0} - \frac{P_s}{\ln P_s} R_s p(R_s). \qquad (4.16)$$

In this relation, there is essentially a single design parameter P_{E0} so that (4.16) can be used to calculate the highest permissible value of the residual error probability. Taking into account Eq. (4.8) and substituting $p(R)$ by its series expansion at $R=0$, the following inequality is obtained:

$$P_{E0} < \left[P + \frac{P_s}{\ln P_s}(S_1 + S_2)\right] D. \qquad (4.16a)$$

Obviously, P_{E0} cannot be arbitrarily small, for example it cannot be a negative number. The following interesting conclusion can thus be drawn from (4.16): S and P cannot be specified independently. If, for instance, we do not strive for high reliability (*i.e., S* is sufficiently large), then high requirements with respect to error probability cannot be met either.

4.2.3 Diversity methods

If the information is transmitted over two or more paths, then the probability of interruption in both (or all) paths will be less than the probability of interruption in a single path. The method of applying more than one path for information transmission in order to counteract fading is called the diversity method. The increase of availability by diversity transmission is brought about through one of two means. According to the first, the fading events on the two paths may not be correlated or may be loosely correlated. In this case, the probability of deep fades in both channels is less than the probability in one channel only. According to the second possibility, there is a strong but favourable correlation between the two paths, sometimes called anticorrelation. A deep fading in one of the channels excludes, with high probability, simultaneous fading in the other channel. The two phenomena have practically identical results but require different design methods. The anticorrelation fading is seldom effective in low and medium capacity systems and will be treated in connection with high capacity systems.

Assuming realized diversity channels, our next task will be the suitable utilization of the signals arriving over two channels. Either the combination of both signals or the prevailing better signal alone can be utilized (diversity combining or diversity switching). According to a modified version of switching diversity, the better channel is not instantly selected, and switch-over only takes place when the performance of the operating channel has become unacceptable. Problems of diversity combining will be treated briefly in connection with high speed systems in Section 4.4.4.

The realization of diversity methods differ in the two frequency ranges above and below 10 GHz. Multipath propagation is the consequence of wave interference with the effect that the fading attenuation due to multipath propagation is a rapidly varying function of both the frequency and the location. Obvious realization methods are therefore the use of different frequencies in the two channels, or placing the receiving antennas of the two channels at a distance (frequency diversity and space diversity). The following relations hold for switching frequency diversity transmission:

$$F(A) = \frac{6 \times 10^{-7} abf^2 D^3}{gA \Delta f} \tag{4.17}$$

where the notations are as in Eq. (4.3), and additionally, Δf is the difference between the two channel frequencies in GHz, and g is a constant depending on the frequency band; its value is 0.5 at 4 GHz, 0.25 at 6 GHz, 0.125 at 8 GHz and 0.095 at 12 GHz. $F(A)$ is now the probability of the fading attenuation exceeding A, but simultaneously in both channels.

The required fade margin is now given by

$$A_f = 7.75 \times 10^{-4} fD \sqrt{\frac{ab}{\Delta f g S_2}} \; ; \tag{4.18}$$

The corresponding values for space diversity transmission are given by

$$F(A) = \frac{5 \times 10^{-4} abD^3}{A^2 Z^2} \tag{4.19}$$

and

$$A_f = 2.2 \times 10^{-2} \frac{D}{Z} \sqrt{\frac{ab}{S_2}} \; . \tag{4.20}$$

Here Z denotes the distance between the two receiving antennas.

Frequency and space diversity transmission have proved to be effective countermeasures against fading due to multipath propagation. However, they are of little use against rainfall outage which is of primary concern above 10 GHz. This is explained by the fact that the rain attenuation is much less dependent on the frequency and is a monotonically increasing function of frequency. Also, the rainfall attenuation depends on the location only as far as the rain rate depends on it. Therefore, the rainfall effect can only be overcome if there is no rainfall over one of the diversity channels. This is accomplished by so-called route diversity transmission which is realized over two or more routes. Rainfall areas are relatively small (at least in the temperate climatic zone), resulting in a low probability of simultaneous deep fading over both routes. This kind of diversity is much more expensive than frequency or space diversity and should only be applied in well justified cases. Note that the route diversity is also ineffective if the heavy rainfall occurs in the vicinity of one of the terminal stations.

4.3 Terrestrial radio relay — high capacity systems

In high capacity systems, the spectrum of the transmitted signals occupies a much larger bandwidth than in medium and low capacity systems. In a channel loaded by multipath fading, this will have the consequence that the fading attenuation which is a rapidly varying function of frequency will not be constant even within the transmission bandwidth of the channel. Consequently, the transmission characteristic of the channel, designed originally to meet for example the Nyquist criteria, will be distorted, and this linear distortion, as shown in Section 2.2, will generate intersymbol interference, thus causing performance degradation. Severe multipath distortion may in itself have the effect of interrupting the channel, even without noise addition. Obviously, this type of interruption cannot be counteracted by increasing the fade margin, *i.e.* the transmitter power. This section deals with the characterization of the selective multipath fading. Effective countermeasures will be presented in Section 4.4.

4.3.1 Characterization of selective multipath fading

In contrast with the transmission characteristic introduced in Section 2.2, the characteristic of a channel loaded by multipath fading is expressed by

$$C(f) = A(f)B(f)H(f)F(f) \tag{4.21}$$

where $A(f)$ and $H(f)$ are the transmit and receive filter characteristics, $B(f)$ is the Fourier transform of the transmitter waveform, and $F(f)$ is the transmission characteristic of the multipath channel. $F(f)$ is not known *a priori* and is also a function of time, and in consequence $C(f)$ cannot be chosen, as usual, to meet the Nyquist criterion. Instead of this, the following questions have to be answered: (*i*) what is the class of practically occurring $F(f)$ functions? and (*ii*) what effect do these functions have on the transmission of digital signals?

Because of the random character of propagation conditions, the transmission characteristic $F(f)$ cannot be accurately determined, so a suitable model has to be chosen providing a good approximation of the practical situation. In the model of the selective fading channel applied in this section, the channel is characterized by a suitable transfer characteristic $F(f)$. This model should meet the following requirements. First of all it should describe the experienced phenomena; it should have a simple analytical form; it should not contain too many parameters, and some statistical description of the parameters should be possible from the measured data. In the models published in the literature, $F(f)$ is given either in polynomial form [11] or as the sum of waves propagated over different paths. The multi-ray model, published in [12, 13], is widely used and will be applied in the following. The so-called simple three-ray model is characterized by the

$$F(\omega) = a[1 - be^{j(\omega - \omega_0)\tau}]. \tag{4.22}$$

It is seen that the attenuation (*i.e.*, $|F|$) has notches and is a periodic function with period of $1/\tau$. The average attenuation is characterized by a, the notch depth

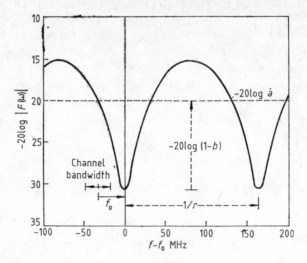

Fig. 4.3. Example for the absolute value of a multipath channel transfer function ($|F|$).

by b, the transmission minimum being $a(1-b)$, and ω_0 the notch frequency. Figure 4.3 shows an example for a transmission function according to Eq. (4.22).

From the transfer function given in Eq. (4.22), the attenuation and group delay functions can be calculated:

$$L(\omega) \overset{\triangle}{=} \frac{1}{|F(\omega)|^2} = \frac{1}{a^2[1+b^2-2b\cos[(\omega-\omega_0)\tau]} \tag{4.23}$$

$$D(\omega) = \tau\,\frac{b\cos[(\omega-\omega_0)\tau]-b^2}{1+b^2-2b\cos[(\omega-\omega_0)\tau]}\,. \tag{4.24}$$

According to the model, the behaviour of the channel with multipath activity is described by Eqs (4.22), (4.23) and (4.24).

It should be noted that the actual transmission function of the channel is not periodic, and the behaviour of the channel will therefore not be correctly described at all frequencies, only in the operating frequency range of the system (*i.e.,* in the Nyquist band or in a somewhat larger band). On the other hand, the channel behaviour will not be correctly described in a band which is larger than $1/\tau$ because of the periodicity mentioned.

Knowing the transmission function of the physical model, our next task will be to determine the statistical properties of the model parameters. The first publication of a physical model originates from Rummler [12] who derived the statistical parameters from measurements carried out on a 26.4 mile radio section in the American state of Georgia, between Atlanta and Palmetto. Strictly speaking, his parameters are applicable only to this section, but they are frequently applied to other radio sections too, and primarily for the laboratory comparison of different equipment types.

According to the Rummler model, the parameter a in (4.22) has a log–normal distribution. In practical calculations, the parameter $A = -20 \log a$ is frequently applied which has thus a Gaussian distribution. Its standard deviation is constant and equal to 5 dB, and its expected value is slightly dependent on a parameter which is defined as $B = -20 \log (1-b)$, thus characterizing the notch depth. This has an exponential distribution and has a constant expected value. Surprisingly, the probability density of ω_0 is a two-step uniform (and not a uniform) function. Finally, the magnitude of τ is constant. The probability density functions are thus given as follows:

$$p(B) = \frac{1}{M_B} \exp\left[-\frac{B}{M_B}\right]; \quad 0 \le B < \infty; \quad M_B = 3.8 \text{ dB} \tag{4.25a}$$

$$p(A|B) = \frac{1}{\sigma\sqrt{2\pi}} \exp\left[-\frac{1}{2}\left(\frac{A - M_A(B)}{\sigma}\right)^2\right]; \quad \sigma = 5 \text{ dB} \tag{4.25b}$$

$$M_A(B) = 24.6 \frac{B^4 + 500}{B^4 + 800}$$

$$p(f_0) = \begin{cases} 5\tau_0/3; & 0 \le |f_0 - f_c| \le \dfrac{1}{4\tau_0} \\[2mm] \tau_0/3; & \dfrac{1}{4\tau_0} < |f_0 - f_c| \le \dfrac{1}{2\tau_0} \end{cases} \tag{4.25c}$$

$$p(\tau) = \frac{1}{2}[\delta(\tau - \tau_0) + \delta(\tau + \tau_0)] \tag{4.25d}$$

where the carrier frequency is denoted by f_c. The sign of τ requires special considerations: the fading is called minimum phase fading if the sign is positive and non-minimum phase fading if the sign is negative. It is seen that the attenuation L does not depend on the sign of τ while the sign of the group delay D is equal to the sign of τ. This will have significance in relation with adaptive equalizers which will be dealt with in Section 4.4.2. In any case, the experimental determination of the sign and thereby a statistical description is extremely difficult because it cannot be performed from attenuation measurements. For lack of alternative, the probabilities of minimum phase and non-minimum phase fadings are usually assumed to be equal, as in Eq. (4.25d). (Note that the original definitions of the two fading types are $b<1$ and $b>1$; however, after rearrangement, we have the definitions of $\tau<0$ and $\tau>0$, with $b<1$ in both cases.)

A slightly modified variant of the Rummler model is presented in Ref. [16]. According to this model, b has a uniform distribution between 0 and 1, and accordingly, B is exponential, but with more pessimistic parameters than in (4.25b); τ is exponential, with different expected values for each fade event, and f_0 is conditionally uniform between $f_c + 1/\tau$ and $f_c - 1/\tau$. The distributions of A and a are not investigated by this model, and primarily for this reason, the Rummler model will be applied in the following.

It is noted that the condition for a model to be acceptable is the following: at a single frequency, the distribution of the attenuation (which is a joint result of A and B) is equal to that given in Eq. (4.3) in Section 4.2.1. Obviously, this requirement is met by the model. Note further that the probability density functions involved are all conditional, the condition being the presence of multipath activity in the channel. Otherwise, the transfer function has the constant value of unity, *i.e.* $a = 1$ and $b = 0$.

In conclusion, note that all investigated models have been physical models, *i.e.* $F(f)$ is given by the model. Further, the models have been static: they do not give any information on the speed of degradation during a fading event.

4.3.2 Equipment response to selective multipath fading

We have seen in the introduction to Section 4.3 that selective fading results in performance degradation. Intersymbol interference will be increased by the linear distortion due to the selective fading, and this will have the effect that the link will be interrupted at less than normal noise level, perhaps even with no noise at all. Obviously, the performance degradation depends on the actual form of the characteristic $F(f)$. Our model has three (variable) parameters, and it is thus feasible to express the equipment response as a function of these parameters. This is accomplished by the so-called system signature, giving the loci of those points in the B–f_0 plane, which correspond to the outage of a noise-free connection. The calculated signatures of 16QAM and 64QAM systems are given in Fig. 4.4, while Fig. 4.5 shows the measured signatures of some equipment published in the literature. The interpretation of the signature curves is the following: if, for a given f_0, B has a value corresponding to the curve, the error probability is 10^{-3}. If B has a higher value, the error probability is also higher, resulting in an interruption. If B has a lower value, the link is operating.

It is seen from the signature curves that the equipment is sensitive to distortions not only within the receiver bandwidth but also somewhat outside this band. Further, the most detrimental effect of the fading is caused by a notch frequency

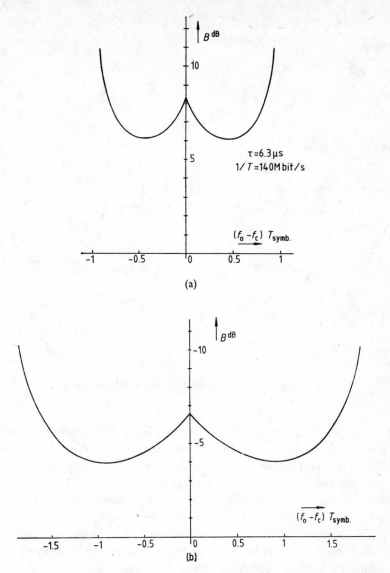

Fig. 4.4. Calculated signatures: (a) 16QAM transmission; (b) 64QAM transmission.

which is slightly detuned from the band centre. The wider the signature and the smaller its minimum value, the more sensitive is the equipment to distortion.

The situation is somewhat different if the connection is loaded not only by distortion but also by thermal noise. If the link performance is jointly determined by noise and distortion, an interruption will be caused by a distortion which is

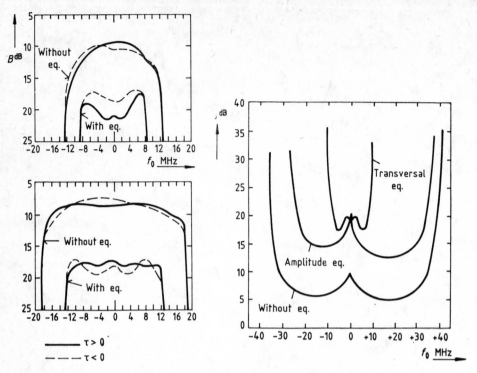

Fig. 4.5. Measured signatures of a few transmission systems.

smaller than in the case in which the distortion is the only performance degradation factor, and a signature will then correspond to each a (or A) parameter. If $A = A_f$, i.e. the flat fade margin, then interruption will already be caused by flat fading. In this case, therefore, no distortion at all is tolerated by the system and the signature is given by the abscissa of the $B\text{-}f_0$ plane. Assume now a sufficiently high fade margin A_f and values of fading decreasing from A_f to 0 dB. During this process, the signature will then change from the constant abscissa value to that corresponding to the noise-free condition according to the original definition. This is shown quantitatively in Fig. 4.6.

The equipment response to selective fading is sometimes characterized by the so-called effective fade margin defined as a flat fade attenuation having a probability of occurrence equal to the probability of outage due to multipath distortion. As a rule of thumb, the effective fade margin is usually assumed to be 20 to 25 dB, more-or-less independently from the flat fade margin. The concept of the effective fade margin is somewhat misleading and will therefore not be applied in this book.

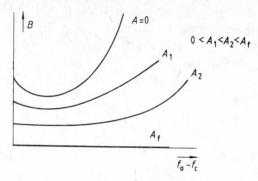

Fig. 4.6. Signatures pertaining to different values of parameter A.

4.3.3 A dynamic phenomenological channel model

Instead of the model discussed in the previous section, one yielding the error probability directly may sometimes be useful. It may also be required to estimate the dynamic behaviour of the channel, *i.e.* the speed of the performance degradation during a fade event. This is for example required in the design of a diversity switch, to be discussed in Chapter 11. In the following, the treatment of such a model, based on Ref. [18], will be presented.

Let us consider Eq. (4.10) as a starting point. If the channel transfer function is time dependent then R will also be time dependent, and can be expressed by a stochastic process. However, the distortion will have the effect of increasing the intersymbol interference, and as a consequence, the signal-to-noise ratio at outage, denoted by R_s, will also be a stochastic process. This can be expressed explicitly as

$$P_E = \exp\left[\ln P_s \frac{R(t)}{R_s(t)}\right]. \tag{4.26}$$

The performance degradation of several systems in which the digital signal is transmitted over a multipath channel have been investigated [12, 17]. According to these investigations, the degradation due to distortion is equivalent to the degradation due to an additional fictitious noise which is additive to the thermal noise. Expression (4.26) can thus be rewritten in the following form:

$$P_E = \exp\left[\frac{\ln P_s}{R_{s0}} \frac{R(t)R_0(t)}{R(t)+R_0(t)}\right] \tag{4.27}$$

where R_0 is the fictitious carrier-to-noise ratio due to the distortion (this is infinite without distortion), and R_{s0} is the carrier-to-noise ratio at outage in a distortion free channel. Knowledge of R and R_s allows calculation of the error probability. On the other hand, the rate of change of R and R_0 has to be estimated to determine the dynamic properties of the channel. This has been accomplished in Ref. [18], based on the Rummler model, the system signature and the dynamic channel behaviour at a single frequency; details of this investigation will not be given here. According to these investigations, the time dependence of the overall signal-to-noise ratio in the exponent of (4.27) can be written in the following form:

$$\frac{RR_0}{R+R_0} = e^{-rt} \tag{4.28}$$

where r is the rate of change of the overall signal-to-noise ratio in dB/s. r is always higher than the rate of change of the fading attenuation at a single frequency. According to a few statistical data, this is seldom higher than 30 dB/s, and probably never exceeds 100 dB/s. Reference [18] gives a method of determining the ratio of the two rates of changes, with the following results: the ratio is 2.21 for unequalized 16QAM, and 10.6 for unequalized 64QAM. This latter can be reduced to 2 by an adaptive equalizer.

4.3.4 Outage probability — transmission without diversity

We have seen that the outage probability is the probability of the parameter B being higher than the value given by the signature. Considering first the noise-free case and denoting the signature as Σ, the outage probability is given by

$$P_m = \int_{f_c-1/2\tau}^{f_c+1/2\tau} \int_{\Sigma(f_0)}^{\infty} p(B)p(f_0)\, dB\, df_0; \tag{4.29}$$

The internal integral can be evaluated by taking into account (4.25a), thus

$$P_m = \int_{f_c-1/2\tau}^{f_c+1/2\tau} \exp\left[\frac{-\Sigma(f_0)}{M_B}\right] p(f_0)\, df_0. \tag{4.30}$$

The integral can easily be calculated numerically if Σ is known in analytical, graphical or tabulated form.

The situation is more complicated if noise has also to be taken into account. (This is the situation if the fade margin of the equipment is not too high, or if the probability of non-flat fading is small but not negligible.) In this case, accurate calculation would require knowledge of signatures corresponding to all operational signal-to-noise ratios, and expression (4.29) should be integrated with respect to the distribution of A. Such a family of signatures is, however, normally not available. The calculation of probability would be extremely difficult even with a knowledge of such a signature family, because the signal-to-noise ratio is a complicated function of A and B. (Instead of going into detail, we refer to [18].) Therefore, lower and upper bounds are given instead of the outage probability. The lower bound is the outage probability calculated from (4.29). In order to calculate the upper bound, the outages due to distortion and noise are separated, and both are substituted by a higher one:

$$P_m = P_{md} + P_{mn} \qquad (4.31)$$

where P_{md} is the outage probability due to distortion, and P_{mn} is the outage probability due to noise. The former is calculated from (4.30), but not with the signature Σ but with a smaller $\tilde{\Sigma}$ signature (pertaining say to 10^{-6}). To this signature corresponds $R_0 = R_{s0} - T$ dB (by substituting $R = \infty$ into Eq. (4.27)). P_{mn} is the probability of having a fading attenuation which is higher than $A - T$ dB, and this can be determined from the distributions of A and B. This is really an upper bound: in the first term of (4.31), the worst signature has been taken instead of the signature which actually lies between Σ and $\tilde{\Sigma}$. In the second term, the signature which is identically zero has been taken, instead of a signature changing continuously between Σ and 0. Let us note two points before concluding this section. In practical cases, the outage is exclusively governed by multipath propagation because the fade margin is high enough not to need to apply Eq. (4.31). (In the model given in Ref. [16] mentioned earlier, the average attenuation A is not included.) On the other hand, the reliability requirements given in Section 4.2.2 are not met at all because of the high value of outage probability calculated from Eq. (4.30). It thus follows that in practical cases, some kind of multipath countermeasure has to be taken, as detailed in the next section.

4.4 Multipath countermeasures

We have seen in Section 4.2 that in low and medium speed systems, a viable countermeasure against flat fading is the increase of the transmitter power, or more accurately, the fade margin. This is not the case in high speed systems because a higher fade margin would only increase the signal-to-noise ratio and would have no effect on the multipath distortion, whereas the signature is interpreted as the outage in the noise-free case. Multipath distortion can be counteracted by the following methods (this list is probably exhaustive):

— decrease in error probability due to distortion: the use of error correction encoding;
— decrease in distortion itself: the use of adaptive equalization;

— relative decrease in distortion: the use of multicarrier system;
— utilization of the undistorted signal in addition to or instead of the distorted
  signal: the use of combining or switching type diversity.

In the following, the basic principles of these methods will be surveyed.

4.4.1 Error correction encoding

This is seldom used as a multipath countermeasure, primarily because of technological difficulties: high speed codecs are difficult to realize even with modern devices. In the recently published references, a high speed rate 12/11 convolutional codec is presented in [19], and a system utilizing this codec is discussed in [20].

4.4.2 Adaptive equalization

This can be regarded as the natural countermeasure of multipath distortion, since it has the function of eliminating the distortion generated by the propagation medium. The adaptive equalizer has a time-dependent transfer function in order to cancel the effect of the changing propagation medium. Adaptive equalization is a specialized subject matter which has a vast literature, being a standard building block of high speed digital transmission systems. However, because of lack of space only basic principles will be given in the following, and any detailed treatment will instead be replaced by ample references also covering recent results.

The adaptive equalization of a channel transmitting digital signals is not a new concept: in most cases, adaptive equalization has long since been applied in high speed data transmission systems having a bit rate in excess of 1200 bit/s. However, its application in microwave transmission is only justified in high speed multilevel systems.

For the realization of adaptive equalizers, two methods of approach can be used. According to the frequency domain method, a linear fourpole is inserted at the output of the channel, and its transfer function is chosen so as to obtain an overall function as close as possible to the ideal (constant) transfer function. According to the time domain method, the distortion products generated by the channel are determined and subtracted from the signal, thus recovering the undistorted signal.

The simplest equalizer is a frequency domain equalizer operating at i.f. and has the basic principle of determining the transfer function of the channel, and generating the inverse function. This method requires a knowledge of the transmitted signal spectrum; the distortion will be the difference between the spectra of the received signal and the (known) transmitted signal. Knowledge of the transmitted spectrum is given if the transmitted signal is random or randomized by a scrambler. At least the latter can certainly be assumed. However, only the absolute value of the channel transfer function can be determined by measuring the spectrum. Consequently, only the amplitude dispersion can be equalized by this kind of equalizer, and there is no possibility of equalizing the group delay distortion as well.

Actually realized frequency domain equalizers are designed to equalize the transfer function (4.22). In spite of the most careful design, the equalizer will not operate perfectly for two reasons: the actual channel transfer function is only approximated by the function (4.22), and the group delay distortion depends considerably on the minimum phase or non-minimum phase character of the fading. Thus if the group delay distortion is decreased for one fading type, it will certainly be increased for the other fading type. A possible method of realizing a frequency domain equalizer is shown in Fig. 4.7.

An example for a time domain adaptive equalizer is the linear transversal filter which has for a long time been applied in wired communication but has also

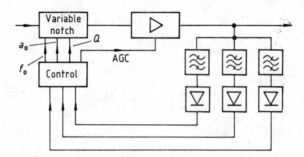

Fig. 4.7. Frequency domain adaptive equalizer.

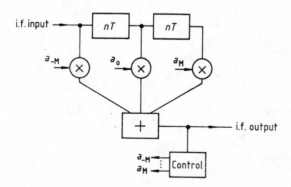

Fig. 4.8. Linear transversal filter.

proved to be suitable for the equalization of multipath distortion. The principle of the transversal filter is shown in Fig. 4.8: it is a tapped delay line, and the branched signals along the line are combined with suitable weights so as to subtract the intersymbol interference products (possibly completely) from the output signal. We thus have the possibility of realizing both feedback and feedforward coupling, compensating the effect of both the preceding and the following symbols. Several optimization algorithms are available for adjusting the weights.

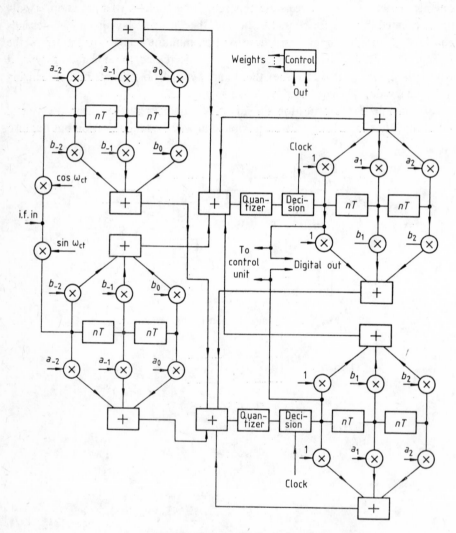

Fig. 4.9. Decision feedback equalizer.

According to one of the conventional methods, the squared error is minimized (LSE — Least Square Error), and according to another one, the zeros of the over-all impulse response are adjusted (ZF — Zero Forcing). The transversal filter can be realized either at i.f., as shown in Fig. 4.8, or in the baseband. In the latter case, two separate delay lines are needed for the two quadrature components.

A disadvantage of linear frequency domain equalizers is the enhancement of noise together with the peaking of a certain frequency band. The time domain decision feedback equalizer, which seems to be one of the best equalizer types, is more favourable in this respect. Its principle, already dealt with in Chapter 2 and shown in its simplest form in Fig. 2.33, is similar to the principle of the transversal filter, but instead of the ISI product, the regenerated digital signal is subtracted from the received signal, following a suitable weighting. In the decision feedback equalizer shown in Fig. 4.9, the crosstalk products due to the symbols following the decision time instant are only available in the form of the received but not yet regenerated signal. Therefore, as shown in the figure, the decision feedback equalizer is combined with a linear feedforward transversal filter.

This short treatment of adaptive equalizers is concluded by giving a few references. A basic treatment of the equalization problem is given in [21], and further references, primarily for applications in digital radio, are [22] to [31].

4.4.3 Multicarrier transmission [32, 33, 34]

A well established method used in short wave communication has recently also been applied in microwave radio under extremely poor propagation conditions. According to this method, the bit stream to be transmitted is converted into several parallel streams which are used to modulate separate carriers. The occupied frequency band is thus divided into several sub-bands, each having a bandwidth which is only a fraction of the total bandwidth. As illustrated in Fig. 4.10, the dispersion within a sub-band has been decreased, though the dispersion within the whole band is unchanged, in contrast with adaptive equalization. The price to be paid for this method is the somewhat wider occupied frequency range because of the finite slope of the filters separating the sub-bands, and the multiplication of several sub-systems. The multicarrier transmission is thus expensive, and is applied in well justified cases only. For instance, multicarrier transmission has been applied in an extremely long over-water span as reported in Ref. [34], and can also be justified for modulations with a very high number of levels, *e.g.* 256QAM or 1024QAM. Multicarrier transmission can, of course, also be applied in combination with adaptive equalization.

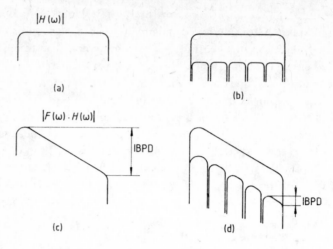

IBPD: in band power difference

Fig. 4.10. Spectrum envelopes illustrating distortion effect in single-carrier and multicarrier transmission. (a) Single-carrier transmission, undistorted; (b) multicarrier transmission, undistorted; (c) single-carrier transmission, distorted by multipath; (d) multicarrier transmission, distorted by multipath.

4.4.4 Space diversity transmission

From the preceding countermeasures, the adaptive equalization is the most plausible. However, because of the limitations of present-day technology (at the time of writing), and of our knowledge about multipath properties, adaptive equalizers in themselves do not provide the required transmission performance (or reliability). For this reason the adaptive equalizer has to be combined with some other countermeasure. For this purpose, space diversity has so far been used in most cases.

We have already seen in Section 4.2.3 that in low and medium capacity systems, the probability of fade depth resulting in outage is decreased by the use of space diversity. It is plausible to expect that the probability of high distortion will also be decreased. This expectation has been perfectly verified by experience.

In order to determine the performance improvement due to space diversity, the diversity improvement has to be defined, and a statistical model has to be given for the parameters of the space diversity channel. The latter task can be accomplished by starting again from the Rummler model. As we now have two channels, the joint density function of eight parameters have to be determined (these are the parameters A, B, f_0 and τ for each of the two channels).

The joint statistics are given, among others, in Ref. [35]. According to this, A_1 and A_2 are correlated Gaussian variables; the correlation coefficient depends on the distance between the antennas; the expected value and the variance are somewhat dependent on the dimension of the antennas. For rough calculations, it can be assumed that the expected value and the variance are the same as for a single channel, see (4.25). B_1 and B_2 are independent exponential variables, f_{01} and f_{02} are also independent; their density function is thus the product of densities applying to a single channel. Finally, τ can again be assumed to be a known constant.

In the following, the diversity improvement ratio will be defined as the ratio of outage parameters corresponding to the transmissions without and with countermeasures. This means that the improvement due to the possible equalization and simultaneous diversity will be investigated.

A few numerical results are presented in Figs 4.11 and 4.12. It is seen that the distance between the antennas may have an optimal value, in contrast with the diversity intended for counteracting the fading attenuation. In this case, the improvement is a monotonic, though flattening function of the distance.

The synergistic effect of combined countermeasures is remarkable: the improvement due to the simultaneous effect of diversity and equalization is much higher than the product of the improvements due to a single countermeasure.

An important question of diversity transmission is the method of utilizing the signals received in the two channels. In digital transmission, two kinds of utili-

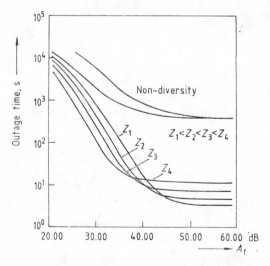

Fig. 4.11. Calculated improvement ratio of a space diversity system, with equal reception levels in the two channels.

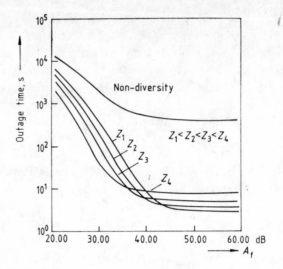

Fig. 4.12. Calculated improvement ratio of a space diversity system, with 5 dB difference between the reception levels in the two channels.

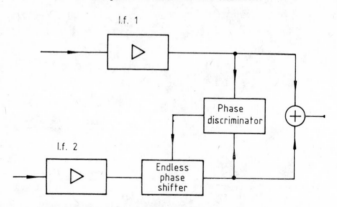

Fig. 4.13. Maximum power diversity combiner.

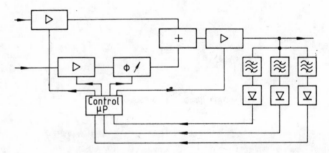

Fig. 4.14. Minimum dispersion diversity combiner.

zations are possible, i.f. combining and baseband switching. Several methods of combining have proved to be effective. In the maximum power combiner, the combining is controlled to yield the highest power of the overall signal due to the two input signals. This is achieved if an in-phase additon is realized, by controlling the phases of either the two local oscillators applied in the two receivers or the two i.f. signals. In the latter case, a single local oscillator is applied for the two receivers, as presented in Ref. [36].

In the minimum dispersion combiner, lowest dispersion within the transmission band is achieved instead of highest power. This can be implemented by either phase control or combined phase and amplitude control of the received signals, as given in Refs [37, 38]. Examples for diversity combiners are shown in Figs 4.13 and 4.14.

4.4.5 Frequency diversity transmission in high capacity digital radio

While space diversity provides an effective countermeasure against multipath propagation, it is not at all an economical solution: two antennas, duplicated microwave receiver circuits and substantial i.f. processing is required for the combination, and frequently, the antenna tower has to be strengthened for carrying two antennas. On the other hand, frequency diversity is extremely economical if (*i*) the radio link has a stand-by channel, (*ii*) this channel provides sufficient improvement when applied as a diversity path.

The first condition is met in almost every (though not every) case. The second condition is also met quite frequently but this was discovered only recently. This probably results from the failure to recognize that the function of diversity is quite different in high capacity digital transmission from analog or low capacity digital transmission. In the latter case, the fading countermeasure has to decrease the noise because the fading results in signal attenuation. On the other hand, in high capacity digital transmission, the countermeasure has to decrease the distortion. We shall see that in this case, the closer the frequencies of the two diversity paths, the higher the diversity improvement. To our knowledge, this interesting result was first pointed out by Frigyes (Ref. [39]).

In the following, the operation of frequency diversity will first be discussed in the case where one diversity route is applied for one operating channel. According to the signature curves discussed in Section 4.3.1, the connection will be interrupted if the notch is within or close to the receiver band. This means that a two-path diversity connection will be interrupted if either the signature occupies a wide band so that a single notch will interrupt both channels, or if there are two notches in the vicinities of both carriers.

The numerical evaluation necessitates a statistical model for the parameters of the transfer function $F(f)$. The number of these parameters is again eight as in space diversity. It can again be assumed that A_1 and A_2 are correlated Gaussian variables, and further that B_1 and B_2 are independent variables with exponential distribution, and τ can again be assumed to be constant. (In any case, a high enough fade margin will be assumed in the following investigations, and the distribution of parameters A then has no significance.)

From the viewpoint of frequency diversity operation, the distributions of f_{01} and f_{02} have the greatest significance, by taking into account our statements on interruption. In our model, the frequency difference between two notches is just $1/\tau$. It has also been stated that the real transfer function is not periodical so $F(f)$ certainly does not hold for frequency bands larger than $1/\tau$. On the other hand, $\tau=6.3$ ns has been assumed for our statistical model, and other investigations have resulted in even smaller values of τ. It can thus rightly be assumed that the probability of two notches within 160 MHz is low. It follows from these considerations that f_{01} and f_{02} are independent variables if there is a great difference between the two carrier frequencies, and a small difference between f_{01} and f_{02} has a very low probability. These properties are summarized in the following probability density function:

$$p(f_{01}, f_{02}) = P_1 p(f_{01})\delta(f_{02}-f_{01})+P_2 p(f_{02})\delta(f_{01}-f_{02})+$$
$$+ h(|f_{02}-f_{01}|)p(f_{01})p(f_{02});$$

$$h(|f_{02}-f_{01}|) = \begin{cases} \dfrac{1}{2}\left[1-\cos\dfrac{\pi}{Q}(f_{02}-f_{01})\right]; & |f_{02}-f_{01}| < Q \\ 1; & |f_{02}-f_{01}| \geqq Q. \end{cases} \qquad (4.32)$$

Here the three terms correspond to the following cases: first term — notch in the vicinity of carrier frequency f_{c1}; second term — notch in the vicinity of carrier frequency f_{c2}; third term — two notches in the vicinities of both carrier frequencies. The raised cosine function in the third term meets the above requirements, and is rather stable against the change in parameters. Of course, this term has not been determined by measurements. Q is an empirical parameter, characterizing the frequency difference at which the minima can be regarded as mutually independent.

The above statistical model can be used to determine the improvement due to frequency diversity. Figure 4.15 shows the results of calculations based on three signature examples. The lowest three curves apply to 16QAM transmission without equalization. The next three apply to an adaptive equalizer of poor quality, and the upper three apply to an adaptive equalizer of intermediate quality. The empirical parameter within a group of three curves is $Q=1.25/\tau$, $1/\tau$ and $0.75/\tau$.

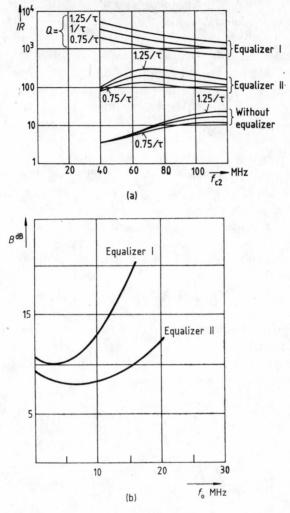

Fig. 4.15. (a) Calculated improvement ratios due to frequency diversity for different Q parameters and different equalizers; (b) signatures taken into account in the calculations.

In Ref. [40], measurement results of frequency diversity improvement are given, and these results approximate the values of the curve $0.75/\tau$ within a factor of two. The model above presented is thus fairly well verified. Also, the Q parameter somewhat smaller than $1/\tau$ corresponds well to the physical insight.

It is noted that in Ref. [41], the assumption of a somewhat different model has given similar results; an advantage of the model discussed above is the fact that it hardly differs from the single channel model.

If a common diversity channel is applied to serve n operating channels then the following can be stated. If any of the operating channels fails, a switch-over to the diversity (stand-by) channel takes place, *i.e.* the information is not lost. However, if multipath distortion results in a loss of a further channel, there is no diversity path left, and the information of this further channel will be lost. This means that the probability of information loss in the kth channel will be equal to the interruption probability of at least another channel. The interruption probability of this system can be calculated because the probability of simultaneous interruption of two channels follows from our previous investigations, provided the difference between carrier frequencies is known.

In the case of two simultaneously interrupted channels, the probability of one of these channels being interrupted earlier is 50 per cent. If one of these channels is just the diversity channel then, from the point of view of information loss, it does not matter which of the channels was interrupted first. In all other cases, the information of the channel interrupted later will be lost. The interruption probability of the kth channel will thus be given by

$$P_0(k) = P_0(k, d) + \frac{1}{2} \sum_{i=1}^{n}{}'' P_0(k, i) \tag{4.33}$$

where $P_0(l, m)$ is the probability of simultaneous interruption of the channels with corresponding serial numbers, and the serial number of the diversity channel is d. The superscript $''$ following Σ indicates that the terms $k=d$ and $k=i$ should be omitted. The information loss probability of any of the channels is given by

$$P_0 = \frac{1}{n} \sum_{k=1}^{n}{}' \left[P_0(k, d) + \frac{1}{2} \sum_{i=1}^{n}{}'' P_0(k, i) \right] \tag{4.34}$$

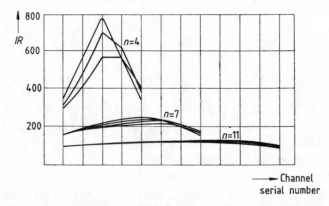

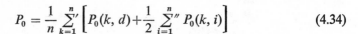

Fig. 4.16. Calculated improvement ratio of $n/1$ channel frequency diversity, for different n values. The signature I in Fig. 4.15b has been assumed in the calculation.

where the superscript ' following the external Σ indicates the omission of the term $k=d$. Figure 4.16 shows the improvement ratio of the 4/1, 7/1 and 11/1 16QAM systems as a function of the channel subscript, by taking into account the signature I of the previous investigations. The individual curves correspond to stand-by channels with different serial numbers.

As a conclusion on frequency diversity, let us investigate the use of the diversity channels. The i.f. combination can only be used in the 1/1 case, and baseband switching has to be used in the $n/1$ case. To obtain the best results, hitless switching has to be applied, *i.e.* switching without canceling or insertion of bits. Because of the frequency difference, the delays of the operating and diversity channels may differ from each other by the time interval of several bits. A delay equalization sub-system has therefore to be applied, rendering the switching procedure somewhat complicated.

4.5 Total probability of outage

We have seen that the probability density functions of our high capacity channel model are conditional density functions, the condition being the presence of multipath activity. The total probability of outage is thus given by the product of the multipath activity probability and the outage probability as determined in the preceding sections. The probability of multipath activity can only be determined empirically, and depends on the radio section involved. As a rough guide, it can generally be assumed that no selective fading occurs if the attenuation of the radio section is less than 19 dB. On this assumption, the probability of multipath activity can be calculated from Eq. (4.3) by substituting $A=19$ dB.

4.6 Satellite systems

The specifications and propagation properties of earth-satellite radio sections are substantially different from those of radio relay sections in which propagation takes place parallel with the surface of the earth, resulting in a different design philosophy. The difference in the specifications originates from the fact that a satellite link comprises only two radio sections [42] so that only those fading attenuations have to be investigated which are allowed in 0.01 per cent of the time. This time percentage is higher by two or three orders of magnitude than the time percentage considered in radio relay systems. Furthermore, only a small part of the propagation takes place in the troposphere, resulting in a much smaller fluctuation of the path attenuation, and this requires a much smaller fade margin.

In contrast with terrestrial radio relay transmission, in which the receiver signal-to-noise ratio is very high for a large fraction of the time, the above properties will result in a relatively low signal-to-noise ratio in satellite systems, and a rather high error probability, 10^{-6} or higher. It could be stated for terrestrial radio relay that only a single point of the $P_E = f(R)$ function, corresponding to the signal-to-noise ratio resulting in outage, has significance. However, this is certainly not applicable to satellite transmission. Without going into detail, it is noted that this will increase the susceptibility to interference because it is seen for example from the curves shown in Fig. 2.36 that the deviation of the signal-to-noise ratio required for a given error probability from the theoretical value is higher for lower error probabilities.

As with our preceding investigations, frequency ranges below 20 GHz only will be considered. In the majority of communication satellite systems, the earth station transmitter operates in the frequency range around 6 GHz, and the earth station receiver in the frequency range around 4 GHz. At these frequencies, level fluctuations are primarily caused by changes in the refractive index (neglecting a few other effects resulting in less than 0.5 dB fluctuation). Obviously, the significant fading attenuations appearing over not too small a time percentage are much less than those in terrestrial radio, and their value is highly dependent on the elevation angle of the antenna. For elevation angles higher than 10 degrees, the fading attenuation will be rather low, and this attenuation does not depend considerably on the frequency. For less than 10 degrees elevation, fading attenuations as high as 10 dB may be present, and below 1 to 1.5°, the fade depth may be as high as 20 to 25 dB, similar to terrestrial propagation. Comparing these numerical values with our previous results shows that in satellite systems, the probability of linear distortion is rather low.

For frequencies above 10 GHz, the additional effect of rainfall will appear, as with terrestrial radio. The two fading types can again be assumed to be independent, so their overall effect can be taken into account by adding the time durations or probabilities. Rain induced attenuation can be given as

$$a = \alpha D_{eff} \text{ dB}$$

where α is the attenuation factor which can be calculated from Eq. (4.4), and D_{eff} is that part of the earth-satellite path which falls within the rainfall area. Its value depends on the extension of the thunderstorm which can vary within wide limits, both in the horizontal and vertical directions, and on the elevation angle of the earth station antenna. The dimensions of the rain cells depend on the rain rate so that the effective section length can be expressed as

$$D_{eff} = f(I, \vartheta)$$

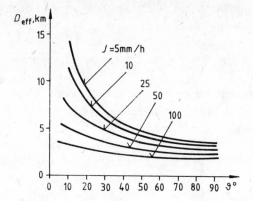

Fig. 4.17. Length of propagation path in the rain area *versus* terrestrial station antenna elevation angle.

where ϑ is the elevation angle. Figure 4.17 shows the effective section length as a function of antenna elevation angle, and is based on Ref. [43]. Below 20 GHz, the actual value of the fading attenuation is generally less than 12 dB for more than 0.01 per cent of the time, though values of 20 to 22 dB have also been measured at earth stations outside Europe.

Above 20 GHz, clouds, water vapour, oxygen and other gaseous media may result in additional fading attenuation which is dealt with in the literature. The majority of present-day systems do not operate in these frequency bands so that in this brief summary, we refer to [42] and its detailed list of references.

4.7 Frequency re-use

The investigations reported in Section 2.4 can be utilized for the determination of the frequency band occupied by a digital radio link. We have seen that this band is 50 to 80 per cent higher than $1/nT$ where T is the bit time and n is the number of bits in a symbol.

The occupied band can be halved if two different transmitters have identical frequencies but orthogonal polarizations. This method is called frequency re-use, and is widely applied in both radio relay and satellite systems.

There may be interference between the two signals of identical frequencies which is called co-channel interference. The level of the interfering signal is primarily determined by the antennas and the transmission medium itself. The discussion of dual polarized antennas is beyond the scope of this book, and we only briefly mention that by using well designed polarization filters, 30 to 40 dB cross polarization discrimination can be obtained between two orthogonally polarized antenna ports. This means that the re-use of the orthogonally polarized signal of the same frequency is not limited by the antennas themselves but by the propagation medium which can generate a substantial cross polarization component causing interference.

Experience shows that the cross polarization discrimination can be decreased both by rainfall and multipath propagation. We have seen in Section 4.2.1 that the rainfall attenuation of horizontally polarized waves is higher than that of vertically polarized waves, and the relative level of vertically polarized waves is thus increased. On the other hand, experience also shows that the polarization plane of the waves may be rotated by the raindrops, even in those frequency ranges in which the rainfall attenuation is not substantial. In certain areas, the cross polarization discrimination due to the transmission medium in the 8 to 20 GHz frequency range can be calculated from the following empirical formula:

$$X^{\mathrm{dB}} = 30 \log f^{\mathrm{GHz}} - 20 \log A^{\mathrm{dB}} + 12 \qquad (4.35)$$

where f is the frequency and A is the fading attenuation [44]. Note that according to other experiences [43], the rainfall has no substantial effect. The cross polarization attenuation is also decreased by multipath propagation, primarily in clear weather when the rotation of the polarization plane is caused by reflections from oblique layers. According to observations on radio sections of longer than 50 km, the cross polarization discrimination may be reduced by the propagation medium to a value for example less than 15 dB over 0.01 per cent of the time.

Summarizing, it may be concluded that the method of frequency re-use can only be applied with modulation systems having a small number of states, thus being not too sensitive to interference.

4.8 References

[1] *CCIR Plenary Assembly.* Vol. V, Geneva 1983.
[2] Livingston, D.: *The Physics of Microwave Propagation.* Prentice-Hall, Englewood-Cliffs 1970.
[3] Schwarz, W., Bennet, G., Stein, S.: *Communication Systems and Techniques.* McGraw-Hill, New York 1966.
[4] Barnett, W.: Multipath propagation at 4, 6, 11 GHz. *Bell Syst. Tech. J.,* Vol. 51, pp. 321–361, Febr. 1972.
[5] Ruthroff, C.: Multipath fading on line-of-sight microwave radio systems. *Bell Syst. Tech. J.,* Vol. 50, pp. 2375–2389, Sept. 1971.
[6] Mie, G.: Contribution to the optics of troubled media (in French). *Ann. Phys.,* No. 25, pp. 337–443, March 1908.

[7] Fedi, F.: Rainfall characteristics across Europe. *Alta Freq.*, Vol. XLVIII, No. 4, pp. 56E–60E, 1979.

[8] *Alta Freq., Special Issue,* Vol. XLVIII, No. 4, 1979.

[9] *CCIR Recommendation 354,* Plenary Assembly, Vol. IX, Geneva 1983.

[10] Frigyes, I.: Design of digital radio-relay systems for reliability, *Budavox Telecommun. Rev.,* 4, pp. 11–21, 1981.

[11] Greenstein, L., Czekaj, B.: Modeling multipath fading response. *Bell Syst. Tech. J.,* Vol. 60, pp. 193–214, Febr. 1981.

[12] Rummler, W.: A new selective fading model: application to propagation data. *Bell Syst. Tech. J.,* Vol. 58, pp. 1037–1071, May–June 1979.

[13] Rummler, W.: More on the multipath fading channel model. *IEEE Trans. Commun.,* Vol. Com-29, pp. 346–352, March 1981.

[14] *CCIR Report 563,* Kyoto 1978.

[15] Boithias, L., Battesti, J.: Method for calculating rain attenuation. In: *Proc. 6th Colloq. Microwave Commun., Budapest,* Vol. 1, pp. III. 5/27. 1–4, 1978.

[16] Campbell, J., Coutts, R.: Outage prediction of digital radio systems. *Electron. Lett.,* Vol. 18, No. 24–25, pp. 1071–1072, 9th Dec. 1982.

[17] Frigyes, I.: *The Microwave Frequency Diversity Channel: High Bit-Rate Digital Transmission* (in Hungarian). Research Report, Technical Univ. Budapest 1985.

[18] Frigyes, I.: A new dynamic multipath channel model. In: *Proc. 8th Colloq. Microwave Commun., Budapest,* pp. 65–66, 1986.

[19] Kavehard, M.: Convolutional coding for high-speed microwave radio communications. *AT&T Tech. J.,* Vol. 64, No. 7, pp. 1625–1637, Sept. 1985.

[20] Witt, J., Barnett, W., Hubbard, J.: 64-QAM digitalization of an analogue microwave channel. In: *Proc. 1986 Europ. Microwave Conf., Dublin,* pp. 53–58, 1986.

[21] Lucky, R., Salz, J., Weldon, E.: *Principles of Data Communications.* McGraw-Hill, New York 1986.

[22] McNally, D., Huang, E.: Propagation protection techniques for high capacity digital radio. *ICC 84, Amsterdam,* paper No. 32.8, 1984.

[23] Moreno, L., Salerno, M.: Adaptive equalization structures. *ICC 84, Amsterdam,* paper No. 32.5, 1984.

[24] Takenaka, S.: An adaptive fading equalizer, *ICC 82, Philadelphia,* paper No. 46.7, 1982.

[25] Fenderson, G., Parker, J., Quigley, P.: Adaptive transversal equalization of multipath propagation. *AT&T Tech. J.,* Vol. 63, No. 8, pp. 1447–1463, Oct. 1984.

[26] Aoki, K., Yamada, H., Ikuta, K., Takenaka, S., Daido, J.: The adaptive transversal equalizer for 90 MBPS 64-QAM radio. *ICC 84, Amsterdam,* paper No. 32.7, 1984.

[27] Giorio, E., Murature, F., Palestini, V.: Design and performance of an adaptive IF equalizer. *ICC 85, Chicago,* paper No. 39.1, 1985.

[28] Wong, W., Williamson, B., Grant, P., Cowan, F.: Adaptive transversal filters for multipath compensation. *ICC 84, Amsterdam,* paper No. 32.1, 1984.

[29] Taylor, D., Shafi, M.: Decision-feedback equalization of multipath. *IEEE Trans. Commun.,* Vol. Com-32, No. 3, pp. 267–279, March 1984.

[30] Shafi, M., Moore, D.: Adaptive equalizer improvement. *ICC 85, Chicago,* paper No. 15.7, 1985.

[31] Bianconi, G., Calandrino, L.: Adaptive baseband equalization. *ICC 85, Chicago,* paper No. 15.7, 1985.

[32] Yoshida, T.: System design and new techniques for an over-water digital radio. *ICC 83, Boston,* paper No. C2.7, 1983.

[33] Frigyes, I., Vanyai, P., Kantor, Cs.: Some points from the activity in Hungary towards high-capacity digital radio. *GLOBECOM 83, San Diego,* paper No. 9.4, 1983.

[34] Araki, M., Ichikawa, H., Hashimoto, A.: 100 km over-water span digital radio system. *ICC 85, Chicago,* paper No. 15.5, 1985.

[35] Rummler, W.: Modeling of diversity performance in digital radio. *ICC 84, Amsterdam,* paper No. 22.6, 1984.

[36] Nichols, R.: An IF combiner for digital and analog radio systems. *ICC 82, Philadelphia,* paper No. 4B.7, 1982.

[37] Komaki, S., Tajima, K., Okamoto, J.: A minimum dispersion combiner for high capacity digital microwave radio. *IEEE Trans. Commun.,* Vol. Com-32, pp. 419–428, April 1984.

[38] Yeh, Y., Greenstein, L.: A new approach to space diversity combining in microwave digital radio. *AT&T Tech. J.,* Vol. 64, pp. 885–935, April 1985.

[39] Frigyes, I.: Frequency diversity in high-capacity digital radio (in Hungarian). *Microwave Seminar,* pp. 16–20, Budapest, Jan. 1985.

[40] Lee, B., Lin, S.: More on frequency diversity in digital microwave radio. *GLOBECOM 85, Atlanta,* paper No. 36, 1985.

[41] Lin, S., Lee, B.: A model of frequency diversity improvement for digital radio. In: *IEEE Int. Symp. Antennas and Propagation, Record, Kyoto* 1985.

[42] *CCIR Recommendation 353–2,* XIII-th Plenary Assembly, Vol. IX, Geneva 1975.

[43] *CCIR Report 564–1,* Vol. IX., Geneva 1975.

[44] Newland, J.: The relationship between rain depolarization and attenuation. *Electron. Lett.,* Vol. 13, 1977.

CHAPTER 5

METHODS OF ENCODING AND DIGITAL TRANSMISSION TASKS

In the previous chapters, the conditions to be met by microwave radio channels transmitting digital signal streams and the obtainable performance have been investigated. In this chapter, the utilization of digital transmission will be discussed, the types and basic properties of digital signal sources will be outlined, together with the digital hierarchical elements. The final performance of digital transmission can be expressed by the error probability of the transmission and the residual jitter in the transmitted digital signal; these two parameters will also be discussed.

5.1 Source and channel encoding

The detailed investigation of primary digital signal sources and the generation of digital signals are presented in the literature [1]; in the following, only the source encoding methods will be outlined from the special aspects of the micro-wave radio channel. By source encoding we mean signal processing where the output signal of the analog information source is transformed into a digital signal stream. On the other hand, the channel encoding procedures are aimed at the optimal utilization of the analog transmission channel. Figure 5.1 shows a simplified model of a digital microwave link, and illustrates that the channel encoding and the source encoding blocks cannot be separated perfectly. According to the model shown, the analog information signal is connected to an analog/digital converter which is the encoder in the conventional sense, and this is followed by so-called baseband signal processing which provides an optimal link between the encoder and the digital modulator. The digital modulator and the digital demodulator are connected by the analog transmission channel, and at the receive side, the inverse operations of the transmit side signal processing are performed on the demodulated signal in order to regenerate the received signal and to recover the carrier and clock signals from the received bit stream. The outcome of the receive side signal processing is connected to the digital/analog converter which is the decoder in the

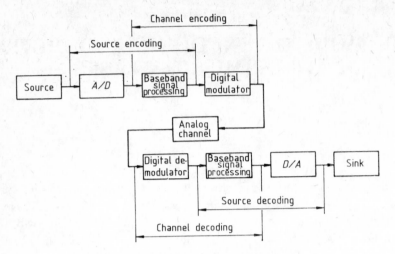

Fig. 5.1. Simplified model of a digital microwave link.

conventional sense, and the signal thus decoded is presented to the sink. Note that the operations performed by the source and channel encoding and decoding blocks cannot be separated in the physical sense, and the method of source encoding even has an effect on the type of channel encoding to be applied. For instance, the "scrambler" transformation, used in many cases, can be regarded as being part of either the source encoding or the channel encoding procedure. In a similar way, the baseband signal processing and the digital modulator/demodulator cannot be separated either. Otherwise, in the case of M-PSK modulation ($M > 2$), the digital modulator usually acts as a Gray encoder too, and in the case of differentially coherent reception, a differential decoding operation can be performed by the digital demodulator.

It should also be borne in mind that the error sensitivity of the source encoding methods may differ and may have different responses to bursts of errors and to various error distributions. The error correction, if applied, can be regarded as the task of either the source or the baseband signal processing, especially in the case of high quality transmission. Similarly, there is no unequivocal decision on whether the reduction of jitter has to be performed by the sink (decoder) or the receive side signal processing. Also, the type of source and channel encoder/decoder applied is important as regards allowable line jitter. Finally, it should be noted that usual encoding methods may sometimes be applied, *e.g.* for transmission of a service channel or to realize private networks, allowing better bandwidth efficiency to be obtained by the reduction of the information source redundancy.

5.2 Conventional source encoding methods

The transmission of digital information is usually realized by pulse code modulation or PCM source encoding [1, 2]. In most PCM applications, time division multiplexing and digital transmission are simultaneously applied. The principle of PCM transmission has originally been developed for the extension of the urban telephone network and thus for voice transmission; however, it is presently also being utilized for the transmission of signals originating from other wideband sources [3]. In the process of PCM signal transmission, the band-limited signal to be transmitted is first sampled at a rate corresponding to the sampling theorems, and the PAM signal stream thus obtained is quantized and transformed into a digital signal stream having discrete values. The quantized signal is transformed, by the use of the binary or other coding rule, into PCM code words by the PCM encoder. In the following discussion, the PCM encoder is defined in a broad sense, comprising all building blocks contributing to the process of transforming the analog signal to be transmitted into the binary bit stream.

The structure of the PCM encoder in this broad sense, together with the structure of the decoder, is shown in Fig. 5.2. At the transmit side, the information signal $s(t)$ is presented, *via* the band limiting filter, to the sampling circuit generating the PAM stream, the sampling frequency being f_s Hz. The sampler is followed by the quantizer and encoder. In voice transmission, quantizing is achieved by logarithmic compression, *i.e.* the absolute value of the quantizing steps is exponentially proportional to the signal level, the relative value remaining thus constant. In contrast with voice transmission, wideband signals are usually encoded by a linear quantizing characteristic. The encoder bit rate is given by

$$f_{bit} = bf_s$$

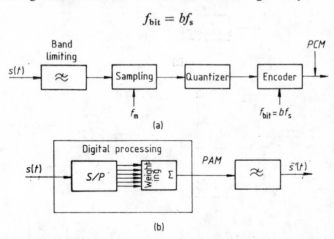

(a)

(b)

Fig. 5.2. General representation of (a) a source encoder and (b) a sink decoder.

where b denotes the number of bits in a code word and f_s is the sampling rate.

At the receive side, the PCM decoder in the broad sense is made up of the decoder itself and of the decoder filter processing the PAM stream. The bit stream arriving at the decoder input is transformed into parallel PCM code words which are presented to a weighting–combining network realizing the inverse of the compression function. The PAM stream is thus recovered at the output of this network and is connected to a band limiting filter in order to generate the replica $s'(t)$ of the signal to be transmitted.

A few realizations of a codec section comprising a PCM encoder and PCM decoder are shown in Fig. 5.3. Conventional transmission is realized by the use of the above encoder and decoder (Fig. 5.3a). Figure 5.3b shows the principle of DPCM transmission during which the PCM encoder receives the difference between subsequent signal amplitudes. This is implemented by a local decoder to

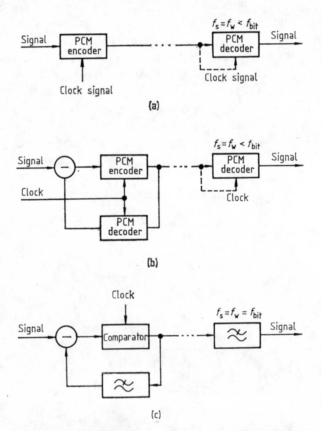

Fig. 5.3. Codec sections: (a) PCM transmission; (b) DPCM transmission; (c) DM transmission. f_s = sampling frequency, f_w = word frequency, f_{bit} = bit frequency.

generate the signal corresponding to the preceding sampling time instant by using a feedback model. At the receive side, decoding is performed by a decoder which is identical with the transmit side local decoder. Finally, Fig. 5.3c shows the delta modulation encoding which is the extreme case of DPCM (differential PCM). Here 1-bit "code words" are generated by the PCM coder, expressing the positive or negative deviation of the actual sample from the previous sample. The coder is realized by a comparator operating at the sampling rate while the decoder is realized by upper band limiting; however, this sampling rate is not determined by the sampling theorems. For PCM and DPCM transmission, $f_s < f_{bit}$ while for DM transmission $f_s = f_{bit}$. The advantage of DPCM transmission is a better bandwidth efficiency by the reduction of redundancy while the advantages of DM transmission are the simple circuit realization and low cost.

Several types of prediction and adaptive encoding methods have been presented to obtain higher bandwidth efficiency for the transmission of wideband signals [4, 5]. The bandwidth requirement is hereby reduced by reducing the redundancy of the information signal. During the process of predictive encoding, an estimate, based on the correlation between elements, sections or samples of the signal to be encoded, is generated for the next sample, and the transmitted signal to be encoded at each sampling instant is only the error of this estimate. In the process of adaptive encoding, the program of the signal processing is changed step-by-step (*e.g.*, by changing the length of the PCM code words) so as to suit the structure of the transmitted bit stream, thus optimizing the transmission.

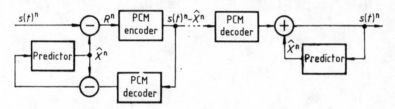

Fig. 5.4. Structure of a prediction codec.

The realization methods of adaptive type encoders are presently in the stage of evolution but predictive encoders have by now a definite model shown in Fig. 5.4. According to this model, a predictor is used to generate the estimate value of the next sample, and the error of the estimate is PCM encoded. The input signal for the predictor is generated by a local PCM decoder. At the receive side, inverse operations are performed, *i.e.* the predictor estimate value is added to the decoder output signal. This means that at the transmit side, the signal at the PCM encoder input is given by

$$R^n = s(t)^n - \hat{x}^n, \tag{5.1}$$

where the superscript n refers to the sampling time instant, $s(t)$ is the signal awaiting transmission, and $\hat{x}^n$ is the value of the estimate. The signal at the output of the predictor is given by

$$x^n = \sum_{i=1}^{z} a_i[s(t)^{n-1} + \hat{x}^{n-1}] \qquad (5.2)$$

where the factors a_i are the weighting factors of the prediction. The most widely used element for realizing the predictor is the transversal filter which is made up of T-element delay line sections and a weighting–combining network as shown in Fig. 5.5. The rate of operation of the signal delay line of the filter depends on the coding rate. The coding structure in Fig. 5.4 shows a predictor for the signal processing of an analog PAM stream. Another coding arrangement can be applied to make the predictor suitable for the processing of binary coded or other bit streams.

It should be noted that there is a price to be paid for the bandwidth reduction obtained by reducing the redundancy: the sensitivity to transmission errors is proportional to the measure of redundancy reduction.

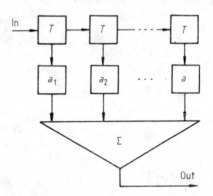

Fig. 5.5. Transversal filter as predictor: $T =$ unit delay; $a =$ weight.

5.3 Digital transmission tasks

It is required that the availability of digital transmission systems should be similar to that of analog systems, and that digital transmission systems should provide the conventional services of data transmission and the transmission of analog signals. In the latter category, the transmission of the analog voice signal (or telephone transmission) is the classical field of PCM encoding. In addition, the digital transmission of sound program signals, FDM signals, videophone signals and television signals is also required. The digital signal to be transmitted is generated either by the encoder driven by an analog signal or by the primary digital signal source. The source may also be of higher hierarchical level, comprising several primary sources. In this section, the encoders appearing as digital signal sources for different types of analog signals to be transmitted will be surveyed. However, the characteristics of data transmission, requiring no encoding, will first be outlined.

5.3.1 Data transmission

In the information traffic, the proportion of data transmission shows a rapidly increasing tendency as a result of the economical development and the widespread use of computer techniques. In most cases, data transmission means the extremely reliable transmission of digital bit streams generated by the digital multiplication of several data signals. The primary data sources may be either synchronous or asynchronous sources, the latter constituting the majority. Data transmission sources are classified according to their transmission speed as follows.

50 Baud

The majority of present telegraph transmission systems operate at this speed according to the start–stop system. This means that the information content between start and stop symbols is made up of isochronous symbols, and there may be an arbitrary time span between start and stop code words.

75 to 100 Baud

This higher transmission speed is used for up-to-date teletypewriter and telex systems.

200 Baud

This is the characteristic speed of data collection systems and simple computer terminals. This speed may also be applied for better quality telegraph transmission.

600/1200 Baud

This is the characteristic speed of transmission between data transmission terminals and computers. This speed was originally intended for data transmission in local and district networks.

2400/4800/9600 Baud

These are typical transmission speeds for *M*-PSK transmission systems connected to data transmission terminals. Recently, these speeds have gained acceptance in local and district networks.

Transmission between computers

At present, there is no recommendation for the transmission speed applicable to this type of transmission. There has been a separate evolution of PCM transmission and data transmission so the utilization of the time slots of 64 kbit/s transmission for data transmission purposes is not yet standardized. In this respect, several proposals have been published [6, 7].

The CCITT, recognizing the importance of the relation between data transmission and the digital hierarchical levels, has produced several recommendations for data transmission multiplex systems. The main characteristics of these recommendations, referring to the 64 kbit/s time slot, are summarized in Table 5.1.

R.101 Recommendations

Refer to the digital multiplication of start–stop primary sources. Output bit rate 2.4 kbit/s. The multiplication of several 2.4 kbit/s sources is covered in Recommendations X.50 and X.51 below.

R.111 Recommendation

Refers to the digital multiplication of telegraph channels with speeds of up to 300 bit/s, and of non-isochronous data signals, independently of the code structure and speed. Output bit rate 64 kbit/s.

X.50 and X.51 Recommendations

Refer to the digital multiplication of synchronous data signals with bit rates of 600, 2400, 4800 and 9600 bit/s. Output bit rate 64 kbit/s.

In order to utilize the digital signal path leading *via* several PCM systems of differing hierarchical levels, special multiplex equipment with input/output bit rates differing from those given in the CCITT Recommendations is also available. Such is the PCM telegraph multiplex equipment TMB 31/32, developed by the Hungarian Research Institute for Telecommunication [8].

Table 5.1. Main data of CCITT Recommendations on data multiplex equipment

Rec.	Bit rate of data stream, bit/s	Bit rate of carrier channel, bit/s	Utilization of carrier channel, %	Highest number of carrier channels	Bit rate of channel driven by the multiplicated bit stream, kbit/s
R.101/A	50 75	51.06 102.12	98 73	46 22	2.4
R.101/B	50 100 200 75 134.5 150 300	51.6 102.12 204.25 76.6 153.19 153.19 306.38	98 98 98 98 98 98 98	46 22 10 30 15 15 7	2.4
R.111	50 50 100 100 200 300	250 500 500 1 000 1 000 1 000	20 10 10 10 20 33	240 120 120 60 60 60	64
X.50	600 2400 4800 9600	800 3 200 6 400 12 800	75 75 75 75	80 20 10 5	64
X.51	600 2400 4800 9600	750 3 000 6 000 12 000	80 80 80 80	80 20 10 5	64

It serves for the multiplication of the following data signals:

— 31 telegraph channels having speeds of 50 Bauds, output bit rate 64 kbit/s;
— 31 telegraph channels having speeds of 100 Bauds, output bit rate 128 kbit/s;
— 31 telegraph channels having speeds of 200 Bauds, output bit rate 256 kbit/s;
— mixed input of telegraph channels having speeds of 50, 100 and 200 Bauds, output bit rate 64, 128 or 256 kbit/s.

For the transmission of data signals, the transmission channel should meet the following additional requirements:

(i) the error multiplication which may be the consequence of the baseband signal processing should not counteract the error correction introduced into the primary bit stream or following the digital multiplication;

(ii) transmission without code restriction should be possible for both the primary data signals and also for the multiplex signal.

5.3.2 Voice transmission

The PCM encoding scheme was originally developed for the transmission of telephone voice signals, and this fact had a decisive effect on the evolution of the digital hierarchical levels and on the typical system parameters. Recently, the digital transmission of non-voice signal types is under consideration but this has in no way diminished the original significance of PCM transmission, *i.e.* the transmission of the multichannel voice signal.

According to CCITT Recommendation G.712, the voice signal is defined as a band limited signal in the range 300 to 3400 Hz, sampled at a rate of 8 kHz. The frame time of a primary PCM multiplex line signal is the reciprocal value of the sampling frequency which is 125 µs. The PAM stream obtained by sampling is converted into 8-bit code words by the PCM encoder. The above frame time was first divided into 24 parts, and recently into 32 parts.* According to the latter division, the primary line transmission speed is $8 \text{ kHz} \times 32 \times 8 \text{ bits} = 2048 \text{ kbit/s}$.

5.3.2.1 Adaptation to the hypothetical reference network

In most cases, the PCM transmission path is part of an international connection. It is then required that the insertion of this digital path into the longest international connection should not introduce a performance degradation with respect to the purely analog transmission. Originally, the noise requirements of a long international connection have been determined by the relations applicable for the reference network defined in CCITT Recommendation G.103. It can be shown that in order to meet these requirements with PCM transmission, the signal-to-quantizing noise ratio of a single codec section should be at least 22 dB in the

---

* The division of the frame time into time slots is intended for the time division multiplication of the voice channels. The frame time of 125 µs is divided into 24 time slots in the United States and in Japan, and into 32 time slots in Europe. This applies to the primary PCM multiplication; the number of time slots for the secondary PCM multiplication is 132, according to the European standard.

"medium" level range of the voice signal at the encoder input (-5 to -25 dB m). At lower voice levels, the signal-to-noise ratio requirements are much lower [9]. By taking into account the reference network mentioned above, the highest number of cascaded codec sections is 14. This means that a single codec section should have a signal-to-quantizing noise ratio which is higher by 10 log 14, $i.e.$ 33.46 dB. It is known that this requirement is just met by an encoder with 7-bit words [1]. In order to accommodate other performance degradation parameters as well, 8-bit code words are generally applied.

5.3.2.2 Quantizing characteristics, code construction

A nonlinear quantizing characteristic has been internationally adapted for the exact transmission of the voice signal dynamics, as detailed in the following.

Linear quantizing means that the quantizing steps Δq_i have constant values, independently of the value of the PAM sample x_i pertaining to the quantizing steps, so

$$\frac{\Delta q_i}{x_i} = \text{constant} \tag{5.3a}$$

if

$$\Delta q_i = \text{constant.} \tag{5.3b}$$

This means that the higher the value of the PAM sample, the more accurate the quantizing. Evidently, if

$$\Delta q_i \neq \text{constant} \tag{5.4a}$$

then the following relation can be approximated:

$$\frac{\Delta q_i}{x_i} \approx \text{constant.} \tag{5.4b}$$

The accuracy of quantizing is thus approximately independent of the PAM sample to be encoded. It thus follows that the compression function $F(x)$, expressing the variation of the quantizing step value as a function of the relative signal level x, should follow some kind of logarithmic characteristic. In this way, the quantizing step will be proportional to the sample height. It is further required that the compression function $F(x)$ should allow easy approximation by segments, and that simple digital circuit elements should be applicable for the realization of the compression and expansion characteristics. An example of the piecewise linear approximation of the compression characteristic is shown in Fig. 5.6.

In CCITT Recommendation G.711, the encoding algorithms are exactly defined for both the "A-law" characteristic used in Europe and for the "μ-law" charac-

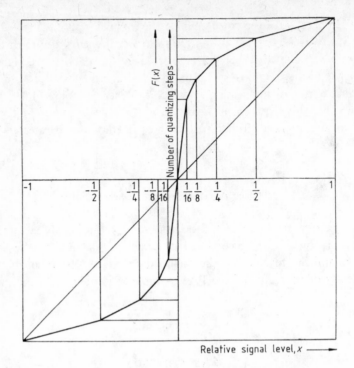

Fig. 5.6. Piecewise linear approximation of a logarithmic quantizing characteristic (*A*-law).

teristic used in America. The analytical formulation of the compression laws is given in the following:

$$F_A(x) = \text{sign}(x)\,\frac{A(x)}{1+\ln A}, \quad \text{if} \quad 0 \leq |x| \leq \frac{1}{A}$$

$$F_A(x) = \text{sign}(x)\,\frac{1+\ln Ax}{1+\ln A}, \quad \text{if} \quad \frac{1}{A} \leq (x) \leq 0 \tag{5.5}$$

and

$$F_\mu(x) = \text{sign}(x)\,\frac{\ln(1+\mu x)}{\ln(1+\mu)}, \quad \text{if} \quad -1 \leq x \leq 1. \tag{5.6}$$

According to CCITT Recommendation G.711, the parameter values to be chosen are $A=87.6$ or $\mu=255$.

According to the segment approximation as shown in Fig. 5.6, a definite number of linear segments are applied: 13 for the *A*-law, 15 for the *μ*-law. Linear quantizing is applied within each segment but the size of the quantizing steps depends on the serial number of the segment. The data of the *A*-law and *μ*-law characteristics are summarized in Tables 5.2 and 5.3, respectively.

Table 5.2. Algorithm of 13-segment approximation of the A-law
quantizing characteristic ($A=87.6$) satisfying Eq. (5.5)

Segment number	Lower limit of segment (linear quantum)	Upper limit of segment (linear quantum)	Code word PXYZABCD	Quantum magnitude (linear quantum)
1	2048	4096	1111ABCD	128
2	1024	2048	1110ABCD	64
3	512	1024	1101ABCD	32
4	256	512	1000ABCD	16
5	128	256	1011ABCD	8
6	64	128	1010ABCD	4
	32	64	1001ABCD	2
7	0	32	1000ABCD	2
	−32	0	0000ABCD	2
	−64	−32	0001ABCD	2
8	−128	−64	0010ABCD	4
9	−256	−128	0011ABCD	8
10	−512	−256	0100ABCD	16
11	−1024	−512	0101ABCD	32
12	−2048	−1024	0110ABCD	64
13	−4096	−2048	0111ABCD	128

Table 5.3. Algorithm of 15-segment approximation of the μ-law quantizing
characteristic ($\mu=255$) satisfying Eq. (5.6)

Segment number	Lower limit of segment (linear quantum)	Upper limit of segment (linear quantum)	Code word PXYZABCD	Quantum magnitude (linear quantum)
1	4063	8159	1000ABCD	256
2	2015	4063	1001ABCD	128
3	991	2015	1010ABCD	64
4	479	991	1011ABCD	32
5	233	479	1100ABCD	16
6	95	233	1101ABCD	8
7	31	95	1110ABCD	4
8	−31	31	1111ABCD	2
9	−95	−31	0110ABCD	4
10	−233	−95	0101ABCD	8
11	−479	−233	0100ABCD	16
12	−991	−479	0110ABCD	32
13	−2015	−991	0010ABCD	64
14	−4063	−2015	0001ABCD	128
15	−8159	−4063	0000ABCD	256

In the case of the A-law characteristic, 12-bit linear encoding is first performed, dividing the input level range into $2^{12} = 4096$ steps, and logic networks are applied to generate the 8-bit code words according to the compression law as given in Table 5.2. In the case of the μ-law characteristic, 13-bit linear encoding is first performed, dividing the input level range into $2^{13} = 8192$ steps, and logic networks are again applied to obtain the compression law as given in Table 5.3.

The binary value of the code word obtained according to the encoding as given in Tables 5.2 and 5.3 is given by

$$M_B = PXYZABCD. \tag{5.7}$$

The sample polarity is expressed by the most significant bit P, the segment applied for the encoding is defined by the next three bits XYZ, and the linear quantizing result within the segment is given by the last four bits $ABCD$.

Using the notation of the code word as given in Eq. (5.7), the sample value is given by

$$M = \alpha(\beta 32 + 2L + 1)2^{s - \beta},$$

where

$$S = 4X + 2Y + Z; \quad L = 8A + 4B + 2C + D; \quad \alpha = 1, \quad \text{if} \quad P = 1$$

and

$$\alpha = -1, \quad \text{if} \quad P = 0; \quad \beta = 0, \quad \text{if} \quad S = 0 \quad \text{and} \quad \beta = 1, \quad \text{if} \quad S \neq 0 \tag{5.8}$$

for the A-law compression, and by

$$M = \alpha[2^s L + 33(2^{s-1} - 1)],$$

where

$$S - 1 = 4\bar{X} + 2\bar{Y} + \bar{Z}; \quad L = 8\bar{A} + 4\bar{B} + 2\bar{C} + \bar{D};$$

$$\alpha = 1, \quad \text{if} \quad P = 1 \quad \text{and} \quad \alpha = -1, \quad \text{if} \quad P = 0 \tag{5.9}$$

for the μ-law compression.

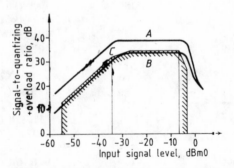

Fig. 5.7. The signal-to-quantizing + overload noise ratio as a function of the input signal level. Quantizing characteristic according to A-law with $A = 87.6$, approximated according to Table 5.2. A — 8-bit encoding, B — 7-bit encoding, C — limit mask according to CCITT Rec. G. 712.

The signal-to-quantizing noise ratio as a function of the input signal level is shown in Fig. 5.7. This figure applies to the A-law compression and also shows the limit mask given in CCITT Recommendation G.712. It is seen that little safety margin would be left by 7-bit encoding and this is why 8-bit encoding is always applied.

5.3.2.3 Comparison of voice encoding methods, bandwidth efficiency

In the summary of source encoding methods given in Section 5.2, only PCM encoding methods have been discussed because of the world-wide acceptance of this encoding method, also shown by the relevant CCITT Recommendations. In spite of this fact, it is feasible to compare PCM with other kinds of encoding methods. This is especially justified considering digital microwave transmission which has a lower bandwidth efficiency than analog microwave transmission of FDM signals. On the other hand, the high degree of redundancy in the voice signal is well known, and this in itself seems to make the research for more economical encoding methods justified. Other encoding methods could also be used to accommodate the service channel voice transmission of microwave links, and to establish connections to special order-wire channels of other international networks or military systems.

Figure 5.8 shows the signal-to-quantizing noise ratio as a function of the bit rate [10] for PCM, DPCM and DM source encoding. The abscissa has a double bit rate scale according to voice bands of 0.3 to 3.4 kHz and 0.3 to 2.2 kHz (sampling rates of 8 and 4.8 kHz, respectively). It can be seen from this figure that up to a bit rate of 28 and 17 kbit/s, respectively, the best bandwidth efficiency is

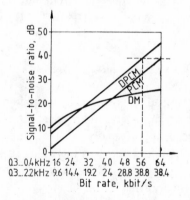

Fig. 5.8. Signal-to-noise ratio due to encoding as a function of the encoder signal bit rate, for different encoders.

offered by DM encoding but the signal-to-noise ratio is only 15 to 16 dB. It can further be ascertained that up to a bit-rate of 42 and 25 kbit/s, respectively, DM has better efficiency than PCM, offering a signal-to-noise ratio of 20 to 21 dB. Another important conclusion is the fact that with DPCM encoding, the line bit rate can be reduced by about 8 kbit/s as compared with the bit rate required by PCM encoding. This means that DPCM encoding allows a 12.5 per cent increase in the transmitted telephone channels.

The predictive, adaptive and hybrid encoding methods, applied for increasing the bandwidth efficiency by decreasing the redundancy of the voice signal, are also significant. In Ref. [5], digital voice transmission with a line speed of 2.4 kbit/s is reported, by applying adaptive type transversal filters and a PCM codec.

It should be noted, however, that a price has to be paid for decreasing the redundancy and thus increasing the bandwidth efficiency. This price is the fact that the signals of diminished redundancy are more susceptible to the channel noise and bit errors. For instance, in the system reported in Ref. [5], the coded voice signal having a bit rate of 2.4 kbit/s is barely understandable at the receive end of a channel having an error probability of 10^{-4} (average time span between two consecutive bit errors of 4 s).

5.3.3 Sound program transmission

From all signals having frequency ranges larger than that of the voice signal, the digital transmission of sound program signals is recently perhaps of the greatest importance. In supplying the radio and television broadcast stations with sound programs and in the exchange of sound programs, the CCITT Series J Recommendations, referring to the transmission requirements of broadcast sound program signals in the frequency range of 0.04 to 15 kHz, are applicable.

5.3.3.1 Main encoding parameters

The choice of these parameters is governed by the requirement that the performance of digital transmission should be not worse than that of the analog transmission. The sampling frequency is usually chosen to be $32 = 4 \times 8$ kHz which is best adapted to the band limiting filters and the frame structure.

The number of required quantizing steps is mainly determined by the idle channel noise requirements. By applying 14 bit linear or 10 bit A-law encoding, the signal-to-idle channel noise ratio of a codec section will be about 81 dB which is about 3 dB higher than the value required for one codec section of the hypothetical reference network. This excess signal-to-noise ratio can be utilized to cover the quantizing noise and the noise due to other encoding errors.

The line transmission speed required for sound program transmission is 448 kbit/s for linear law encoding and 320 kbit/s for logarithmic law encoding. Both of these encoding laws have merits and faults, and the optimal solution has not yet been found.

5.3.3.2 Effect of error probability

For the digital transmission of sound programs, the transmission performance has to be much higher than for voice transmission. This is explained by the fact that errors have not only a random noise background effect but also give rise to strong clicks during the program which certainly cannot be tolerated. There are three possibilities of achieving higher performance:

— reduce the error probability,
— apply error correction,
— reduce the effect of errors.

It was mentioned in Section 5.3.2 that the method of PCM encoding was originally developed for the transmission of voice telephone signals, and to this end, an error probability around 10^{-6} is guaranteed between terminal stations of the PCM network. However, this means that in a sound program channel having a speed of 420 kbit/s, the average time span between consecutive bit errors is about 2.3 s (see Table 5.4), in contrast with the average time span of about 15 s for 64 kbit/s

Table 5.4. Average time span between two consecutive bit errors
at a transmission speed of 420 kbit/s

Error probability	10^{-10}	10^{-9}	10^{-8}	10^{-6}
Time span	6.7 h	0.7 h	4 min	2.3 s

voice transmission. This latter frequency of clicks will have no disturbing effect during a telephone conversation but clicks every 2.3 seconds may render a sound program transmission unacceptable, especially during very quiet periods of music programs. Experience has shown [12] that the time span between consecutive bit errors should be at least 40 minutes, and this would require an error probability of less than 10^{-9} for sound program transmission. As this low error probability cannot be obtained with existing PCM networks, suitable measures have to be taken for the correction of errors or for the diminishing of the error effects.

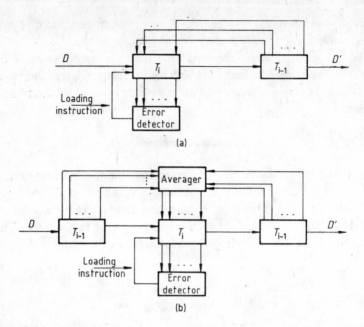

Fig. 5.9. Error correction possibilities preceding the decoding of the digital sound program signal by recognition of the code word error: (a) repetition of the preceding code word T_{i-1}; (b) substitution of the errored code word T_i by a code word representing the average value of T_{i-1} and T_{i+1}.

Two error correction methods applicable to the transmission of binary code words are shown in Fig. 5.9. In both methods, some kind of error detection is applied. As soon as a code word error is detected, the errored code word is cleared and either the preceding code word is repeated or the average of the preceding and the following code words is built.

The error detection is not necessarily extended for the complete code word but less significant bits may be excluded. The error detection can be realized either by parity bit detection or by the application of UD and ZD codes [14]. According to the method of parity bits, the number of symbols within code words is extended to be either an even or an odd number. According to the method of unity disparity (UD) or zero disparity (ZD) codes, the number of zeros may be different from the number of ones by unity, or may not differ at all.

5.3.3.3 Effect of jitter

Because of the effect of jitter accumulated during digital transmission, the PAM samples after decoding will exhibit unwanted phase modulation [13], giving rise to an intermodulation-like noise in the received analog signal. For instance, subjective sound program transmission tests carried out by the BBC have shown that the jitter amplitude of the digital bit stream at the decoder input should be less than 50 ns p–p [14].

The task of reducing the jitter to a tolerable value can be assigned either to the source decoder or to the channel. However, the considerations on error probability given in Section 5.3.3.2 also apply to jitter, and it is not feasible to reduce the jitter by modifying channel parameters because the channel may be shared by both voice and sound programs, and reduced jitter would be worthless for voice transmission. Therefore, the jitter reduction task is usually assigned to the digital sink of the user. The primary function of the transmission path is the reduction of the error probability (making definite decision) and the allowance of transmission without any code restriction.

5.3.4 PCM transmission of FDM signals

There is a choice between two kinds of multichannel transmission systems, FDM and TDM. FDM systems have been well established for many years, and FDM signals are extensively transmitted by analog means (*e.g.*, FDM–FM transmission). However, because of the advent of digital transmission, the compatible digital transmission of both TDM and FDM signals is frequently required. One possibility of achieving this compatibility is the PCM encoding of FDM signals [15].

Assume as an example that we require the digital transmission of an FDM supergroup signal, made up of 60 voice channels in the frequency band of 312 to 552 kHz. This digital transmission can be realized by two methods shown in Fig. 5.10. According to Fig. 5.10a, the FDM supergroup signal is demultiplied, thus generating 60 voice signals in the voice frequency band. These are then PCM encoded by two pieces of primary PCM multiplex equipment and multiplied by time division; the line bit rate of the digital transmission is thus 2×2048 kbit/s. At the far end, the inverse operations are performed. According to Fig. 5.10b, the FDM supergroup signal is directly PCM encoded, resulting in a digital bit stream having a bit rate of about 6 Mbit/s. At the far end, a PCM decoder is applied to recover the original FDM supergroup signal.

Comparison of the above two methods shows that the arrangement of Fig. 5.10b has the advantage of requiring only a wideband PCM coder and decoder

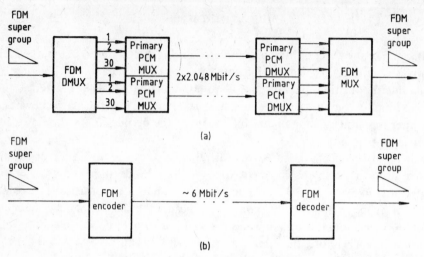

Fig. 5.10. Digital transmission of an FDM supergroup signal: (a) method of elementary demultiplication; (b) direct PCM encoding of the FDM supergroup.

(two printed circuit boards) while the arrangement shown in Fig. 5.10a requires a 60-channel FDM multiplex and demultiplex equipment, and further two items of primary PCM multiplex and demultiplex equipment. The noise performance is also improved according to Fig. 5.10b as no demultiplication and remultiplication of the supergroup takes place. One drawback of the direct PCM encoder/ decoder method is a bit-rate increase of about 50 per cent.

5.3.4.1 Adaptation to the analog network*

FDM transmission requirements are summarized in CCITT Recommendation G.332 which is applicable to a hypothetical reference network of 2500 km length. This hypothetical network is made up of three 833 km sections, each having demultiplied voice frequency terminals. According to this Recommendation, the one-hour average power of the noise should not exceed 10 000 pWOp at the zero relative level point of the reference circuit, even in the worst voice frequency channel. From this allowable noise power, 2500 pWOp is allotted to the modem equipment performing the multiplication and demultiplication, and 7500 pWOp to the transmission path itself. The allowable noise contribution of a digital transmission path thus depends on the insertion points of the path in the reference network.

* The compatibility considerations will be presented from the FDM aspect since FDM is presently the most commonly used transmission method. This means that the digital transmission path should be "transparent" for the transmitted FDM signals.

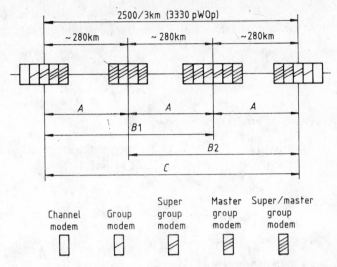

Fig. 5.11. Composition of part of a hypothetical reference network according to CCITT Rec. G. 332.

Figure 5.11 shows a 833 km section of an FDM hypothetical reference network between voice frequency terminals, with subsections between modem terminals denoted by A, $B1$, $B2$ and C. The frequency ranges and channel numbers of the multichannel groups as defined in CCITT Rec. G.211 are summarized in Table 5.5, together with the above subsections over which the multichannel groups can be transmitted.

Table 5.5. FDM group frequency ranges, channel numbers and subsections over which transmission is possible

FDM group	Frequency range limits		Channel number	Subsection according to Fig. 5.11
	f_{min}, kHz	f_{max}, kHz		
Basic group A	12	60	12	C
Basic group B	60	108	12	C
Basic supergroup	312	552	60	$A, B1, C$
Basic mastergroup	812	2 044	300	$A, B1, C$
Basic supermaster group	8516	12 388	900	$B, B1, B2, C$

According to CCITT Rec. G.232, the following highest allowable noise contributions of FDM modems (FDM multiplex–demultiplex equipment) are specified:

channel modem	330 pWOp
group modem	90 pWOp
supergroup modem	80 pWOp
mastergroup modem	25 pWOp
supermaster group modem	25 pWOp

These noise contributions can be used to specify the noise requirements of PCM encoded FDM modems (see Section 2.5.2 of Ref. [1]).

5.3.4.2 Main encoding parameters

For the PCM encoding of FDM signals, linear quantizing is generally applied because nonlinear quantizing would give rise to intermodulation distortion, with the effect of intelligible crosstalk between telephone voice channels. The length of code words is determined by the noise requirements mentioned earlier, about 50 per cent of the total permitted noise being allotted to the quantizing and overload distortion. The number of quantizing steps is chosen by taking into account the statistical properties of the FDM signal [1, 15]: usually, 11 to 12 bit code words are used for the PCM encoding of the 12 channel FDM group, 9 to 10 bit code words for the 60 channel FDM supergroup, and 8 to 10 bit code words for the 300 channel FDM mastergroup and 900 channel FDM supermaster group.

In the following, the choice of the sampling frequency f_s for the PCM encoding of FDM signals will be investigated. This choice should be governed by the requirements of the sampling theorems and by the practical realization requirements of the band limiting filters.

For voice frequency signals possessing only an upper band limiting frequency f_{max}, the sampling frequency range is bounded by the well known inequality

$$f_s > 2f_{max}.$$

However, FDM signals have both an upper band limiting frequency f_{max} and a lower band limiting frequency f_{min}, and in this case, the sampling frequency range is bounded by the following inequality:

$$\frac{2}{n} f_{max} \leqq f_s \leqq \frac{2}{n-1} f_{min}, \tag{5.10}$$

where n is the highest integer which is still less than $f_{max}/(f_{max}-f_{min})$. According to these considerations, the sampling frequency ranges for the PCM encoding of FDM signal groups are summarized in Table 5.6.

Table 5.6. Sampling frequency ranges for the PCM encoding of FDM signal groups
according to CCITT Rec. G.211, calculated by Eq. (5.10)

FDM signal group	n	$f_{s\,min}$, kHz	$f_{s\,max}$, kHz
Basic group, A	1	120	∞
Basic group, B	2	106	120
Basic supergroup	2	552	624
Basic mastergroup	1	4088	∞
Basic supermaster group	3	8222	8516

By taking into account the above values of code word lengths and sampling
frequency ranges, the PCM encoding of FDM signal groups can be realized at
the following hierarchical levels:

1 or 2 12-channel FDM groups	1st level
60 channel FDM supergroup	2nd level
300 channel FDM mastergroup	4th level

5.3.4.3 Code word construction

For the PCM encoding of FDM signals, it is feasible to convert the binary
coded words into Gray coded form (see Chapter 10). This is advantageous because the encoding of an analog signal having normal distribution results in code
words in which the second significant digit is a "1", with a probability of almost
100 per cent [15]. This peculiar statistical property of the Gray encoding can be
well utilized for self-contained frame synchronization.

5.3.5 Videophone and television transmission

The digital transmission of video signals has recently gained importance. Two
groups of transmission tasks are normally considered:

(*i*) transmission of videophone signals of about 1 MHz bandwidth;
(*ii*) transmission of black-and-white or colour television signals of about 5 MHz
bandwidth.

The following video encoding methods are possible:

— linear PCM encoding without redundancy reduction;
— encoding based on direct analog or direct digital signal processing;

— predictive encoding;
— adaptive encoding;
— combined encoding.

The bit rate required for the digital transmission of video signals without redundancy reduction is rather high. On the other hand, the redundancy in video signals is the highest of all previously discussed signal types, resulting in intense research to develop video signal encoding methods based on redundancy reduction [4, 17, 18, 22].

5.3.5.1 Videophone transmission

It is expected that videophone, a new service for telephone subscribers, will show a rapid expansion around the end of this century. This is why the transmission requirements of digital videophone are initially taken into account in the design of digital networks now in progress.

For the linear PCM transmission of a videophone signal with 2 MHz sampling frequency and 8 to 9 bit code words, a line bit rate of 16 to 18 Mbit/s would have to be applied which is available at the third or higher PCM hierarchical levels. On the other hand, videophone signals have a low dynamic range and a high redundancy; an encoding method based on redundancy reduction will therefore probably be applied worldwide, thus decreasing the above high line speed which would render videophone transmission uneconomical. Previous experiments have shown that DPCM source encoding will probably be applied. In the Picturephone® system developed at Bell Laboratories [23], linear DPCM encoding with 3-bit code words is applied, and digital transmission takes place at the second PCM hierarchical level which is 6.312 Mbit/s in the United States. DPCM encoding should be applied also according to an Italian PTT proposal [24], but with a non-linear quantizing characteristic and 4-bit code words, in accordance with the 8448 kbit/s bit rate at the second PCM hierarchical level in Europe. It can be foreseen that in future videophone transmission systems, the line bit rate requirement will be reduced further to around 2 Mbit/s by more efficient redundancy reduction, probably with DPCM and predictive encoding.

5.3.5.2 Television transmission

It is expected that the requirement for digital transmission of television video signals will be emphasized by the wide-spread use of high capacity multichannel digital radio relay systems. This would result in a better exploitation of the transmission system, *e.g.* multichannel transmission during day hours and television transmission during evening hours. Further, in developing countries, investments for establishing up-to-date infrastructures are frequently preferred to catch up with modern technologies, and this may result in exclusively digital transmission networks which would have to be suitable for television transmission too.

Linear PCM transmission of television video signals requires 7 to 11 bit code words and a sampling rate of 11 to 13 MHz which may be possible at the fourth hierarchical level, at the bit rate of 139.264 Mbit/s. In spite of this high bit rate, the standardization of linear PCM encoding is preferred by several PTT administrations [24] to meet the stringent quality requirements of television transmission. It can be expected that in the distant future, the transmission bit rate will be decreased to around 30 Mbit/s which would be suitable for the third hierarchical level. This reduction will probably be made possible by the application of "interframe" and "intraframe" type predictive encoding [20, 21] according to which the correlation between consecutive pictures is also utilized, by the use of adaptive signal processing in which the encoding algorithm is continuously varied [4], and by high performance error correction methods.

5.4 References

[1] Lajtha, Gy., Lajkó, S.: *PCM Transmission* (in Hungarian). Műszaki Könyvkiadó, Budapest 1978.
[2] Cattermole, K. W.: *Principles of Pulse Code Modulation.* Iliffe Books, London 1969.
[3] Bylanski, P., Ingram, D. G. W.: *Digital Transmission Systems.* Peter Peregrinus, London 1976.
[4] Millard, J. B., Maunsell, H. I.: Digital encoding of the video signal. *Bell. Syst. Tech. J.,* Febr. pp. 459–479, 1971.
[5] Moye, L. S.: Digital transmission of speech at low bit rates. *Electr. Commun.,* Vol. 47, No. 4, 1972.
[6] Dinglinger, H. C.: Digital multiplex signals for data transmission. *Electr. Commun.,* Vol. 52, No. 1, 1977.
[7] Williams, M. B.: Developments in data communications. *Post Office Electr. Eng. J.,* pp. 70–80, 1971.
[8] *PCM telegraph multiplex equipment TMB-31/32* (in Hungarian). TKI product report 1978.
[9] Richards, D. L.: Transmission performance of telephone networks containing p.c.m. links. *Proc. IEE,* 115, (15) pp. 1284–1292, 1986.
[10] Rao, V. G.: Comparison of PCM, DPCM and DM for coding speech signals. *Electro-Technol.,* March–April, pp. 98–93, 1972.
[11] CCITT Com. Sp. D. No. 50, 1973–76.
[12] Hessenmüller, H.: The transmission of broadcasting programmes in a digital integrated network. In: *Proc. Int. Seminar on Integrated Systems for Speech, Video and Data Communications, Zürich,* 1972.
[13] Ványai, P.: Some jitter problems of PCM transmission. *Budavox Telecommun. Rev.,* 1, 1976.

[14] CCITT Com. Sp. D. No. 127, 1973–76.

[15] Ványai, P.: *Considerations on PCM encoding of FDM signals* (in Hungarian). TKI reports, 4, 1971.

[16] Sakashita, T., Murata, T.: Direct encoding for FDM basic supergroup signal. *Jpn Telecommun.*, Oct. 1970.

[17] Geissler, H.: Planning of higher order PCM systems and broadband signal transmission. *Telefunken Yearbook,* 1973.

[18] Bauch, H. H., Häberle, H. G.: Picture Coding. *IEEE Trans. Commun.*, Vol. Com-22, No. 9, pp. 1158–1167, 1974.

[19] Bostelmann, G.: A simple high quality DPCM-coder for video telephony using 8 Mbit/s. *NTZ*, Vol. 3, pp. 115–117, 1974.

[20] Connor, D. J., Brainhard, R. C.: Intraframe coding for picture transmission. *Proc. IEE,* Vol. 60, July, pp. 779–791, 1972.

[21] Haskell, B. G., Mounts, P. W.: Interframe coding of videophone pictures. *Proc. IEE,* Vol. 60, July, pp. 792–800, 1972.

[22] Ristenbatt, M. P.: Alternatives in digital communications. *Proc. IEE,* Vol. 61, June, pp. 703–721, 1973.

[23] Millard, J. B., Maunsell, H. T.: Digital encoding of video signals. *Bell. Syst. Tech. J.,* Febr., pp. 459–479, 1971.

[24] CCITT Com. Sp. D. No. 89, 1973–76.

[25] CCITT Com. Sp. D. No. 17, 1973–76.

[26] Ványai, P.: The second PCM hierarchical level: the secondary multiplex (in Hungarian). *Híradástechnika,* 2, pp. 33–47, 1973.

[27] Aratani, T.: Jitter effect on the error rate of PCM repeaters. *Electron. Commun. Jpn,* 11, 1966.

[28] Pospishil, R.: Jitter problems in PCM transmission. *NTZ,* 11, 1971.

[29] Bennett, W. R.: Statistics of regenerative digital transmission. *Bell Syst. Tech. J.,* Nov., 1958.

[30] Proakis, J. G.: *Digital Communications.* McGraw-Hill, New York 1983.

[31] Bassinet, B., Madec, B.: Error performance of digital transmission networks. *Proc. 3rd Int. Conf. Telecommun. Transmission, London,* pp. 23–26, 1985.

CHAPTER 6

DIGITAL MICROWAVE TRANSMISSION SYSTEMS

6.1 Radio relay transmission

6.1.1 Network configuration

The CCIR Recommendations applying for digital microwave radio relay systems [3] apply for specific frequency bands of the 0.4 to 20 GHz frequency range, presenting recommended values of the radio channel frequency arrangements and parameters of the hypothetical reference network.

Figure 6.1 shows the structure of stations involved in one channel of a digital microwave radio relay connection. If more than one radio channel is applied, frequency division multiplexing of the channels is used by utilizing microwave components. Transmission of all channels is realized by common antennas, with the possibility of using both polarizations. As shown in Fig. 6.1, most stations in a long radio relay connection are repeater stations. Switched sections of the link are terminated by main stations while at the end points of the link, terminal stations are applied to interface with other transmission media.

An essential parameter of a radio relay link is the distance between consecutive repeater stations which depends on terrain conditions and on the frequency range applied. For instance, the length of radio sections is 30 to 40 km in the 2 to 4 GHz range, and 1.5 to 3.5 km in the 20 to 40 GHz range. The highest number of repeater stations within a section between main stations is 20 to 50 in the higher frequency ranges, and 5 to 10 in the lower frequency ranges.

Digital radio relay systems provide the same services as analog systems such as supervisory equipment, service telephone channels, continuous performance monitoring and stand-by channel switching. The information required by these services

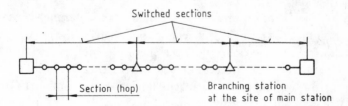

Fig. 6.1. Station configuration of a digital microwave radio relay link: ○ repeater station; Δ main station; □ terminal station.

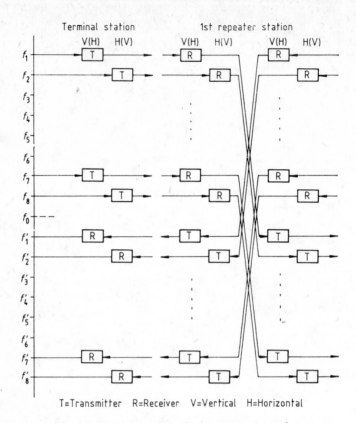

Fig. 6.2. Multichannel terminal and repeater station.

is usually transmitted by the additional modulation of one or more operating radio channels, without requiring a separate channel. The main information may be transmitted by 4PSK modulation, for example, and the additional information by some kind of FSK modulation. It is normally required that the supplementary information should be accessible without the frame structure recognition of the main information bit stream.

Figure 6.2 shows the arrangement of a multichannel terminal station and repeater station together with the frequency designations of the radio channels.

The radio frequency channel arrangements for digital systems, as for analog systems, are given by the CCIR Recommendations. Thus, for instance, the channel arrangement for low and medium capacity links operating in the 11 GHz band are given in Rec. 387, and for medium capacity links operating in the 13 GHz band in Rec. 497 [3]. According to the channel arrangement, lower and upper halfbands are designated around the centre frequency f_0; and the channel frequencies f_1,

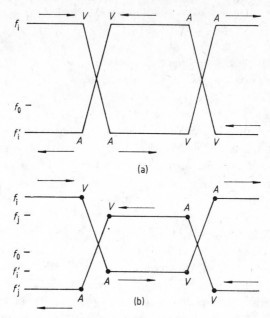

Fig. 6.3. Frequency arrangements: (a) two-frequency arrangement; (b) four-frequency arrangement.

$f_2, ...,$ and $f_1', f_2', ...,$ are evenly distributed in these two halfbands. The channel frequencies with identical subscripts are related to each other: at repeater stations, a channel received at frequency f_i is transmitted at frequency f_i'.

In Fig. 6.2, the polarization of the radio channels is shown by the designations $V(H)$ and $H(V)$ (vertical — horizontal). It is seen that adjacent channels have opposite polarizations. It is also possible to apply two simultaneous co-channel transmissions at any channel frequency, one with vertical and one with horizontal polarization. In principle, this would allow us to double the radio frequency channel capacity shown in Fig. 6.2.

Frequently, the full capacity of the channel arrangement is not utilized, and only two radio channels in each direction are applied for $1+1$ transmission (one operating and one stand-by channel). In this case, a two-frequency or a four-frequency channel arrangement can be used as shown in Fig. 6.3. In the case of two-frequency arrangement, reception from the two directions at a repeater station is on identical channel frequencies, and the same is true for transmissions in the two directions.

The two-frequency arrangement allows economical utilization of the available spectrum; however, interference requirements are more easily met by the use of the four-frequency arrangement shown in Fig. 6.3b. Only rarely are more than four carrier frequencies used for a two-way radio channel.

6.1.2 Radio frequency equipment

Radio frequency equipment includes passive microwave components (polarization filters, channel branching filters, *etc.*), microwave transmitters and receivers, carrier supplies, intermediate frequency circuits, and also digital modulator, demodulator and carrier recovery circuits. These last three are discussed in detail in Chapters 7, 8 and 9.

In the following, possibilities of realizing the carrier supply will be presented. At a repeater station, two high-level carrier supplies for the two transmitters and two low-level carrier supplies for the two receivers are needed in each two-way radio channel. In conventional systems, these carrier supplies are provided by separate units. In these units, crystal controlled oscillators operating around 100 MHz and having a frequency stability of 10^{-5} to 10^{-6} are applied, providing the specified carrier signal after frequency multiplication and amplification.

In up-to-date systems, the required carrier signals are derived from a single crystal controlled oscillator by frequency synthesis using phase-locked loops (see Fig. 6.4). In the PLL circuit, a frequency divider with a division ratio of modulo N is used to provide the control voltage of the phase comparator, and the desired frequency is provided by a q-times frequency multiplier following the voltage controlled oscillator (VCO) of the PLL. In the lower microwave frequency ranges, there is a possibility for $q=1$. A suitable value for the reference frequency f_k feeding the phase comparator is 1 MHz, and the output signal frequency of the PLL circuit is given by

$$f_i = qNf_k \tag{6.1}$$

where $0 \leq N \leq 2^n$, 2^n being the highest division ratio of the frequency divider.

The frequency division ratio N is adjusted by suitable programming of the programmable frequency divider. This allows application of identical circuits for the generation of signals f_i and $f_i \pm f_{i.f.}$ by different programming, and all signals thus generated will have frequency stability of the common crystal oscillator. Several variants can be used to implement this principle.

There is another inexpensive carrier supply principle which can be applied if the frequency difference between the two-way radio channels can be chosen to be equal to the intermediate frequency. In this case, a series of identical carrier supplies, driven by a single crystal oscillator, can provide all carrier frequencies needed at a repeater station. It should be noted that upper and lower frequency conversion has to be applied at consecutive repeater stations. The simple carrier supply system shown in Fig. 6.5 has been applied in the MIDAS microwave data transmission system [5].

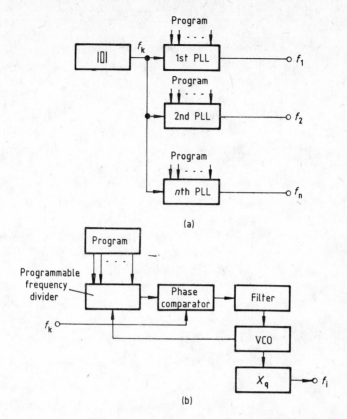

(a)

(b)

Fig. 6.4. Synthesizer carrier supply: (a) distribution of n carriers; (b) details of a single synthesizer.

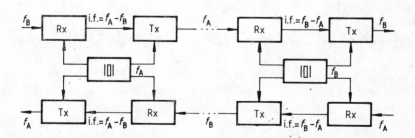

Fig. 6.5. One-channel transmission, with a common carrier supply at each station.

6.1.3 Baseband signal processing equipment

The baseband signal processing equipment provides an optimal interface between the digital bit stream to be transmitted and the digital modem. At the transmit side, the functions of the baseband signal processing equipment are the channel coding (*e.g.*, scrambling, differential encoding, Gray encoding *etc.*), the generation of parallel (*e.g.* dibit, tribit *etc.*) bit streams and possibly the insertion of additional information. At the receive side, the inverse functions are provided: channel decoding, and further the recovery of the series bit stream and the additional information signal (if present). The receive side signal processing equipment also has the functions of equalizing the intersymbol interference due to the normally strong band limiting, and providing the bit time recovery and regeneration of the digital bit stream.

The functions of the baseband signal processing equipment actually performed depend on the station type. For instance, the channel encoding and decoding, the insertion and recovery of the additional information, and the series–parallel and parallel–series converters are only needed at terminal and branching stations. On the other hand, the bit time recovery and the regeneration of the received digital bit stream is implemented at every regenerating repeater station, in some cases together with encoding and decoding of some orders of wire channels. A detailed discussion of baseband signal processing is presented in Chapter 10.

6.1.4 Stand-by channel equipment

Multichannel radio relay links are divided into switched sections by the main stations. Within these sections, a faulty operating channel can be substituted by a stand-by channel in order to meet the transmission requirements. Usually, a single stand-by channel is applied to replace all operating channels of a multichannel link. The switch-over process is carried out in three steps:

(*i*) the performance, *i.e.* the error probability of all radio channels is continuously monitored;

(*ii*) in the case of a channel failure, a decision is taken at the receive side to replace the operating channel by the stand-by channel;

(*iii*) simultaneously with this decision, the switch-over at the output side is signaled back to the preceding main station at the transmit end of the switched section, and the above substitution is performed at the transmit side too.

The situation is simplified if a $1+1$ (operating$+$stand-by) system, *i.e.* a diversity system is applied, with two radio channels driven in parallel at the transmit side.

In this case, switch-over is only required at the receive side, and no other operations are needed.

After the failed channel is back in normal operation, it may be switched back into the transmission path. This requires a switch-back with a hysteresis ratio in the range of 100 to 1000, *e.g.* a switch-over with an error probability of 10^{-4} is followed by a switch-back with 10^{-6} error probability. However, the number of switch-over operations may be halved if no switch-back to the original channel takes place, and the prevailing free radio channel is taken as the stand-by channel. This latter method is always employed in $1+1$ systems.

Two methods may be used for continuous monitoring of the error probability. In low capacity systems, the transmission of the auxiliary channel implemented by additional modulation is monitored while in high capacity, multichannel systems, the monitoring of the channel generated by bit insertion may take place. An example of the latter is the Japanese 10 radio channel relay system 20 G—400 M [1, 2] in which 4PSK modulation is used. The bit rate of the two dibit channels is increased by bit insertion from 198.6 to 198.989 Mbit/s, thus providing two additional channels at 389 kbit/s. The transmission of a low periodicity bit sequence over these channels provides a means of continuous monitoring of the radio channel performance.

6.1.5 Interface equipment

The interconnection of the radio relay link with the PCM multiplex equipment at terminal or branching stations is governed by the relevant CCITT Recommendations. The required line codes and signal levels to be provided by the interface equipment of the radio relay link are summarized in Table 6.1.

6.1.6 Supervisory and service channel equipment

This kind of equipment is applied at main stations, terminal stations and branching stations. The main functions of this equipment are the following:

— control and fault location of the repeater stations pertaining to a main station;
— transmission of switch-over instructions;
— provision of an omnibus type service telephone channel within the line section between main stations;
— provision of an express service channel along the complete microwave link;
— operations to provide status signaling and fault location over the complete microwave link.

Table 6.1. Interconnection data of digital transmission

Hierarchical level	Bit rate, Mbit/s		Line code		Signal level
	Europe	USA	Europe	USA	
1	2.048	1.544	HDB$_3$	B3ZS	3.0 V
2	8.448	6.312	HDB$_3$ or AMI+ clock on sepa- rate wire	B6ZS	1.05 V
3	34.368	44.736	HDB$_3$	B3ZS	0.7 dBm
4	139.264	274.178	CMI	B3ZS	0.7 dBm

The service channel is generally implemented independently from the main digital bit stream by applying additional modulation. In multichannel systems, the additional modulation is realized in one radio channel only. In the case of 4PSK transmission, the additional modulation is generally 2FSK or 3FSK (partial response transmission) if digital modulation is used in the auxiliary channel. Time division multiplexing is normally used for the transmission of the individual information signals. The bit rate of the service channel of microwave radio links is in the range of 100 to 300 kbit/s. This is utilized for the transmission of telephone channels over short and long distances, and of data to provide the above signaling.

6.1.7 Station types

The block diagrams of stations utilizing the equipment discussed in the preceding sections are shown in Fig. 6.6.

Terminal station

The terminal station comprises one transmitter and one receiver, both connected to the baseband signal processing equipment. The terminal station provides one terminal of a switched section, thus comprising also the switching unit in the case of multichannel links. The interface equipment provides the connection to the

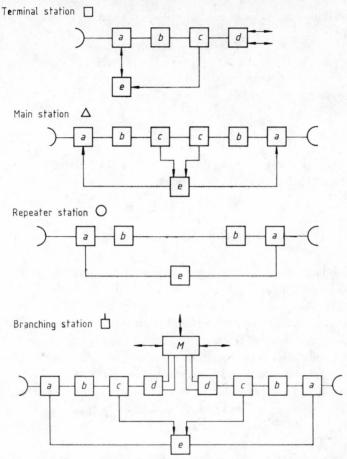

Fig. 6.6. Configuration of station types. *a* — radio frequency equipment; *b* — baseband signal processing equipment; *c* — switching equipment; *d* — interface equipment; *e* — supervisory and service channel equipment; *M* — multiplex equipment.

external equipment. Finally, the supervisory and service channel equipment is connected to the radio frequency equipment, utilizing the service channel operating with the additional modulation system.

Main station

The main station is located between switched sections of a radio relay link, comprising radio frequency equipment, baseband signal processing equipment, interfacing equipment and service channel equipment. The service channel is used to transmit the supervisory signals and service channel voice signals, and also the switching signal information.

Repeater station

This is the simplest type of station of the radio relay link. It comprises, in addition to two sets of radio frequency transmitter and receiver, the baseband circuits required for the regeneration of the bit streams. Service channel equipment is only partially implemented at repeater stations.

Branching station

The branching station is composed of two terminal stations and a multiplex equipment suitable for the transmitted bit stream.

6.2 Waveguide transmission

The advantage of waveguide transmission is the absence of absorption effects characterizing free-space transmission and thus the elimination of instabilities. However, the waveguide attenuation and dispersion may have adverse effects on high capacity transmission. The waveguide attenuation is generally high and is measurable in short waveguide sections of a few meters length. An exception to this is the TE_{01}^0 mode wave propagation which has small transmission losses for specific geometrical dimensions and frequencies [6].

The realization of a waveguide transmission system requires high investment and is therefore justified for only extremely high capacity transmission. Accordingly, the channels should be suitable for the transmission of bit rates in the range of 400 to 1600 Mbit/s. This is possible through the use of millimeter wavelength signals in the 40 to 100 GHz frequency range. The TE_{01}^0 mode in the circular waveguide is not a basic mode, and the geometrical relations also allow the propagation of other modes. Inhomogeneities of the waveguide may have the effect of converting part of the signal energy into another mode of higher loss, and *vice versa*. A special waveguide type can be used to reduce the possibility of mode conversion and re-conversion.

One of the application fields of waveguide transmission is the trunk-like interconnection of extremely large cities which are reasonably close. A typical realization is a Japanese waveguide transmission system [4] applying a circular waveguide of 51 mm diameter as the transmission medium, propagating the TE_{01}^0 mode. The waveguide link is capable of accommodating up to 24 duplex channels over which bit streams at 800 Mbit/s are transmitted by 4PSK modulation. The intermediate frequency of the receivers is 1400 MHz. Within the frequency range of 40 to 80 GHz, 12 duplex channels are transmitted in the upper half band, and 12 further

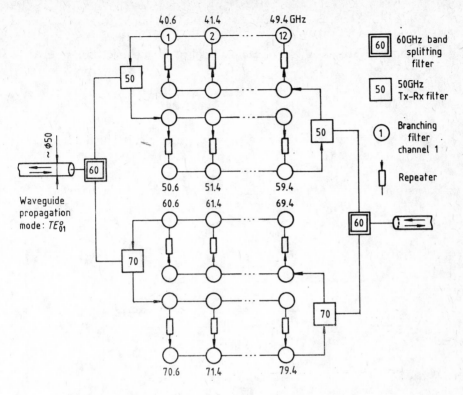

Fig. 6.7. Repeater station of a digital waveguide transmission system.

duplex channels in the lower half band. The full capacity of the waveguide transmission system shown in Fig. 6.7 corresponds to the transmission of 288 000 duplex telephone channels. The waveguide attenuation is about 3 dB/km, and the distance between the regenerating repeater stations is in the range 15 to 20 km.

6.3 Satellite transmission

In addition to terrestrial line-of-sight radio relay transmission and waveguide transmission, satellite transmission is also applied in digital microwave systems, primarily for intercontinental connections. According to the International Radio Regulations, the following frequency ranges are provided for fixed satellite services: for the up-link direction, 4400 to 4700 MHz, 5725 to 6425 MHz, 7900 to 8400 MHz; for the down-link direction, 3400 to 4200 MHz and 7250 to 7750 MHz. The main advantage of satellite communication is the fact that a single satellite can be simultaneously accessed by several earth stations, as implied in the designation "multiple access." Either frequency division multiple access (FDMA) or time division multiple access (TDMA) can be applied in multiple access networks used in practice.

6.3.1 TDMA system

According to the time division multiple access principle, the digital bit streams of several earth stations are sequentially transmitted by a single radio frequency carrier in the satellite radio channel. The TDMA system is flexible in the sense that a re-routing of the transmission path according to traffic requirements can be implemented by programming. This is in contrast with the FDMA system which requires a complicated and expensive operation for this purpose.

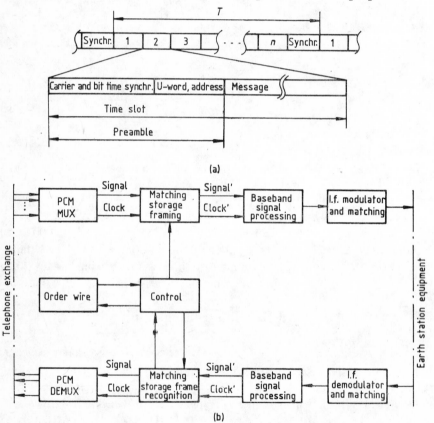

Fig. 6.8. TDMA system: (a) frame structure characterizing the operation; (b) block diagram of TDMA equipment.

In a typical TDMA system, PCM transmission is applied, and the frame time interval of the system is generally an integral multiple of 125 µs, corresponding to the 8 kHz sampling frequency of the PCM multiplex. A typical TDMA frame structure is shown in Fig. 6.8a. In the INTELSAT network application [7], the frame time is 750 µs, the bit rate of the completed digital bit stream is 60.032 Mbit/s, and time slots for 60 earth stations are provided. This means that the average number of bits in a time slot is 750, corresponding to a bit rate of about 900 kbit/s, *i.e.* about 14 telephone channels per station can be transmitted. The digital signals of TDMA earth stations are transmitted in short bursts which is in contrast with the even signal distribution in terrestrial transmission systems. The TDMA system is asynchronous *i.e.* there is only a plesiochronous relation between the earth station transmitter carriers and the clock signals of the digital signal sources. Correspondingly, the series of information bits within the time slots is preceded by a so-called preamble which contains the address of the receive station and contributes to a rapid carrier and bit time recovery at the receive side.

One time slot denoted by X in Fig. 6.8a is intended for housekeeping purposes, determining the time sequence of the individual transmitters and making possible service channel transmission. Figure 6.8b shows a simplified block diagram of a TDMA earth station. The interface-storage-framing block has the function of transforming the PCM multiplex signal into a format corresponding to the TDMA repetition rate, according to the instructions of the central control. The baseband signal processing unit is driven by the bursts comprising the preambles. The applied digital modulation is 4PSK. In the receive path, inverse operations are carried out.

6.3.2 SPADE system

In FM–FDM–FDMA systems, only complete FDM groups of voice channels can be branched at individual earth stations. However, traffic demands at some earth stations may call only for a few telephone channels, and the TDMA system is not economical in meeting this requirement because the number of bits in the preamble and in the information part of the burst are of the same order of magnitude. This problem is solved by SCPC (Single Channel Per Carrier) systems in which individual voice channels are transmitted by separate carriers. One implementation of the SCPC principle is the SPADE system (Single Channel Per Carrier PCM Multiple Access Demand Assignment Equipment). This is an FDM system in which the capacity of the radio channels is a single voice channel, and the carrier is transmitted by voice activation. 4PSK modulation of the individual carriers by the PCM encoded voice signals is used, and the carriers are radiated after suitable up-conversion. The spectrum and simplified block diagram of the

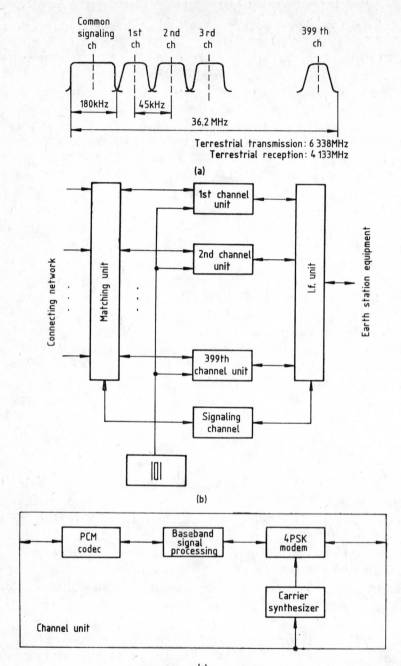

Fig. 6.9. SPADE system: (a) carrier spectrum; (b) block diagram of SPADE equipment; (c) block diagram of a channel unit.

SPADE system are shown in Fig. 6.9 [8]. The capacity of the system is 400 telephone channels, and from the available 36 MHz frequency band, approximately 180 kHz is utilized for the transmission of the common signaling. After baseband signal processing, the PCM signal streams are used to drive 4PSK modulators at a speed of 32 kB. A bandwidth of 45 kHz is available for the transmission of the modulated signals.

The average utilization of a duplex telephone channel is only around 40 per cent, so that the demand assignment feature of the SPADE system approximately doubles the transmission capacity.

6.4 References

[1] Juichiya, Y., Yoshikawa, T.: 20G–400M digital radio system. *Rev. Electr. Commun. Lab.*, Vol. 24, No. 7–8, pp. 521–551, 1976.
[2] Kuramoto, M., Omori, H.: 20G–400M terminal equipments. *Rev. Electr. Commun. Lab.*, Vol. 24, No. 7–8, pp. 573–600, 1976.
[3] *CCIR Vol. IX–1, Fixed service using radio relay systems.* Plenary assembly, Dubrovnik, 1986.
[4] Shimada, S., Ohtomo, I.: An experimental channel multiplexing and demultiplexing network for 4-phase PSK millimeter waveguide transmission system. *Rev. Electr. Commun. Lab.*, No. 7–8, pp. 507–520, 1973.
[5] Ványai, P.: Digital microwave radio relay equipment for high speed data transmission. *Budavox Telecommun. Rev.*, pp. 1–11, 1977.
[6] Karbowiak, K. E.: *Trunk Waveguide Communication.* Chapman and Hall, London 1965.
[7] Pontano, B.: The Intelsat TDMA/DSI system. *Int. J. Satellite Commun.*, Vol. 3, No. 1–2, Jan.–June, 1985.
[8] Hall, G. C.: Single-channel-per carrier, pulse code modulation, multiple access, demand assignment equipment SPADE for satellite communication. *Post Office Electr. Eng. J.*, April, pp. 42–48, 1974.

CHAPTER 7

MODULATORS AND TRANSMITTERS

7.1 General considerations

The function of the digital modulator is the generation of a modulated carrier from the digital bit sequence carrying the information. The digital bit sequence is generally a two-level NRZ stream which is directly suited for generating a two-state carrier. However, a multilevel modulation scheme is frequently applied for better utilization of the frequency band. By forming groups of n bits, a modulated signal with $M = 2^n$ states can be generated. The bit sequences thus formed have a repetition rate which is equal to the bit rate/n and is called the symbol rate. Most frequently utilized are the 4-state, 8-state and 16-state modulations, derived by forming $n = 2, 3$ and 4 bit groups. The conversion of the bit sequence by the coder will be dealt with in Chapter 10.

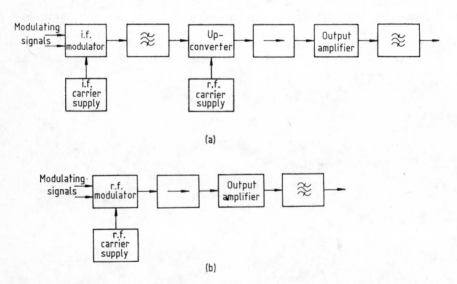

(a)

(b)

Fig. 7.1. Block diagram of a microwave transmitter: (a) transmitter with i.f. modulator; (b) transmitter with r.f. modulator.

I. f. and r. f. modulators. The digital modulation process is carried out either in the i.f. part or r.f. part of the transmitter. The i.f. implementation is shown in Fig. 7.1a; according to this method, the i.f. modulator is fed by an i.f. carrier supply, and the modulated i.f. signal is shifted to the microwave range by an up-converter. Figure 7.1b shows the microwave implementation: modulation is accomplished directly in the microwave frequency band, thus eliminating up-conversion. In both variants, a microwave power amplifier can be applied if this is justified by the higher output power required.

Modulation methods. In most digital transmission systems, phase modulation is applied (PSK — Phase Shift Keying). This can be relatively simply and accurately realized both in the i.f. and r.f. frequency ranges [1, 2, 3]. PSK modulation also has the advantage of allowing a simple realization of a coherent demodulation.

The PSK signal has 100 per cent amplitude modulation which can be reduced by applying O-QPSK (Offset Quadrature PSK) or MSK (Minimum Shift Keying). This may be feasible for satellite communication when amplitude modulation gives rise to side lobe regeneration by the nonlinear power amplifier of the satellite transponder, resulting in a spectrum spread and thus in adjacent channel interference.

In the case of high speed transmission, 16 or 64 state QAM (Quadrature Amplitude Modulation) is frequently applied for better utilization of the frequency band. For low speed transmission, amplitude modulation (ASK — Amplitude Shift Keying) is normally utilized. It was shown in Chapter 2 that this has a much lower efficiency than PSK but the realization is quite simple. Digital frequency modulation (FSK — Frequency Shift Keying) is midway between PSK and ASK from the point of view of efficiency, and is applied in transmission systems of lower speed. PSK, FSK and ASK modulators can be realized both in the intermediate frequency and microwave ranges; on the other hand, QAM and MSK modulators are applied in the intermediate frequency range only because linear multipliers, needed in these modulator types, cannot be realized in the microwave range.

Spectrum limiting. The spectrum of the modulated carrier has to be limited in order to eliminate adjacent channel interference. Band limiting can be accomplished either by a low-pass filter, inserted at the digital input of the modulator, or a bandpass filter, inserted at the carrier output of the modulator. A low-pass filter can be applied if the band limiting effect of the filter is not offset by a nonlinear circuit following the filter (nonlinear modulator or nonlinear power amplifier, the latter for example in satellite communication systems [4, 5]). QAM and MSK can be realized by linear multipliers so that a low-pass filter preceding the modulator can be used for spectrum limiting. The same method is also applicable for FSK modulators.

A carrier frequency bandpass filter following the modulator results in an effective band limiting, independently from the modulator linearity. However, in the case of low-speed microwave modulators, the required narrow-band filter may be difficult to realize at microwave frequencies. In certain cases, the correct placement of the carrier frequency bandpass filter may be of importance. Suppose, for instance, that a phase modulator is applied, and the unwanted amplitude modulation is reduced by a limiter following the phase modulator. The limiter then has the effect of spreading the spectrum, so the bandpass filter for band limiting should follow the limiter.

7.2 Phase modulators (PSK)

7.2.1 I.f. phase modulator circuits

The frequency of the i.f. phase modulators is 35 MHz, 70 MHz or 140 MHz, corresponding to the intermediate frequency of the digital transmitter. In most cases, the i.f. modulator is realized by a ring modulator circuit. In some cases, modulators utilizing monolithic IC's are also applied.

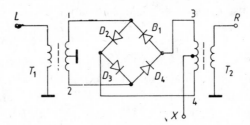

Fig. 7.2. Ring modulator.

Two-state phase modulator (2PSK or BPSK). This is shown in Fig. 7.2. Assume that there is a positive voltage at the x input of the modulator. Diodes D_1 and D_3 are then conducting, thus connecting terminal 1 of transformer T_1 to terminal 3 of transformer T_2, and terminal 2 of T_1 to terminal 4 of T_2. In the period of negative voltage at input x, diodes D_2 and D_4 are conducting, thus connecting terminal 1 of T_1 to terminal 4 of T_2, and terminal 2 of T_1 to terminal 3 of T_2. This means that the output signal polarity is changed for each change of the modulating signal polarity, *i.e.*, we have a 180° phase change. The above process is realized if the modulating signal at input x is of higher amplitude than the carrier signal amplitude at input L, *i.e.* the modulating signal will then switch over the carrier signal.

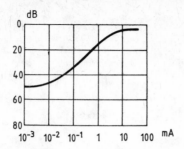

Fig. 7.3. Attenuation of a ring modulator.

Let the carrier signal be

$$V_c(t) = \cos \omega_0 t, \tag{7.1}$$

then the output signal is given by

$$V_0(t) = a(t) \cos \omega_0 t \tag{7.2}$$

where $a(t) = \pm 1$. Figure 7.3 shows the insertion loss of a typical ring modulator as a function of the diode current. It is seen that at least 10 mA driving current is required for correct operation. Typical operational parameters for ring modulators: diode driving current 20 to 40 mA, carrier signal amplitude 0.5 V r.m.s. on 50 ohms, corresponding to 5 mW [6].

The ring modulator can also be applied as an amplitude modulator. In this case, the carrier signal amplitude is higher than the modulating singal amplitude, and the diodes are switched by the carrier signal. $a(t)$ may have an arbitrary waveform. This kind of modulator is used for generating a double sideband suppressed carrier signal. The ring modulator is also called a multiplying type modulator because the output signal is the product of the modulating signal and the carrier signal. For an amplitude modulator, the carrier frequency level is the same as above, the modulating signal being 150 mV r.m.s. The TKI R110 modulator has similar parameters.

Four-state phase modulator (4PSK or QPSK). Two ring modulators can be connected to form a four-state modulator as shown in Fig. 7.4. Let the carrier signal be given by $V_c = \cos \omega_0 t$. Similarly to the two-state phase modulator, the two multipliers supply the two following output signals:

$$V_{01}(t) = a \cos \omega_0 t \tag{7.3}$$

$$V_{02}(t) = -b \sin \omega_0 t. \tag{7.4}$$

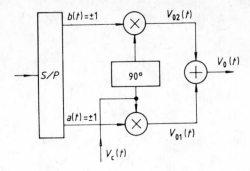

Fig. 7.4. 4PSK modulator.

The signal at the summing circuit output will be

$$V_0(t) = a \cos \omega_0 t - b \sin \omega_0 t = \sqrt{a^2 + b^2} \cos(\omega_0 t + \varphi)$$

$$\varphi = \tan^{-1} \frac{b}{a} \tag{7.5}$$

where $a = \pm 1$ and $b = \pm 1$. The S/P converter, fed by the modulating bit sequence, has the function of series–parallel conversion by generating dibits of TTL level, and further of converting the TTL signals into ± 1 signals. In the case of current driven multipliers, the S/P converter may also comprise the current generator, made up of transistors or monolithic IC's.

The vector diagram of a 4PSK signal is made up of two vectors which are perpendicular to each other (Figs 7.5b and c). During the modulation process, the vector is shifted either to the neighbouring position (90° phase change) or to the opposite direction (180° phase change). In the latter case, the signal amplitude goes through zero during the phase transition as shown in Figs 7.5a and 7.5e. This results in a 100 per cent amplitude modulation. A 90° phase change brings about an amplitude change of only 3 dB.

Eight-state phase modulator (8PSK). The output signal of this modulator is given by

$$V_0(t) = \cos\left(\omega_0 t + \frac{2\pi n}{8} + \frac{\pi}{8}\right), \quad n = 0, 1, 2, \dots, 7. \tag{7.6}$$

The eight states are defined by three binary numbers represented by three modulating signals a, b, c.

One method of generating a 8PSK signal is to combine the signals of a 4PSK modulator and a 2PSK modulator, fed by carriers which are 45° out of phase. However, two amplitude levels with a difference of 6 dB will appear. This could be

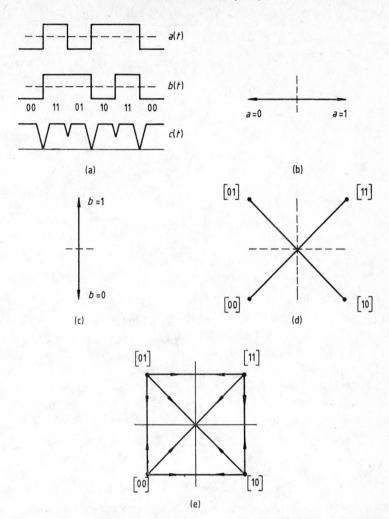

Fig. 7.5. Characteristics of 4PSK modulation: (a) waveforms of dibits $a(t)$ and $b(t)$ and envelope $c(t)$ of the modulator output signal; (b) vector diagram due to dibit $a(t)$; (c) vector diagram due to dibit $b(t)$; (d) vector diagram of the complete 4PSK modulator; (e) possible phase transitions.

reduced by a limiter but this usually has AM-to-PM conversion, increasing the phase error.

Another method is to convert the a, b, c modulating signals into two four-level signals having amplitudes $\pm\sin \pi/8$ and $\pm\cos \pi/8$. The 8PSK signal is obtained by modulating a quadrature amplitude modulator with these signals, providing the multipliers of the quadrature amplitude modulator are linear. The signal of this modulator has a constant level in all modulation states [7].

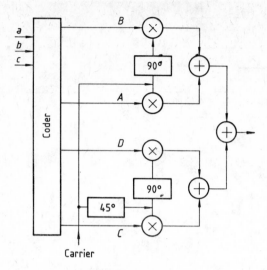

Fig. 7.6. 8PSK modulator.

A third method of generating an 8PSK signal, most frequently applied, is to combine the signals of two 4PSK modulators [8, 9]. In this case (see Fig. 7.6), the carriers of the two modulators should have a phase difference of 45 degrees. A coder is applied to convert the incoming signals a, b, c into the modulating signals A, B, C, D. These would generate a 16PSK signal having two different amplitude levels at alternate states. An additional function of the coder is the suppression of every second signal in order to obtain an 8PSK signal.

7.2.2 I.f. phase modulator errors

In the preceding, several types of i.f. modulators were investigated, of which only the errors of the 4PSK modulator will be studied. The output signal of a 4PSK modulator is given by Eq. (7.5). The error sources can be classified as follows.

The phase difference between the two modulator carriers is not exactly 90°. This can be the result of the inaccurate adjustment of the phase shifter or of temperature effects.

The signal amplitudes in the two branches are different. This may result from the difference between the input signal amplitudes or from the differing insertion losses of the ring modulators.

One of the ring modulators is asymmetrical, the output signals for the two modulation states have different amplitudes.

It can be shown that a deviation from 90° in the phase shifter results in amplitude modulation and does not affect the $n90°$ phase difference between the modulation states. Let the deviation from 90° be $2\Delta\Phi$; the output signal can then be expressed in the following form:

$$V(t) = a(t) \cos (\omega_0 t + \Delta\Phi) - b(t) \sin (\omega_0 t - \Delta\Phi) \qquad (7.7)$$

where $a = \pm 1$ and $b = \pm 1$. The amplitude ratio is calculated to be

$$\frac{A_2}{A_1} = \frac{1 + \tan \Delta\Phi}{1 - \tan \Delta\Phi}. \qquad (7.8)$$

According to Eq. (7.8), a deviation of 10° results in an amplitude variation of 1.4 dB while 5° introduce a variation of 0.8 dB (Fig. 7.7a).

On the other hand, an amplitude difference between the multiplier output carriers results in a phase error. Let $a = a_0(1 + \delta)$ and $b = b_0(1 - \delta)$. Substituting these amplitudes into Eq. (7.5), we have the phase error of

$$\varphi = \tan^{-1}\left(\frac{b_0}{a_0} \frac{1 - \delta}{1 + \delta}\right), \qquad a_0 = \pm 1, \quad b_0 = \pm 1. \qquad (7.9)$$

According to Eq. (7.9), $\delta = 0.1$ yields 6.5° and $\delta = 0.05$ yields 3° (Fig. 7.7b).

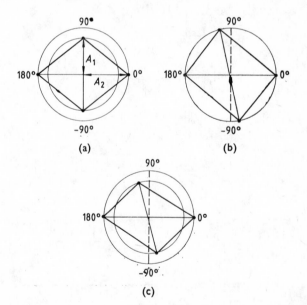

Fig. 7.7. Vector diagrams showing errors of 4PSK modulator: (a) the phase difference between the two quadrature carriers is not exactly 90°; (b) the attenuation values of the two branches are not exactly identical; (c) asymmetry between the two multipliers.

An asymmetrical ring modulator introduces both an amplitude variation and a phase error (Fig. 7.7c).

In practical phase modulators, the three error types are present simultaneously. A 4PSK modulator having a phase error of $\pm 2°$ and an amplitude variation of 0.3 dB can be considered as a good modulator. The transition between phases is extremely short in the case of ring modulators (around 1 ns) so transient phenomena need not be considered.

7.2.3 Up-conversion of the i.f. signal into the microwave frequency range

There are three basic methods of converting the modulated i.f. signal into the microwave frequency range. The widely applied method is the use of the well known up-converter realized by a high-level mixer (Fig. 7.8a). Either the sum-frequency or the difference frequency output signal is filtered out and utilized as a modulated microwave signal [1, 10].

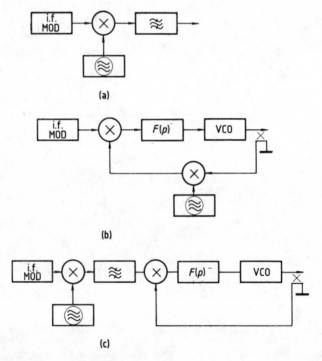

Fig. 7.8. Methods of up-converting the i.f. signal into the r.f. range: (a) conventional up-converter; (b) PLL type up-converter; (c) conventional up-converter and PLL amplifier.

Figure 7.8b shows a special up-conversion arrangement utilizing a PLL [11]. A voltage controlled oscillator operates at microwave frequencies, and its signal is down-converted to the i.f. range by a microwave oscillator. The down-converted signal is then compared with the modulated i.f. signal in an i.f. phase discriminator which supplies, after suitable filtering and amplification, the error signal controling the VCO. This control has the effect that the modulation phase at i.f. is followed by the VCO phase at r.f.

In the arrangement of Fig. 7.8c, a conventional up-converter is followed by a PLL acting as a microwave amplifier. The phase discriminator operates at microwave frequencies.

An important property of the PLL arrangements of Figs 7.8b and 7.8c is the limiting action eliminating the unwanted AM of the PSK signal. An upper limit of the modulating signal frequency range is given by the need to meet the stability criterion. The PLL arrangements are therefore primarily applied in low and medium capacity PSK systems.

7.2.4 Phase modulator — multiplier arrangement

In many cases, the microwave signal is generated by frequency multiplication of a lower frequency signal. This system allows the phase modulation to be carried out at some intermediate frequency at which only a small fraction of the specified phase deviation has to be achieved. Let ω_0 be the output frequency, ω_1 be the frequency at which modulation is performed, and $n = \omega_0/\omega_1$. If the phase deviation at the output is $\Delta\Phi_0$ then the phase deviation of the modulator is only $\Delta\Phi_0/n$. This deviation being small, a linear phase modulator can be realized.

Figure 7.9a shows the block diagram of such an arrangement while Fig. 7.9b is a circuit diagram of a transistor phase modulator. The latter shows a tuned amplifier stage with a resonant circuit comprising a varactor diode. Modulation is achieved by applying the modulating signal as a varactor bias, resulting in a capacitance variation and thus a detuning of the resonant circuit, with a corresponding change of the amplifier phase shift.

Let the current driving the resonant circuit be given by

$$i(t) = I_0 \cos \omega_0 t. \tag{7.9a}$$

The voltage across the resonant circuit is then

$$V(t) = Zi(t) = |Z| I_0 \cos (\omega_0 t - \varphi) \tag{7.10}$$

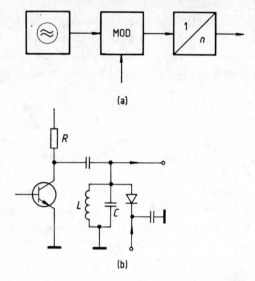

(a)

(b)

Fig. 7.9. (a) PSK modulator with frequency multiplier; (b) circuit of i.f. PSK modulator with varactor modulation.

where

$$Z = \cfrac{1}{pC + \cfrac{1}{R} + \cfrac{1}{pL}} = R\,\cfrac{1}{1 + jQ\left(\cfrac{\omega}{\omega_0} - \cfrac{\omega_0}{\omega}\right)}$$

(7.11)

$$\omega_0 = \frac{1}{\sqrt{LC}}, \quad Q = \omega_0 RC.$$

Introducing the notation $\eta = \dfrac{\omega}{\omega_0} - \dfrac{\omega_0}{\omega}$ we have

$$|Z| = \frac{R}{\sqrt{1 + Q^2\eta^2}}$$

(7.12)

and the modulator phase characteristic is given by

$$\varphi = \tan^{-1}Q\eta.$$

(7.13)

For small angles of phase modulation, $\varphi \approx KV_m$, i.e. the phase modulation can be regarded as being linear. For small detuning, $|Z|$ shows small changes only so the amplitude modulation can be neglected.

A shortcoming of the above system is the fact that the phase error of the modulator is multiplied by n at the output, requiring a modulator of extremely low phase error and high frequency stability. The multiplying chain has to transmit a modulated signal so a suitable bandwidth is required which, in turn, reduces the efficiency of multiplication. The multiplier arrangement is therefore applied only in low capacity systems.

7.2.5 Microwave phase modulator circuits

Varactor diode phase modulator. According to the operating principle of such a modulator, the carrier signal is used to drive the first port of a circulator while the second port is connected to a transmission line section terminated by the reactance of a varactor (see Fig. 7.10a). The varactor is driven by the modulation signal, thus varying its reactance and thus the phase of the carrier reflected back to the circulator at the modulation rate; the varactor phase modulator is therefore also called a reflection type modulator. The phase modulated carrier appears at the third port of the circulator.

Let us assume a lossless varactor, and consider the impedance diagram shown in Fig. 7.10b. The two points corresponding to the capacitive impedances in the two modulation states will appear on the circle of total reflection (points A and B). However, the phase difference between points A and B is not sufficient with capacitance changes due to conventional varactors. In order to increase the phase difference, let us find plane X of the transmission line section at which the transmission line impedance is zero at the varactor impedance corresponding to point A; in this plane, the two phases will be A_1 and B_1 instead of A and B. Let us connect in this plane a second transmission line section, short circuited at the far end, in parallel with the section terminated by the varactor. Because of the zero impedance, the impedance as seen from the circulator will not change, because of the

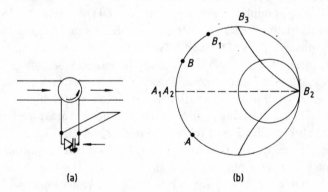

(a)   (b)

Fig. 7.10. (a) R.f. varactor modulator; (b) impedance diagram.

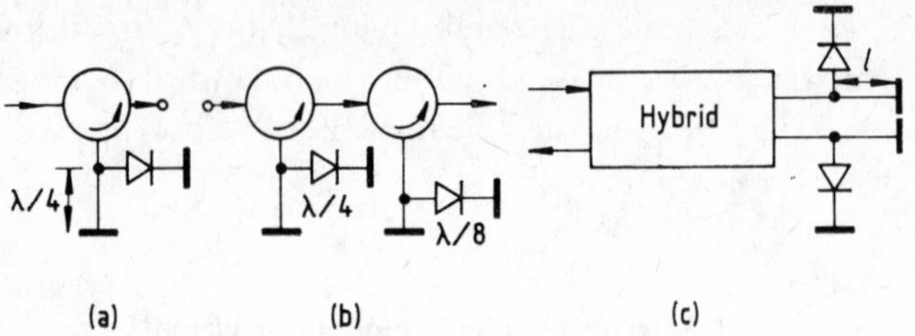

Fig. 7.11. PIN diode path-length modulators: (a) 2PSK modulator with circulator; (b) 4PSK modulator with circulator; (c) 2PSK modulator with hybrid.

parallel connection. A suitable length of the section with the short circuits has the effect of shifting point B_1 into B_3 or B_2, resulting in a phase change of $A_1 - B_3$ (90°) or $A_1 - B_2$ (180°), respectively. Point A_1 is not shifted, independently of the length of the transmission line (because of the zero impedance).

PIN diode phase modulator. This type of phase modulator, applied more frequently in the microwave frequency range, has two variants. The first is the *path-length modulator;* in this modulator, the length of a transmission line section is made to vary at the modulation rate, the variation being implemented by a PIN diode, driven by the modulation signal and coupled to an appropriate point of the line. A circulator realization of the path-length modulator is shown in Figs 7.11a and 7.11b. The signal propagating from the circulator is reflected from the short circuit plane in the case of a non-conducting diode and from the diode plane in the case of a conducting diode. Let the distance between the diode plane and the short circuit plane be l; with the choice of $l = \lambda/4$ and $\lambda/8$, the phase difference between the two reflected signals will be 180° and 90°, respectively. Figure 7.11a shows a 2PSK modulator, with $l = \lambda/4$ (180°) while in Fig. 7.11b, a 4PSK modulator is shown, implemented by the cascade connection of 180° and 90° modulators. The two PIN diodes are driven by two suitable dibit sequences, similarly to the 4PSK i.f. ring modulator presented in Fig. 7.4 [14, 15, 16, 17, 18]. A hybrid realization of the path length modulator, for 2PSK, is seen in Fig. 7.11c; here the modulation input terminals of the two PIN diodes are connected in parallel.

The second variant of the PIN diode phase modulator is the *reflection type modulator* which has a principle of operation similar to the varactor phase modulator shown in Fig. 7.10a. Phase modulation is provided by a changing susceptance but this is realized by driving a PIN diode and not a varactor. The most frequently used hybrid realization of this modulator is shown in Fig. 7.12a for 2PSK; the two susceptances denoted by b are parallel driven by the modulating signal. Two examples for susceptance realization are shown. The arrangement in Fig. 7.12b is suitable for 90° 2PSK modulation, the terminating capacitance being C_1 for a non-conducting diode and $C_1 + C_2$ for a conducting diode. The arrange

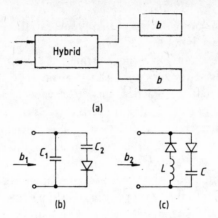

Fig. 7.12. (a) PIN diode reflection type modulator; (b) circuit for 90° phase modulation; (c) circuit for 180° phase modulation.

ment in Fig. 7.12c is suitable for 180° 2PSK modulation; here the two diodes are push–pull driven, the terminating susceptance thus being alternately an inductance L and a capacitance C.

In the preceding, we have only presented the operation principle of microwave modulators, with the aid of simple equivalent circuits. The tuning out of diode stray parameters or the realization of hybrid and circulator circuits is a microwave circuit design task not covered here. Also, for simplicity, the modulating signals driving the diodes have not been shown.

7.2.6 Microwave phase modulator errors

The phases corresponding to the modulation states are adjusted at a specific frequency, so a change of frequency normally results in a phase error which limits the useful frequency range of the modulator. In the case of a varactor modulator, the plane of the susceptance given by the varactor capacitance and the transmission line section electrical length are changed with frequency. The change of temperature will also cause a phase error by changing the diode capacitance and the transmission line dimensions. Considering the PIN diode modulators, the frequency range of the reflection type modulators is higher than that of the path-length type modulator [31]. Further, amplitude modulation will be generated by PIN diode modulators because the insertion loss of the PIN diode is changed at the modulation rate. To a lesser extent, the amplitude modulation is also present with varactor modulators. The insertion loss of PIN diode modulators is 2 to 3 dB, depending on the operating frequency. A typical amplitude variation is about 0.3 dB, and a typical phase accuracy is about $\pm 3°$ [19].

Transient phenomena. For increasing bit rates, the switch-over time cannot be neglected, and the phase changes during the transitions will affect the waveform of the transmitted signal. In practice, this effect can be essential for $t_f/T > 0.2$ where t_f is the transition time and T is the bit time. The finite switch-over time results in waveform distortion and jitter, which have the effect of reducing the eye pattern opening, increasing the distortion caused by other transmission effects (intersymbol interference, band limiting).

Considering the switch-over time, two groups of microwave modulators can be distinguished according to the type of phase transition. In conductance type (PIN diode) modulators, the signal amplitude is substantially changed during the phase transition while the amplitude change is less for susceptance type (varactor diode) modulators. It is thus feasible to apply a susceptance type modulator if unwanted amplitude modulation has to be reduced.

Figures 7.13a and 7.13b show the vector diagrams of ideal modulators, also presenting vectors **A**, **B** of the reference signals, needed for coherent demodulation. In Figs 7.13c and 7.13d, the demodulated signals given by ideal demodulators are shown, derived from the projection of the PSK signal vector to the reference signal vector (the demodulated signal is the product of the PSK signal and the reference signal). These signals have been drawn on the assumption that the phase transition has a constant speed; in this case, the distance covered by the vector

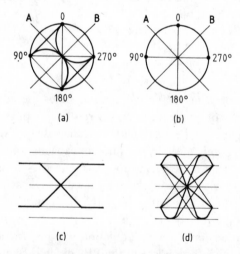

(a)   (b)

(c)   (d)

Fig. 7.13. Characteristics of 4PSK modulator: (a) vector diagram of ideal conductance modulator; (b) vector diagram of ideal susceptance modulator; (c) demodulated signal of conductance modulator; (d) demodulated signal of susceptance modulator.

during the phase transition is proportional to time. It is seen that the overshoot and jitter due to the finite phase transition time is higher for the susceptance modulator. Figures 7.13a and 7.13c are applicable to the 4PSK ring modulator operating at intermediate frequency and treated in Section 7.2.1, as this has a phase transition similar to that of the microwave PIN diode modulator.

7.3 Quadrature amplitude modulator (QAM)

It was mentioned in Section 7.1 that in order to make better use of the available frequency band, multi-state combined amplitude and phase modulation, called M-ary quadrature amplitude modulation (M-QAM), is applied. The term is explained by the method of generating the modulated signal by adding two amplitude modulated signals in quadrature (*i.e.*, having carriers which are 90° out-of-phase), each being an $\sqrt{M}$-ary amplitude modulated signal.

The output signal of an M-QAM modulator can be expressed in the following form:

$$V_0(t) = a(t) \cos \omega_0 t - b(t) \sin \omega_0 t$$

$$\begin{aligned} a(t) &= \pm 1, \pm 3 \\ b(t) &= \pm 1, \pm 3 \end{aligned} \quad \text{for} \quad M = 16 \tag{7.14}$$

$$\begin{aligned} a(t) &= \pm 1, \pm 3, \pm 5, \pm 7 \\ b(t) &= \pm 1, \pm 3, \pm 5, \pm 7 \end{aligned} \quad \text{for} \quad M = 64.$$

A possible realization of an M-QAM modulator is shown in Fig. 7.14a. The series–parallel converter converts the incoming modulating signal into two parallel NRZ sequences, while the $2/L$ converters have the function of converting each NRZ sequence into a sequence having $L=\sqrt{M}$ states. $L=4$ for $M=16$, so in a 16QAM modulator, both low-pass filters are driven by four-state sequences. The modulator itself is made up of two suppressed carrier amplitude modulators (DSB-SC).

The building blocks of the quadrature amplitude modulator are similar to those of the 4PSK modulator discussed in Section 7.2.1. However, in the 4PSK modulator, the multipliers have switching functions only (multiplication by 1 or -1) while in the quadrature amplitude modulator, real multipliers are applied to provide linear amplitude modulation. For this purpose, so-called four-quadrant multiplier IC's can be utilized but ring modulators are also suitable. In the latter case, the modulating signal level should be lower than the carrier level. The modulator

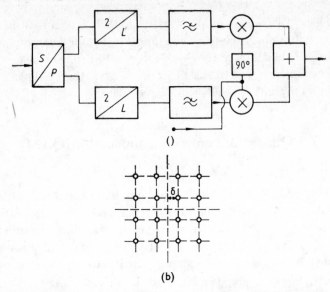

(b)

Fig. 7.14. M-QAM modulator: (a) block diagram; (b) vector diagram for $M=16$.

being linear, the output spectrum can be formed by the low-pass filter preceding the modulator.

The most straightforward quadrature amplitude modulator is the four-state modulator which is simply called a QAM modulator. This is identical with the 4PSK modulator because of the identity of the vector diagrams; however, for QAM, the spectrum of the modulated carrier can be limited by the filters preceding the modulator.

Figure 7.14b shows the vector diagram for 16QAM which is widely used for high speed digital transmission, though 64QAM is also gaining in importance.

If a QAM modulator is followed by a carrier amplifier it is important to know the peak power, the average power and the minimum power in order to adjust a correct driving level. The following expressions are the results of a simple calculation:

$$P_{peak} = c(\sqrt{M}-1)^2\delta^2$$

$$P_{av} = c\,\frac{M-1}{3}\,\delta^2 \qquad\qquad (7.14a)$$

$$P_{min} = c\delta^2$$

where δ is half the distance between two vectors and c is a constant. For 16QAM,

$$P_{peak} = 9c\delta^2,$$

$$P_{av} = 5c\delta^2$$

and for 64QAM,

$$P_{peak} = 49c\delta^2,$$

$$P_{av} = 21c\delta^2.$$

Another variant of the M-QAM modulator is based on the addition of signals generated by a suitable number of 4PSK modulators. Preceding this addition, the 4PSK signals can be amplified by nonlinear amplifiers; this is why this variant is also called NLA-QAM (nonlinearly amplified QAM). A 16QAM signal can be generated by adding two QPSK signals as shown in Fig. 7.15a [24] while a 64QAM signal requires the addition of three QPSK signals according to Fig. 7.16a [25]. A 64QAM signal can be considered as the sum of a 16QAM signal

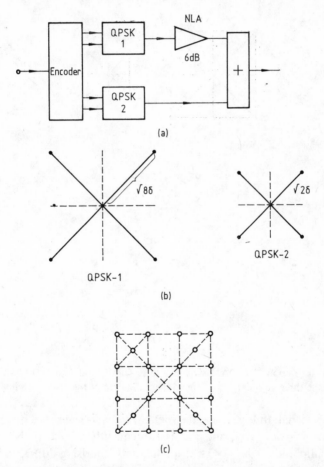

Fig. 7.15. (a) Generation of 16QAM signal by two QPSK modulators; (b) vector diagrams of the two modulators; (c) signal space diagram of the resultant signal.

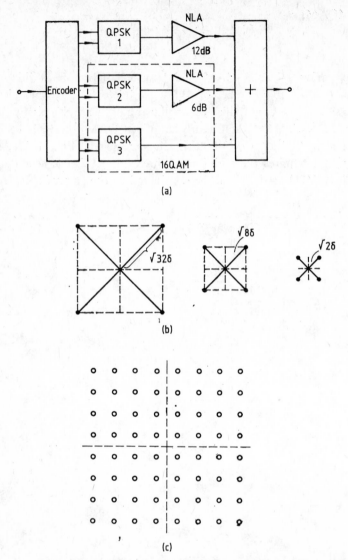

Fig. 7.16 (a) Generation of 64QAM signal by three QPSK modulators; (b) vector diagrams of the three modulators; (c) signal space diagram of the resultant signal.

and a QPSK signal. In Fig. 7.16a, the 16QAM signal is generated by the two QPSK modulators of lower level, and this is subsequently added to the QPSK signal of highest amplitude. The M-QAM modulator comprising QPSK building blocks can be realized both in the intermediate and microwave frequency ranges.

7.4 Partial response system modulator (PRS)

From several variants of the partial response systems, the quadrature partial response system (QPRS) was investigated in Section 2.2.4, so in the following, the modulator of this system will be dealt with. The building blocks of the QPRS signal are similar to those of the QAM signal but an additional encoder is also utilized (see Fig. 7.17a). The incoming bit sequence is first converted into two parallel sequences, and each of these is then converted into three-state sequences; both multipliers are thus driven by sequences which have levels of −1, 0 and +1. The QPRS signal is generated by the addition of the two modulators fed by carriers which are 90° out-of-phase; the vector diagram is made up of 9 vectors as shown in Fig. 7.17b [26].

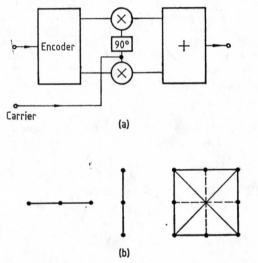

(a)

(b)

Fig. 7.17. QPRS modulator: (a) block diagram; (b) vector diagram.

7.5 O-QPSK and MSK modulator

The offset QPSK (O-QPSK) is aimed at reducing the amplitude modulation of the normal QPSK modulator. This is realized by delaying in one branch of the QPSK modulator the modulating bit sequence by the time of half a symbol, *i.e.* by the time of one bit. In this way, a phase jump occurs only in one of the branches at a time, and as a consequence, the phase jumps in the output signal have a magnitude of only 90° [27]. The O-QPSK modulator can be realized, similarly to the conventional QPSK modulator, both in the intermediate and microwave frequency ranges (Fig. 7.18).

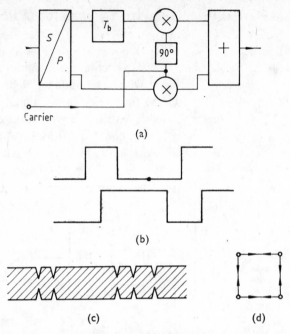

Fig. 7.18. O-QPSK modulator: (a) block diagram; (b) waveforms of the two modulating signals; (c) modulated signal showing the envelope; (d) vector diagram.

MSK modulation is a special type of phase modulation, equivalent to frequency modulation with low deviation (minimum shift keying). It has the advantage of showing no amplitude modulation, and of having a spectrum envelope which diminishes rapidly at frequencies remote from the carrier frequency. The modulation index is given by

$$h = \frac{2\Delta f}{f_b} = 0.5$$

where Δf is the peak deviation and f_b is the bit frequency. In transmission systems with coherent detection, MSK can be realized with a quadrature amplitude modulator [27, 28]. However, the waveforms of the two modulating sequences have a cosine shape instead of a square-wave shape. This is necessary for achieving a linear phase transition in order to obtain a rapid spectrum envelope reduction at frequencies remote from the carrier (see Section 2.3.1):

$$a(t) = \pm\cos\frac{\pi t}{2T_b}$$

$$b(t) = \pm\sin\frac{\pi t}{2T_b}.$$

(7.15)

Substituting these into Eq. (7.14) we have the following expression for the quadrature amplitude modulated output signal:

$$V_0(t) = \pm\cos\frac{\pi t}{2T_b}\cos\omega_0 t \mp \sin\frac{\pi t}{2T_b}\sin\omega_0 t. \tag{7.15a}$$

This can be written in the form

$$V_0(t) = A(t)\cos\left(\omega_0 t + \varphi(t)\right) \tag{7.16}$$

where

$$A(t) = \sqrt{\cos^2\frac{\pi t}{2T_b} + \sin^2\frac{\pi t}{2T_b}} = 1$$

$$\tag{7.16a}$$

$$\varphi(t) = \tan^{-1}\left(\pm\frac{\sin\dfrac{\pi t}{2T_b}}{\cos\dfrac{\pi t}{2T_b}}\right) = \pm\frac{\pi t}{2T_b}.$$

It is thus seen that the phase/time function is indeed linear, the phase changing by $\pi/2$ during the bit interval T_b, and the amplitude is constant. Noting that

$$\cos\frac{\pi(t-T_b)}{2T_b} = \sin\frac{\pi t}{2T_b} \tag{7.16b}$$

only one kind of pulse shaping network is needed, generating a waveform of $\cos\pi t/2T_b$. Delaying this pulse by a time interval of T_b we obtain the waveform of sine $\pi t/2T_b$. The block diagram, the carrier envelope and the vector diagram of this modulator are shown in Fig. 7.19. The multiplier applied in the modulator circuit has to be linear in order to allow true modulation with a specified waveform. MSK modulators can preferably be realized in the intermediate frequency range.

The spectrum of the MSK modulator is given in Section 2.3.1. Should the spectrum be confined to a smaller frequency range, then additional filtering is required. The filter should either be inserted between the pulse forming network and the multiplier [29], or else follow the MSK modulator. MSK can also be implemented by frequency modulation having a modulation index of $h=0.5$. However, there are difficulties in applying this kind of MSK in the case of coherent demodulation because the phase accuracy needed for coherent demodulation cannot be provided with conventional frequency accuracy.

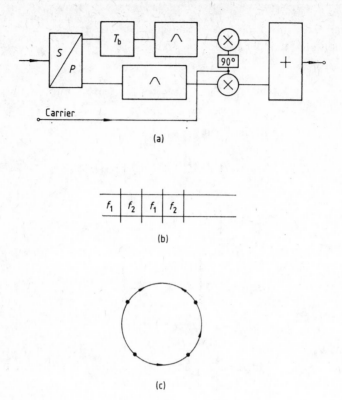

Fig. 7.19. MSK modulator: (a) block diagram; (b) modulated signal envelope with carrier
frequency designations; (c) vector diagram.

7.6 Amplitude modulators (ASK)

In digital microwave equipment, switching type amplitude modulators are
applied (ASK — Amplitude Shift Keying). This means that for an ideal ASK
modulator the output power is either zero (the modulator has infinite insertion
loss) or equal to the input power (the modulator has zero insertion loss); switch-
over between these two modulation states is fast compared with the bit time.
Thus low insertion loss and high isolation should be provided by real modulators. However, there are no requirements for a linear transition.

The ASK modulator can be realized either at microwave or at intermediate
frequencies. A more straightforward digital transmitter can be realized by using
microwave ASK, so this is more conventional. I.f. ASK is used if filtering of the
ASK spectrum is necessary. Filtering is then also implemented at i.f. because the
filter realization would be much more difficult in the microwave frequency range.

Any microwave diode with a resistance which can be varied in a wide range can be applied as a switch. Point-contact, Schottky and PIN diodes are normally used; in some cases, tunnel diodes and varactor diodes are also applied [31]. Diode selection is governed by several parameters. Most important is the output power transmitted. For low powers up to about 1 mW, all diode types are suitable. Up to several hundred milliwatts, Schottky diodes, varactor diodes and PIN diodes are equally applicable. For higher powers, the PIN diode is normally applied; PIN diodes for switching powers as high as 100 watts are commercially available. In earlier days, higher switching speeds could only be obtained by using Schottky and tunnel diodes because PIN diodes were too slow [31]. However, high speed PIN diodes are now available and can be applied for transmission speeds of several hundred Mbit/s. Compared with other diode types, the PIN diode has the advantage of providing a resistance which can be linearly changed with the current flowing through the diode, within wide limits.

In the following, a PIN diode ASK modulator will be investigated. However, since the equivalent circuit and other properties of Schottky diodes are similar to those of the PIN diode, the results also will be applicable to Schottky diode modulators.

The diode mount, together with the tuning elements, can be arranged to provide either a series impedance or a parallel admittance in the transmission line. The arrangement actually used will depend on the realization possibilities of the transmission line applied (waveguide, coaxial line, stripline). At the resonant frequency at which the stray diode parameters are tuned out, an ideal modulator with zero insertion loss and infinite isolation could be realized on the assumption that the diode has no loss. The actual parameters for the parallel diode arrangement shown in Fig. 7.20 can be calculated by taking into account the following parameters:

— conductance G and susceptance B of the non-conducting diode and tuning elements,
— series resistance R and reactance X of the conducting diode and tuning elements.

According to the calculation, the insertion loss is given by

$$a_l = 10 \lg \left[\left(1 + \frac{GZ}{2} \right)^2 + \left(\frac{BZ_0}{2} \right)^2 \right] \tag{7.17}$$

and the isolation is given by

$$a_i = 10 \lg \left\{ \left[1 + \frac{RZ_0}{2(R^2 + X^2)} \right]^2 + \left[\frac{XZ_0}{2(R^2 + X^2)} \right]^2 \right\}. \tag{7.18}$$

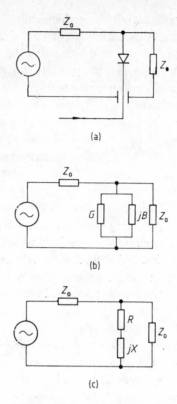

Fig. 7.20. Parallel PIN diode ASK modulator: (a) circuit principle; (b) equivalent circuit with non-conducting diode; (c) equivalent circuit with conducting diode.

In a practically realized ASK modulator, $G=1/R_r$ where R_r is the parallel resistance of the non-conducting diode, and $R=R_s$ where R_s is the series resistance of the conducting diode, we further have that $B=X=0$ as the susceptance and reactance are tuned out at the frequency of the modulator. The insertion loss is then given by

$$a_l = 20 \lg\left[1+\frac{Z_0}{2R_r}\right]$$

(7.19)

and the isolation is given by

$$a_i = 20 \lg\left[1+\frac{Z_0}{2R_s}\right].$$

(7.20)

A similar calculation can be performed for the series diode arrangement; in this case, the conducting diode results in transmission and the non-conducting diode in isolation. In the microwave range, typical values for ASK modulators are an insertion loss of 0.1 to 0.3 dB and an isolation around 30 dB [32]. In the intermediate frequency range, ASK can be implemented by a simple transistor or diode circuit but ring modulators are also applicable.

7.7 Frequency modulators (FSK)

Switching type frequency modulators (FSK — Frequency Shift Keying) are widely applied in recent low-capacity digital transmission systems. Two types of FSK can be distinguished, CPFSK (Continuous Phase FSK) and non-continuous phase FSK. The latter is accomplished by switching the signals of two oscillators at the modulation rate. This type of FSK is not applied in microwave digital equipment because of the need for two oscillators, and also because of the phase transients generated, giving rise to overshoot in the demodulated signal.

Continuous phase FSK can be generated by the modulation of voltage controlled oscillators which are also applied in analog equipment (Fig. 7.21). In most cases, two-state FSK is applied, but M-FSK is also possible in order to obtain better bandwidth efficiency.

An important parameter of the two-state FSK is the modulation index:

$$h = 2\Delta fT$$

where $2\Delta f$ is the peak-to-peak frequency deviation and T is the bit-time. Here h can also be considered as the normalized frequency deviation, *i.e.* the ratio of the peak-to-peak deviation to the bit frequency. It has been shown in Section 7.5 that for MSK, $h=0.5$. MSK is frequently implemented by frequency modulation when a simple demodulator has to be used, and no coherent reception is needed. In the case of orthogonal frequency modulation (see Section 2.1.1), $h=1$. It can be shown that the lowest error probability of NRZ transmission is obtained around $h=0.7$ if a limiter–discriminator is applied for demodulation.

The spectrum of the modulated FSK signal depends strongly on the value of h. The most rapidly decreasing spectrum envelope is obtained for $h=0.5$. At

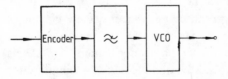

Fig. 7.21. Frequency modulator.

$h=1$, peaking is present at frequencies $f_1=f_0-\Delta f$ and $f_2=f_0+\Delta f$. The spectrum envelope at distant frequencies can be reduced by a low-pass filter inserted in the modulating signal branch.

In some cases, the frequency stability of the modulated oscillator does not allow DC coupling of the modulating signal. An encoder has then to be applied eliminating the DC component of the modulating signal, thus allowing the use of AC coupling. This is provided by the biphase, the AMI–NRZ and AMI–RZ encoding.

FSK can be realized both at intermediate and microwave frequencies. The latter realization eliminates the up-converter and results in a more straightforward transmitter. Principally, the FSK signal has no amplitude modulation so it can be amplified, if necessary, by a nonlinear amplifier.

7.8 Microwave carrier supplies

Microwave carrier supplies in digital transmitters are usually unmodulated, with the exception of the microwave frequency modulator, implemented by FSK of the carrier supply. Another case of carrier supply modulation is the transmission of the service channel information by frequency modulation of the supply while the main information is transmitted for example by PSK.

The frequency stability of the microwave carrier supply is usually a few times 10^{-5}, most often realized by the application of a crystal oscillator. One method is to use a crystal oscillator around 100 MHz, and to apply a frequency multiplier to obtain a desired frequency f_0. However, harmful sidebands will also appear, and a selective filter is then required to suppress the components $f_0 \pm f_{cr}$. Higher purity of the spectrum can be obtained by two variants of the PLL loop. According to Fig. 7.22a, a voltage controlled oscillator is applied at frequency f_L, and the crystal oscillator frequency is multiplied to f_L and compared with the VCO signal frequency in a phase discriminator. The output signal is either the VCO signal directly, or else the VCO signal is further multiplied by N_x. According to Fig. 7.22b, the frequency of the VCO signal is divided down to the crystal oscillator frequency f_{cr}, and the phase comparison is achieved at this frequency. Further variants can be obtained by the combination of dividing and multiplying circuits. The output frequency can easily be adjusted by changing the division number.

The PLL arrangement has high attenuation at frequencies $f_L \pm k f_{cr}$, thus resulting in high spectral purity. If the arrangement in Fig. 7.22a is used, the suppression of components $f_0 \pm f_L$ can be accomplished by simple filters (f_L is the VCO frequency).

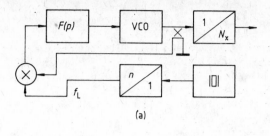

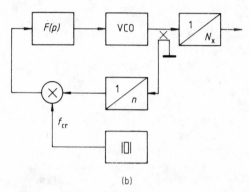

(a)

(b)

Fig. 7.22. PLL type carrier supply: (a) realization with frequency multiplier; (b) realization with frequency divider.

A transistor, a Gunn diode or an IMPATT diode can be applied in the voltage controlled oscillator. PLL carrier supplies are widely used in both digital and analog equipment [15, 33, 34, 35]. Another realization of the carrier supply is possible using a frequency multiplier, followed by an injection locked Gunn diode or IMPATT diode oscillator [10].

7.9 Microwave power amplifiers

In the majority of digital transmitters, no microwave power amplifiers are applied [1, 10, 14], and the power output of the carrier supply is chosen so that the required power is available after modulation. Should amplification of the microwave modulated signal be necessary, then two kinds of microwave amplifiers can be applied. In early equipment, the traveling wave tube was used extensively, but this is now replaced by solid state devices such as Gunn and IMPATT diode or transistor amplifiers. These may be either linear amplifiers or operated as injection locked amplifiers [37, 38, 39]. PLL-type amplifiers can also be applied [36].

7.10 References

[1] Morita, T., Hayasi, M., Ogawa, K.: 11/15 GHz band digital radio equipment. *FUJITSU, Sci. Tech. J.*, June, pp. 41–64, 1976.

[2] Kobayasi, T., Maruyama, H.: New 100 Mb/s digital radio-relay system in 11 GHz band. *Jpn Telecommun. Rev.*, April, pp. 157–162, 1982.

[3] Sato, M., Yamazaki, F., Iwamoto, M.: Commercial test on 20 GHz digital radio-relay system. *Jpn Telecommun. Rev.*, July, pp. 234–242, 1978.

[4] Divsalar, D., Simon, M. K.: The power spectral density of digital modulation transmitted over nonlinear channels. *IEEE Trans. Commun.*, Vol. Com-30, No. 1, pp. 142–151, 1982.

[5] Dewieux, C.: Analysis of PSK signal power spectrum spread with a Markov chains model. *COMSAT Tech. Rev.*, Vol. 5, No. 2, Fall, pp. 225–249, 1974.

[6] *Operating and service manual.* Model 10514 A, B HP mixers, Jan., 1967.

[7] Benett, W. R., Davey, J. R.: *Data Transmission.* McGraw-Hill, New York 1965.

[8] Lindsey, W. C., Simon, M. K.: Carrier synchronization and detection of polyphase signals. *IEEE Trans. Commun.*, Vol. Com-20., No. 3, pp. 441–454, 1972.

[9] Lindsey, W. C., Simon, M. K.: *Telecommunication System Engineering.* Prentice-Hall, Englewood-Cliffs 1973.

[10] Nishino, K.: 11 and 15 GHz PCM radio-relay systems. In: *Proc. 6th Europ. Microwave Conf., Rome*, pp. 445–459, 1976.

[11] Schertler, K. E.: Microwave power up-converters for angle modulated signals using phase-locked microwave oscillators. In: *Proc. 5th Europ. Microwave Conf., Hamburg* 1975.

[12] Bors, L.: Linearity analysis of reflection type phase modulators. In: *Proc. 5th Colloq. Microwave Commun., Budapest* 1974.

[13] Eisman, Z., Rak, B.: Integrated digital phase modulator for the 11 GHz band. In: *Proc. 6th Colloq. Microwave Commun., Budapest*, Vol. II, pp. 551–556, 1978.

[14] Kohiyama, K., Horikawa, I., Momma, K., Morita, K.: 20 G–400 M radio-relay system repeater. *Rev. Electr. Commun. Lab.*, Vol. 24, No. 7–8, pp. 552–572, 1976.

[15] Frenstadt, P.: 11 GHz PCM link. In: *Proc. 4th Microwave Conf., Montreux*, pp. 567–571, 1974.

[16] Clementson, W. I., Kenyon, N. D., Kurokowa, K.: An experimental MM-wave path length modulator. *Bell Syst. Tech. J.*, Nov., pp. 2917–2946, 1971.

[17] Ohm, G., Alberty, M.: 11 GHz MIC QPSK modulator for regenerative satellite repeater. *IEEE Trans. Microwave Theory Techn.*, Vol. 30, No. 11, pp. 1921–1926, 1982.

[18] Sasaki, S., Hayasi, M., Hamada, S.: 17 Mbit/s 2 GHz digital radio system. *FUJITSU Sci. Tech., J.*, Sept., pp. 1–28, 1980.

[19] Junghaus, H.: A KU-band hybrid-coupled 4-phase modulator in MIC technology. In: *Proc. 5th Europ. Microwave Conf., Hamburg*, pp. 133–137, 1975.

[20] Kurematsu, H., Dooi, Y., Saikowa, T.: Error rate caused by transient response of microwave PSK modulator. *FUJITSU Sci. Tech. J.*, Vol. 8, No. 3, Sept., pp. 83–107, 1972.

[21] Szabó, L., Schünemann, K., Sporleder, F.: PIN diode transients in high speed switching applications. In: *Proc. 5th Europ. Microwave Conf., Hamburg*, pp. 138–142, 1975.

[22] Simon, M. K., Smith, J. G.: Carrier synchronization and detection of QPSK signal sets. *IEEE Trans. Commun.*, Vol. Com-22, No. 2, pp. 98–105, 1974.

[23] Aghvami, A. H.: Performance analysis of 16-ary QAM signaling through two-link nonlinear channels in additive Gaussian noise. *IEE Proc.*, Vol. 131, No. 4, July, pp. 403–406, 1984.

[24] Miyauchi, M., Seiki, S., Ishio, H.: New technique for generating and detecting multilevel signal format. *IEEE Trans. Commun.*, Vol. Com-24, No. 2, pp. 263–267, 1976.

[25] Demerest, K., Plourde, J. K.: An optimum combination of power levels and combiner weighting for generating 64–QAM from three 4–PSK signals. *IEEE Trans. Commun.*, Vol. Com-32, No. 3, pp. 320–322, 1984.

[26] Kurematsu, H.: The QAM 2G–10R digital radio equipment using partial response systems. *FUJITSU Sci. Tech. J.*, June, pp. 27–48, 1977.

[27] Gronemeyer, S. A., McBride, A. L.: MSK and offset QPSK modulation. *IEEE Trans. Commun.*, Vol. Com-24, No. 8, pp. 809–810, 1976.

[28] Austin, M. C., Chang, M. U., Horwood, D. F., Maslov, R. A.: QPSK, staggered QPSK, and MSK, a comparative evaluation. *IEEE Trans. Commun.*, Vol. Com-31, No. 2, pp. 171–182, 1983.

[29] Morais, D. H., Feher, K.: Bandwidth efficiency and probability of error performance of MSK and offset QPSK systems. *IEEE Trans. Commun.*, Vol. Com-27, No. 12, pp. 1794–1801, 1979.

[30] Murota, K., Hirade, K.: GMSK modulation for digital mobile radio telephony. *IEEE Trans. Commun.*, Vol. Com–29, No. 7, pp. 1044–1050, 1981.

[31] Wattson, H. A.: *Microwave Semiconductor Devices and their Circuit Application.* McGraw-Hill, New York 1969.

[32] Reiter, G., Béres, V.: A PIN diode switch with high isolation and low loss. In: *Proc. 6th Colloq. Microwave Commun., Budapest,* pp. 40.1–40.4, 1978.

[33] Noesen, P.: 4 GHz radio link. In: *Proc. 4th Europ. Microwave Conf., Montreux,* pp. 562–566, 1974.

[34] Myrseth, E.: An economical microwave link. In: *Proc. 6th Europ. Microwave Conf., Rome,* pp. 475–479, 1976.

[35] Payne, J. B.: Recent advances in solid state phase-locked microwave signal sources. *Microwave Syst. News,* Febr.–March, pp. 79–87, 1976.

[36] Hines, M. E., Posner, R. S., Sweet, A. A.: Power amplification of microwave FM communication signals using a phase-locked voltage-tuned oscillator. *IEEE Trans. Microwave Theory Techn.,* July, pp. 393–404, 1976.

[37] Blix, R., Thaler, H. J.: Dynamic behaviour of stable and injection-locked IMPATT diode amplifiers for PSK microwave communication systems. In: *Proc. 4th Europ. Microwave Conf., Montreux,* pp. 310–317, 1974.

[38] Kuno, H. J.: Analysis of nonlinear characteristics and transient response of IMPATT amplifiers. *IEEE Trans. Microwave Theory Techn.,* Nov., pp. 694–702, 1973.

[39] Berceli, T.: Transfer properties of injection-locked diode oscillators. In: *Proc. 5th Colloq. Microwave Commun., Budapest,* pp. 694–702, 1974.

CHAPTER 8

DEMODULATORS AND RECEIVERS

8.1 General considerations

Figure 8.1 shows the block diagram of a digital receiver. The microwave and i.f. filters are primarily intended to provide suitable selectivity. The design and construction of microwave filters are covered in the literature [1, 2, 3] so they will not be dealt with in this chapter. It should be noted, however, that the stop-band attenuation of the microwave filters at the frequencies of interfering carriers should be high enough to ensure linear operation of the mixer at the level of the required signal. Otherwise, intermodulation products would result in signal suppression and the selectivity contribution of the i.f. filter would be reduced.

In the case of linear mixer operation, receiver selectivity is determined by the overall response of the microwave and i.f. filters. The effect of filters and the filter characteristic which provides ideal transmission have been dealt with in Section 2.2.1. The filters actually applied in digital equipment will be treated in Section 8.7.1.

The mixers in digital receivers are similar to those applied in analog receivers and are dealt with in the literature [4, 5, 6, 7, 8].

The value of the intermediate frequency depends on the bit rate to be transmitted. It is feasible to satisfy the relation i.f. $\geq 3f_s$ where $f_s = 1/T_s$ is the symbol frequency [9]. In low speed equipment, the intermediate frequency is 35 MHz while 70 MHz is applied for medium bit rates. 140 and 350 MHz are also applied, and for high bit rates, the intermediate frequency 1.7 GHz is preferred.

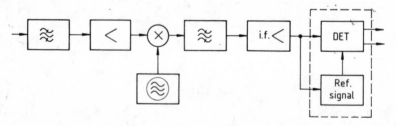

Fig. 8.1. Receiver block diagram.

At low bit rates, a limiting type i.f. amplifier may be applied but normally, i.f. amplifiers utilizing automatic gain control (AGC) are used. The design methods of i.f. amplifiers are similar to those applied for analog equipment, with the exception of i.f. amplifiers for high bit rates for which special considerations apply.

The considerations which apply for microwave carrier supplies of transmitters are also applicable for receiver local oscillators, the only difference being the low power requirement.

8.2 Demodulation of PSK signals

In digital receivers, two kinds of phase demodulators, coherent and non-coherent, are applied. In the majority of cases, coherent phase demodulators are utilized because of the better signal-to-noise ratio obtainable. These demodulators require a reference signal which is generated by the so-called carrier recovery process. The carrier recovery circuits will be treated separately in Chapter 9.

8.2.1 M-PSK demodulator

In principle, an M-ary PSK signal can be demodulated by M multipliers parallel fed by the M-PSK signal. The reference signal is split into M branches, the signal of each branch being shifted by $360/M$ degrees. The reference signal phase at each multiplier is thus equal to one of the possible phases of the incoming M-ary PSK signal. The multipliers are connected to a signal processing circuit selecting the highest output level. This method is not actually applied because of difficulties in the circuit realization.

Methods used in practice for demodulating an M-ary PSK signal are dealt with in the literature [10, 11, 12]. One of the most frequently applied methods is the use of $M/2$ multipliers as shown in Fig. 8.2. In the following, the voltage at the output of the ith low-pass filter will be calculated.

Let the incoming signal be given by

$$V(t) = \sqrt{2} \cos [\omega_0 t + \Phi_0 + \varphi(t)]$$

$$\varphi(t) = \frac{2\pi}{M} n, \quad n = 0, 1, ..., M-1.$$

(8.1)

Any value of Φ_0 can be assumed: in most cases $\Phi_0 = \pi/M$.

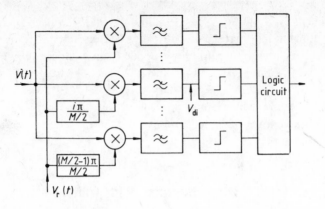

Fig. 8.2. Block diagram of M-PSK demodulator.

The reference signal voltage is given by

$$V_r(t) = \sqrt{2} \cos (\omega_0 t + \Phi_R). \tag{8.2}$$

This signal is passed through $M/2$ phase shifters, and the output signal voltage of these phase shifters is given by

$$V_{ri}(t) = \sqrt{2} \cos \left[\omega_0 t + \Phi_R + \frac{i\pi}{M/2} \right], \quad i = 0, 1, ..., (M/2-1). \tag{8.3}$$

In the multiplier circuits, difference and sum frequency components are generated but the latter are attenuated by the low-pass filters. Utilizing the relation

$$\sqrt{2} \cos \alpha \sqrt{2} \cos \beta = \cos (\alpha - \beta) + \cos (\alpha + \beta) \tag{8.3a}$$

and taking into account only the first term, the output signal of the i-th filter will be given by

$$V_{di}(t) = \cos \left[\varphi(t) - \frac{i\pi}{M/2} + (\Phi_0 - \Phi_R) \right]. \tag{8.4}$$

The filters are followed by zero comparators which are connected to a digital decision circuit which has the function of sampling and parallel-to-series conversion.

In Eq. (8.4), $\Phi_0 - \Phi_R = \pi/M$. Φ_R may have M different values differing by $2\pi/M$. This phenomenon introduces phase ambiguity in the demodulated signal which is eliminated by signal processing (see Chapter 10).

In another variant of an M-PSK demodulator, only two multipliers are utilized (see Fig. 8.3), fed by reference signals which are in quadrature (90° phase dif-

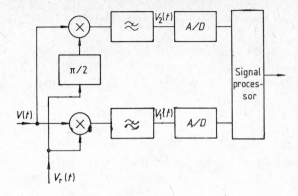

Fig. 8.3. Demodulation of M-PSK signal by quadrature demodulator.

ference). This circuit is called a quadrature demodulator or quadrature amplitude demodulator since it is equally applicable for the demodulation of M-PSK and QAM signals. The preferred expression for the incoming signal is now given by

$$V(t) = a(t) \cos \omega_0 t - b(t) \sin \omega_0 t \qquad (8.5)$$

while the two reference signals are given by

$$V_{r1}(t) = 2 \cos \left[\omega_0 t + n \frac{2\pi}{M} \right]$$

$$V_{r2}(t) = -2 \sin \left[\omega_0 t + n \frac{2\pi}{M} \right] \qquad (8.6)$$

$$n = 0, 1, \ldots, M-1.$$

The phase of the incoming signal will be given by

$$\varphi(t) = \tan^{-1} \frac{b(t)}{a(t)} = \tan^{-1} \frac{V_2}{V_1}. \qquad (8.7)$$

The low-pass filters are followed by analog-to-digital converters which are connected to a signal processing circuit (see Chapter 10).

The phase of the reference signal with respect to the incoming PSK signal phase is not definite, resulting in a phase ambiguity of the demodulated signal. Several methods of eliminating this phase ambiguity are known and will be covered in Chapter 10.

8.2.2 2PSK demodulator

The expression of the demodulated signal in the case of 2PSK transmission can easily be derived from the M-ary case (Fig. 8.4). The phase of the reference signal is equal to one of the phases of the received signal, so if the received signal is given by

$$V(t) = \sqrt{2} \cos\left(\omega_0 t \pm \frac{\pi}{2}\right) = \sqrt{2}\, a(t) \sin \omega_0 t, \quad a(t) = \pm 1 \tag{8.8}$$

then the reference signal can be expressed as

$$V_r(t) = \pm\sqrt{2} \sin \omega_0 t \tag{8.9}$$

and the demodulated signal is thus given by

$$V_d(t) = \pm a(t). \tag{8.9a}$$

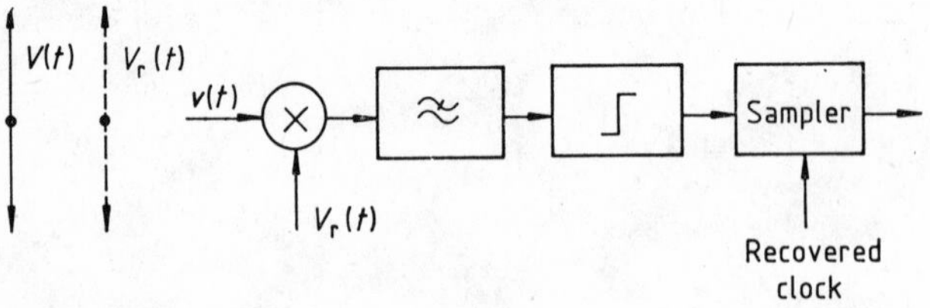

Fig. 8.4. Block diagram of a 2PSK demodulator.

8.2.3 4PSK demodulator

The received signal can be expressed by either of two equivalent expressions. According to the cosine expression

$$V(t) = \sqrt{2} \cos\left[\omega_0 t + \frac{\pi}{4} + \varphi(t)\right], \quad \varphi(t) = \frac{n\pi}{2}, \quad n = 0, 1, 2, 3 \tag{8.10}$$

and according to the quadrature expression

$$V(t) = a(t) \cos \omega_0 t - b(t) \sin \omega_0 t, \quad a(t) = \pm 1, \quad b(t) = \pm 1. \tag{8.10a}$$

The reference signals feeding the two multipliers are given by

$$V_{r1}(t) = \sqrt{2} \cos\left(\omega_0 t + n\frac{\pi}{2}\right)$$

$$V_{r2}(t) = -\sqrt{2} \sin\left(\omega_0 t + n\frac{\pi}{2}\right). \tag{8.11}$$

For the cosine expression, the demodulated signals are given by

$$V_{d1}(t) = \cos\left[\varphi(t) + \frac{\pi}{4} - n\frac{\pi}{2}\right] = \pm 1/\sqrt{2}$$

$$V_{d2}(t) = \sin\left[\varphi(t) + \frac{\pi}{4} - n\frac{\pi}{2}\right] = \pm 1/\sqrt{2}$$

(8.12)

and for the quadrature expression, by

$$V_{d1}(t) = \frac{a(t)}{\sqrt{2}} = \pm 1/\sqrt{2}$$

$$V_{d2}(t) = \frac{b(t)}{\sqrt{2}} = \pm 1/\sqrt{2}.$$

(8.13)

The circuit of the 4PSK demodulator is identical to the circuit of the QAM demodulator (Fig. 8.5), similarly to the 4PSK modulator and QAM modulator case.

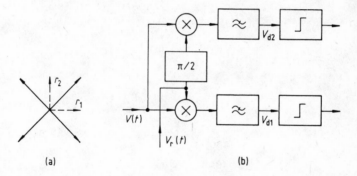

(a)   (b)

Fig. 8.5. 4PSK demodulator: (a) vector diagram of received signal; (b) block diagram.

8.2.4 8PSK demodulator

The input signal is given by

$$V(t) = \sqrt{2}\cos\left[\omega_0 t + \frac{\pi}{8} + \varphi(t)\right]; \quad \varphi(t) = n\frac{2\pi}{8}; \quad n = 0, 1, ..., 7 \quad (8.14)$$

and the reference signal is expressed by

$$V_r(t) = \sqrt{2}\cos(\omega_0 t + \Phi_R); \quad \Phi_R = n\frac{2\pi}{8}. \quad (8.15)$$

The expressions for the demodulated signals at the output terminals of the low-pass filters are as follows:

$$V_{d1}(t) = \cos\left[\varphi(t) + \frac{\pi}{8} - \Phi_R - \frac{3}{4}\pi\right]$$

$$V_{d2}(t) = \cos\left[\varphi(t) + \frac{\pi}{8} - \Phi_R - \frac{\pi}{2}\right]$$

(8.16)

$$V_{d3}(t) = \cos\left[\varphi(t) + \frac{\pi}{8} - \Phi_R - \frac{\pi}{4}\right]$$

$$V_{d4}(t) = \cos\left[\varphi(t) + \frac{\pi}{8} - \Phi_R\right].$$

The low-pass filters are followed by zero comparators supplying four binary signals. However, the eight phases are determined by three binary numbers. The logic circuit in Fig. 8.6 has the function of providing the three required signals by suitable logic operations.

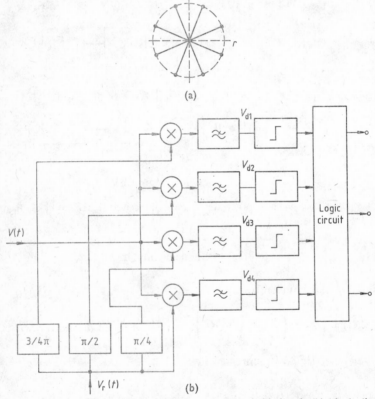

Fig. 8.6. 8PSK demodulator: (a) vector diagram of received signal; (b) block diagram.

8.2.5 Differentially coherent PSK demodulators

The circuit of these demodulators is identical to that of the coherent PSK demodulators with the exception that the reference signal is not generated by carrier recovery but by the incoming i.f. signal after a delay of one unit time. Thus no carrier recovery circuit is required. This means that if the received signal is given by

$$V(t) = \sqrt{2} \cos [\omega_0 t + \varphi(t)] \qquad (8.17)$$

then the expression of the reference signal is

$$V_r(t) = \sqrt{2} \cos [\omega_0(t-T) + \varphi(t-T)] \qquad (8.18)$$

where T is unit time.

A 2PSK differentially coherent demodulator is shown in Fig. 8.7a. The signal at the multiplier output terminal is given by

$$V_0(t) = \cos [2\omega_0 t - \omega_0 T + \varphi(t) - \varphi(t-T)] +$$
$$+ \cos [\varphi(t) - \varphi(t-T) + \omega_0 T]. \qquad (8.19)$$

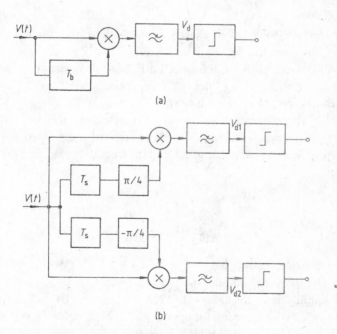

(a)

(b)

Fig. 8.7. Differentially coherent demodulator; (a) block diagram for 2PSK signals; (b) block diagram for 4PSK signals.

The component of frequency $2\omega_0$ is attenuated by the low-pass filter. Assume that $\omega_0 T = k2\pi$; the demodulated signal at the filter output terminal is then given by

$$V_d(t) = \cos[\varphi(t) - \varphi(t-T)]. \tag{8.20}$$

For a 2PSK input signal, the phase difference between two consecutive phases will be either 0 or 180° so that the demodulated signal value will be either $+1$ or -1.

A 4PSK differentially coherent demodulator is shown in Fig. 8.7b. Here the reference path is split into two branches, the reference delay being T_s, symbol time, with an additional phase shift of $\pi/4$ and $-\pi/4$ inserted into the two branches.

The two demodulated signals at the output terminals of the low-pass filters are given by

$$V_{d1}(t) = \cos\left[\varphi(t) - \varphi(t-T_s) - \frac{\pi}{4}\right]$$

$$\tag{8.21}$$

$$V_{d2}(t) = \cos\left[\varphi(t) - \varphi(t-T_s) + \frac{\pi}{4}\right].$$

Here the phase difference between consecutive phases may be $n\dfrac{\pi}{2}$ where $n = 0, 1, 2, 3$. Substituting these values into Eq. (8.21), the demodulated signal values may be $\pm 1/\sqrt{2}$. This is exactly equal to the values for the coherent demodulator; however, the demodulated signal will be differentially decoded [13]. The realization method of the reference path delay depends on the bit rate. For small bit rates, acoustic delay lines may be applied while filters or simply pieces of cables can be used for higher bit rates [14].

In differentially coherent demodulators, it is difficult to achieve a sufficient timing stability of the delayed signal. The phase error of the delayed signal has two contributions, the delay time change ΔT and the frequency change Δf of the incoming signal:

$$\Delta\Phi = \Delta\Phi_1 + \Delta\Phi_2 \tag{8.22}$$

where

$$\Delta\Phi_1^\circ = \Delta T f_0\, 360$$

$$\tag{8.22a}$$

$$\Delta\Phi_2^\circ = T_s \Delta f\, 360.$$

Let us now evaluate the phase error for an 8 Mbit/s QPSK system operating at 8 GHz. The frequency stability is typically given by $\Delta f/f_0 = \pm 10^{-5}$ while the delay line stability is in the range of $\Delta T/T_s = 5 \times 10^{-4} \ldots 10^{-3}$. The intermediate frequency at which the demodulator operates is 70 MHz while the symbol time corresponding to a bit rate of 8 Mbit/s is given by $T_s = 250$ ns. Substituting these values into Eqs (8.23) we have $\Delta\Phi_1 = 3.1 \ldots 6.3°$ and $\Delta\Phi_2 = 14°$. These phase errors would result in a perceptible degradation of the error probability so the delay line stability has to meet stringent requirements. In some cases, automatic control of the phase shift is applied.

The error probability of differentially coherent demodulators has been investigated in Section 2.1 and it has been shown that it is higher than for coherent demodulators. At high signal-to-noise ratios, the S/N ratio difference pertaining to a given error probability is 0.5 dB for 2PSK, 2.5 dB for 4PSK and 3 dB for 8PSK [15].

The jitter introduced by differentially coherent demodulators is higher than for coherent demodulators because not only is the incoming signal noisy, but also the delayed signal used for reference, and the transitions of the demodulated signal are influenced by both signals [13]. In spite of all these shortcomings, the differentially coherent demodulator is still applied because of the advantage that there is no need for carrier recovery which results in a simplified receiver structure.

Differentially coherent demodulators are frequently applied in FH (frequency hopping) systems in which the phase of the incoming i.f. signal may show rapid changes. The carrier recovery circuit would follow these changes with a certain delay during which the transmitted information would be lost. The application of differentially coherent demodulators may thus be of advantage in FH systems.

8.3 Demodulation of QAM signals

QAM signals can be demodulated by so-called quadrature demodulators [16, 17, 18] which will be analyzed in the following (Fig. 8.8).

The expression of an M-QAM signal is given by

$$V(t) = a(t) \cos \omega_0 t - b(t) \sin \omega_0 t. \tag{8.23}$$

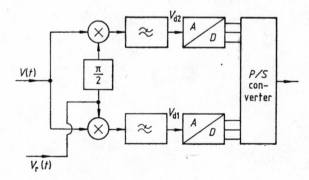

Fig. 8.8. Block diagram of QAM demodulator.

In the ideal case, the expressions of the reference signals are given by

$$V_{r1}(t) = 2 \cos \omega_0 t$$
$$V_{r2}(t) = -2 \sin \omega_0 t.$$

(8.23a)

After performing the multiplication and suppression of the signal component of frequency $2\omega_0$, the two demodulated signals will be

$$V_{d1}(t) = a(t)$$
$$V_{d2}(t) = b(t).$$

(8.24)

The amplitudes of the demodulated signals may have $L = \sqrt{M}$ values where M is the number of signal states. The signal with L levels is converted by the analog–digital converter into $n/2$ two-level signals where $L = 2^{n/2}$ and $M = 2^n$. For 16QAM, $n = 4$ and $L = 4$, for 64QAM $n = 6$, $L = 8$. From these n signals, a parallel–series converter generates the demodulated output signal.

Assume now the practical case in which the reference signals have phase errors of Φ_1 and Φ_2:

$$V_{r1}(t) = 2 \cos (\omega_0 t + \Phi_1)$$
$$V_{r2}(t) = -2 \sin (\omega_0 t + \Phi_2).$$

(8.25)

The demodulated signals are then given by

$$V_{d1}(t) = a(t) \cos \Phi_1 + b(t) \sin \Phi_1$$
$$V_{d2}(t) = b(t) \cos \Phi_2 - a(t) \sin \Phi_2.$$

(8.25a)

These phase errors result in an amplitude reduction of the demodulated signals and in a crosstalk between the quadrature channels. Figure 8.9 shows the signal-to-noise ratio increment needed to obtain a given error probability as a function of

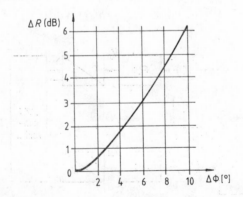

Fig. 8.9. Signal-to-noise ratio degradation due to the static phase error of the reference signal for the case of 16QAM transmission.

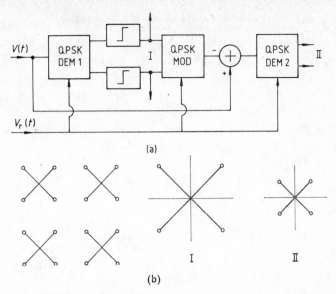

Fig. 8.10. 16QAM demodulator implemented by two QPSK demodulators: (a) block diagram; (b) decomposition of the 16QAM vector diagram.

the phase error. For the case of 16QAM, $a=1$ and $b=-3$ may take place, for example. If the phase error is given by $\Phi_1=10°$, the demodulated a-signal may be attenuated by as much as 6 dB.

The QAM demodulator can be realized by QPSK demodulators, just as a QAM modulator can be realized by QPSK modulators [19]. Figure 8.10 shows such a 16QAM demodulator which is explained in the following. The first QPSK demodulator yields two four-level signals which are used to drive two zero comparators giving the QPSK signal denoted by I. This is used to modulate a QPSK modulator, and the modulated signal is subtracted from the incoming 16QAM signal. The resultant is shown as the QPSK signal denoted by II, and this is demodulated by the second QPSK demodulator. This arrangement has the advantage of not needing linear multipliers: the four-level eye diagram may be distorted because decision takes place only at mid-level. However, the relative amplitudes and phases of the incoming and locally generated signals are critical.

Differentially coherent demodulation of QAM signals is also possible [20]. It is primarily applied in FH systems for reasons given in Section 8.2.5, but results in a degradation of the error probability. The signal-to-noise ratio pertaining to a given error probability is equal to that applying to coherent 16PSK demodulation at low S/N ratios, and to differentially coherent 16PSK demodulation at high S/N ratios.

8.4 Demodulation of QPRS signals

In Section 7.4, the generation of 9QPRS signals was discussed. In the following, the demodulation of these signals will be dealt with. A QAM demodulator, shown in Fig. 8.11, can be applied [21]. In both branches of the demodulator, a three-level signal will be demodulated. The vector diagram shows the decision levels of the analog–digital converters which yield two 2-level signals in each branch. These are subsequently processed by a baseband signal processor.

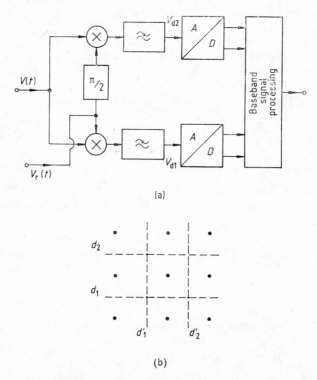

(a)

(b)

Fig. 8.11. QPRS demodulator: (a) block diagram; (b) vector diagram.

To obtain a given error probability, the S/N ratio needed for QPRS signals is 2.5 dB higher than that needed for QPSK signals, and is equal to the S/N ratio obtained with a differentially coherent QPSK demodulator. The main advantage of QPRS transmission is the smaller bandwidth required for a given bit rate; the bandwidth efficiency is 2.25 bit/s/Hz [15].

8.5 Demodulation of O-QPSK and MSK signals

8.5.1 Coherent demodulation

The demodulation of offset-QPSK signals is similar to the demodulation of QPSK signals [22, 23] shown in Fig. 8.5 with the difference that the demodulated signal in one of the demodulator branches is shifted by the bit time T_b with respect to the signal in the other branch. Figure 8.12 shows the structure of the demodulator which has to be followed by the receive side baseband signal processor. This has the function of resolving the phase ambiguity of the recovered carrier.

In principle, differentially coherent demodulation can also be used [15]. However, this is not applied in practice because the baseband signal processing would be more complicated.

MSK signals can also be demodulated by a QAM demodulator [22, 23, 24]. The structure of this demodulator is identical with that shown in Fig. 8.12 for O-QPSK signals.

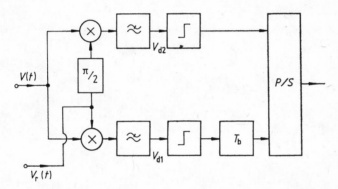

Fig. 8.12. Block diagram of coherent demodulator for MSK and O-QPSK signals.

8.5.2 Non-coherent demodulation

Non-coherent demodulation can be applied only to MSK signals. It was shown in Section 7.5 that an MSK signal is really an FSK signal having small deviation, and the modulation can be realized by continuous phase frequency modulation. Coherent demodulation of MSK signals, though possible [25], is seldom applied because of the insufficient transmit side frequency stability of the modulator. Therefore, an non-coherent demodulator, realized by either a differential detector or a limiter–discriminator is applied in most cases, resulting in a simplified receiver structure. The same type of demodulator is also applied for the demodulation of FSK signals, so their discussion will be presented in connection with FSK signal demodulation (Section 8.7).

8.6 Demodulation of ASK signals

Before discussing the demodulation process itself, a property of the i.f. amplifier will be outlined. During the transmission of digital bit streams, long breaks may be present during which the AGC control voltage should not increase the amplification as this would also increase the noise level. To avoid this phenomenon, the AGC detector should be a peak detector with a large time constant, and the number of consecutive zeros should be limited in the transmitted signal. This problem can be solved by the application of biphase coding in which ones and zeros are equally distributed. However, the transmission of biphase signals requires double bandwidth.

The demodulation of ASK signals can be realized by either envelope detectors or synchronous detectors. ASK transmission is normally used in low speed, inexpensive equipment so that the synchronous demodulation (Fig. 8.13a), requiring the generation of a reference signal by carrier recovery, is seldom applied.

The envelope detector can have a series or parallel arrangement, with one or two diodes. Figure 8.13b shows a parallel detector utilizing two diodes. The low-pass filter should be designed to transmit the demodulated signal and suppress the i.f. signal. The post detection low-pass filter may have the additional function

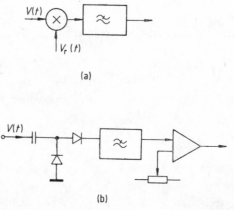

(a)

(b)

Fig. 8.13. ASK demodulator variants: (a) synchronous demodulator; (b) envelope detector.

of noise suppression if the i.f. bandpass filter is not sufficiently narrow (*e.g.*, in low bit-rate systems with relatively poor frequency stability).

For high signal-to-noise ratios, the threshold level of the comparator following the demodulator can be chosen to equal half of the demodulated signal peak voltage ($V_0/2$). In the following, the dependence of the error probability on the decision threshold will be calculated for small signal-to-noise ratios [26].

Let the noisy input signal be given by

$$V(t) = A \cos \omega_0 t + n_{\rm c}(t) \cos \omega_0 t - n_{\rm s}(t) \sin \omega_0 t \qquad (8.26)$$

where A is the signal amplitude, while $n_{\rm c}(t)$ and $n_{\rm s}(t)$ are the cosine and sine noise component amplitudes. Equation (8.26) can also be expressed in the following form:

$$V(t) = E \cos [\omega_0 t + \varphi(t)] \qquad (8.27)$$

where the following notation has been applied:

$$E = [(A+n_{\rm c})^2 + n_{\rm s}^2]^{1/2}$$

$$\varphi(t) = \tan^{-1} \frac{n_{\rm s}(t)}{A+n_{\rm c}(t)}. \qquad (8.27a)$$

The noise power of the receiver input signal is given by

$$\sigma^2 = \frac{n_{\rm c}^2}{2} + \frac{n_{\rm s}^2}{2}. \qquad (8.28)$$

The probability density function of the noisy signal amplitude is given by

$$p_{\rm s}(E) = \frac{E}{\sigma^2} I_0\left(\frac{AE}{\sigma^2}\right) \exp\left[-\frac{E^2+A^2}{2\sigma^2}\right]. \qquad (8.29)$$

If only noise is present ($A=0$), then

$$p_{\rm n}(E) = \frac{E}{\sigma^2} \exp\left[-\frac{E^2}{2\sigma^2}\right]. \qquad (8.30)$$

The corresponding distribution functions are

$$Q_{\rm s}(E) = \exp\left[-\frac{A^2+E^2}{2\sigma^2}\right] \sum_{m=1}^{\infty} \left(\frac{E}{A}\right)^m I_m\left(\frac{EA}{\sigma^2}\right) \qquad (8.31)$$

$$Q_{\rm n}(E) = 1 - \exp\left[-\frac{E^2}{2\sigma^2}\right] \qquad (8.32)$$

where I_m is the modified Bessel function of mth order. Utilizing these expressions, the error probability can be determined as explained in the following. If the decision threshold is chosen to be kA then the error probabilities will be given by

$$P(1/0) = \exp\left[-\frac{k^2A^2}{2\sigma^2}\right] = e^{-k^2R} \tag{8.33}$$

$$P(0/1) = \frac{e^{-(1-k)^2R}}{2(1-k)}\sqrt{\frac{k}{\pi R}}\left[1-\frac{3+6k-k^2}{16k(1-k)^2R}\right] \tag{8.33a}$$

where $P(1/0)$ and $P(0/1)$ are the error probabilities when transmitting zero and one (space and mark), respectively.

Figure 8.14a shows the dependence of these error probabilities on the decision threshold for a few signal-to-noise ratios. The dependence of the optimal threshold on the signal-to-noise ratio is plotted in Fig. 8.14b. In practice, the decision threshold is not shifted, with dependence on the signal-to-noise ratio, but has a fixed value. It is seen that this introduces a small degradation.

Biphase coding eliminates the DC component of the bit stream so a capacitive coupling between the demodulator and the comparator can be applied.

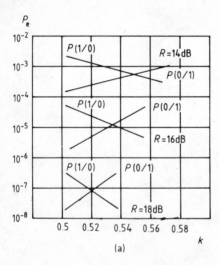
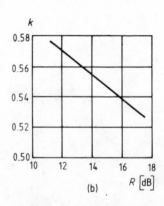

Fig. 8.14. (a) Error probability as a function of the decision threshold; (b) optimal decision threshold as a function of signal-to-noise ratio.

8.7 Degradations due to the transmit–receive section

8.7.1 Intersymbol interference due to bandpass filters

The receiver filter has to meet the requirement of intersymbol-free transmission and interference suppression while the transmitter filter should have a response providing a specified radiated power density spectrum. Furthermore, both filters should be easily realizable. These requirements can only be approximated and met with certain compromises.

In Section 2.2, the effect of intersymbol interference on the error probability has been calculated but these calculations can only be carried out numerically, with the aid of a computer. However, a simpler expression can be derived relating the error probability to the reduction of the so-called eye diagram opening [27]. In the following, the concept of the eye diagram will be explained.

The eye diagram is an oscilloscope display showing the demodulated bit stream as appearing at the input of the comparator with a horizontal deflection time nT where T is the bit time. A single eye opening is obtained with $n=1$ (see Fig. 8.15), while several openings are obtained with $n=2, 3$, etc. A noise and distortion free transmission results in equal heights of all consecutive bits, and thus the superposition on the screen results in a single envelope (see Fig. 8.15a). The presence of noise and distortion has the effect of receiving differing bit heights and of the reduction of the eye opening (Fig. 8.15b). The peak reduction can be calculated from the relative deviation of the consecutive bits from their nominal value:

$$D = \frac{\sum_{n}' x_n}{x_0} \tag{8.34}$$

where x_0 is the nominal value of the bits and x_n are the deviations which may be either positive or negative. The prime in the above expression indicates that $n=0$ does not appear in the addition. The eye diagram can also be characterized by the squared average value of the eye diagram reduction, given by

$$\varepsilon^2 = \frac{1}{x_0^2} \sum_{n}' x_n^2. \tag{8.35}$$

Assuming an ideal eye height of unity, the smallest eye opening will be given by

$$\eta_m = 1 - D. \tag{8.36}$$

Evidently, this smallest eye opening cannot be used for the direct calculation of the error probability as it appears only with a certain probability. However, it can be applied for the calculation of error probabilities less than about 10^{-6} [29].

The eye opening is reduced by the group delay and amplitude response of the

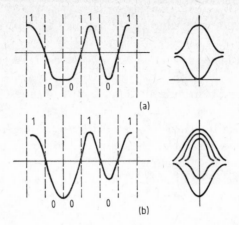

(a)

(b)

Fig. 8.15. Eye diagram variants: (a) ideal transmission; (b) distorted transmission.

receiver bandpass filter. Let us assume a raised cosine signal spectrum at the demodulator input, showing that the amplitude response of the filter meets the requirement of intersymbol interference-free transmission. In this case, the eye opening is reduced only by the group delay distortion of the filter. This reduction can be calculated and is plotted against the group delay distortion in Fig. 8.16 for linear and parabolic group delay responses [29].

The amplitude response of the receiver filter should have arithmetic symmetry to avoid the reduction of the eye opening. Otherwise, unwanted quadrature components will be generated which give rise to crosstalk between the demodulated component signals [30]. However, filter design based on conventional low-pass–bandpass transformation results in geometrical symmetry. The correct filter response can then be realized by suitable shift of the pole frequencies [31].

Transmission performance is, in principle, determined by the overall responses of the transmit and receive bandpass filters. In most cases, however, the transmit

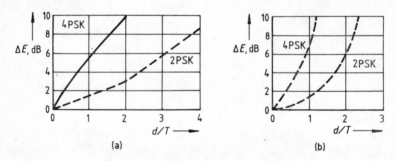

(a)   (b)

Fig. 8.16. Decrease of eye diagram opening due to group delay distortion: (a) parabolic group delay distortion; (b) linear group delay distortion.

filters have only the function of limiting the radiated spectrum bandwidth and thus provide suitable selectivity, their bandwidth being much wider than the receive filter bandwidth. As a rough guide, the transmit and receive filter bandwidths are chosen to be $1.5...2/T_s$ and $1...1.3/T_s$, respectively, where T_s stands for symbol time [18, 32]. In most cases, Gaussian filters, Thomson filters or 2-pole Butterworth filters are applied. These have low group delay distortion so there is no need for group delay equalizers in the case of 2PSK or 4PSK transmission [33, 34, 35, 36]. In some cases, special filters resulting in lowest intersymbol interference are applied.

For higher-order Butterworth or Chebyshev filters, equalizers have to be applied. In the simplest case, only group delay equalizers are used, and the group delay response in the $1/T_s$ frequency range is equalized to a value of $\Delta\tau/T_s \sim 0.05$ to 0.1 [31]. However, gain slope equalizers may also be used to minimize the slope which would introduce unwanted quadrature components. In a given item of equipment for QPSK transmission [37], the group delay has been set equal to $0.1T_s$ and the gain slope to 0.6 dB in the frequency range of $1/T_s$. Each of these residual values gives rise to a signal-to-noise degradation of 0.2 dB.

16QAM transmission is more susceptible to distortions. In 16QAM equipment for 140 Mbit/s transmission, group delay is equalized to $\Delta\tau \sim 1$ ns corresponding to $\Delta\tau/T_s = 3.5 \times 10^{-2}$ [17], resulting in a signal-to-noise degradation of 0.2 dB.

The signal-to-noise ratio degradation due to intersymbol interference can be further reduced by higher-order equalizers applied both at the transmit and receive side, resulting in a response meeting the Nyquist criteria [31]. However, these more complicated equalizers are not usually justified because the signal-to-noise ratio improvement is only a few tenth of a dB.

8.7.2 Phase error of the reference signal

The phase error of the reference signal at the receive side is due to synchronization errors resulting from carrier frequency instability at the transmit and receive sides, and also circuit adjustment inaccuracies. The effect of this error on the error probability should be considered together with the phase error of the modulator at the transmit side. Four different additions of these two errors, each having a probability of 1/4, may take place at 4PSK:

$$\Delta\Phi = \begin{cases} -\Delta\Phi_r - \Delta\Phi_m \\ -\Delta\Phi_r + \Delta\Phi_m \\ \Delta\Phi_r - \Delta\Phi_m \\ \Delta\Phi_r + \Delta\Phi_m \end{cases} \qquad (8.37)$$

where the modulator and reference signal phase errors are denoted by $\Delta\Phi_m$ and $\Delta\Phi_r$, respectively. According to Ref. [34], the error probability due to these errors is calculated to be

$$P_E = \sum_1^4 \frac{1}{4}\,\mathrm{erfc}\left[\sqrt{R}\,\sin\left(\frac{\pi}{4}+\Delta\Phi\right)\right] \qquad (8.38)$$

where R is the signal-to-noise ratio at the demodulator input. With good approximation, the error probability will be determined by the term comprising the highest $\Delta\Phi$ value:

$$P_E \cong \frac{1}{4}\,\mathrm{erfc}\left\{\sqrt{R}\,\sin\left[\frac{\pi}{4}-(\Delta\Phi_r+\Delta\Phi_m)\right]\right\}. \qquad (8.39)$$

Figure 8.17 shows the signal-to-noise ratio degradation due to the phase error for 2PSK and 4PSK modulation. The degradation for QAM modulation is given in Fig. 8.9.

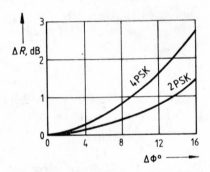

Fig. 8.17. Signal-to-noise ratio degradation as a function of the static phase error.

8.7.3 Decision threshold shift

The oscilloscope displaying the eye diagram is connected to the output of the low-pass filter following the detector of the demodulator. This point is followed by a comparator for decision making, the reference voltage of the comparator being adjusted to the centre level of the eye diagram. In the following, an expression is presented [34] relating, in the case of 4PSK transmission, the error probability to the decision threshold shift from the centre level ΔV_{th}, the demodulator offset ΔV_D and the comparator hysteresis ΔV_H:

$$
\begin{aligned}
P_\mathrm{E} = &\frac{1}{4}\,\mathrm{erfc}\left\{\sqrt{\frac{R}{2}}\left[1+\left(\Delta V_\mathrm{th}+\Delta V_\mathrm{D}+\frac{1}{2}\,\Delta V_\mathrm{H}\right)\middle/V_\mathrm{D}\right]\right\}+ \\
&+\frac{1}{4}\,\mathrm{erfc}\left\{\sqrt{\frac{R}{2}}\left[1+\left(\Delta V_\mathrm{th}+\Delta V_\mathrm{D}-\frac{1}{2}\,\Delta V_\mathrm{H}\right)\middle/V_\mathrm{D}\right]\right\}+ \\
&+\frac{1}{4}\,\mathrm{erfc}\left\{\sqrt{\frac{R}{2}}\left[1-\left(\Delta V_\mathrm{th}+\Delta V_\mathrm{D}+\frac{1}{2}\,\Delta V_\mathrm{H}\right)\middle/V_\mathrm{D}\right]\right\}+ \\
&+\frac{1}{4}\,\mathrm{erfc}\left\{\sqrt{\frac{R}{2}}\left[1-\left(\Delta V_\mathrm{th}+\Delta V_\mathrm{D}-\frac{1}{2}\,\Delta V_\mathrm{H}\right)\middle/V_\mathrm{D}\right]\right\}
\end{aligned}
\tag{8.40}
$$

where V_D is the amplitude of the demodulated voltage and R is the signal-to-noise ratio at the demodulator input. A useful upper limit is given by the following simplified expression:

$$
P_\mathrm{E} \leqq \frac{1}{2}\,\mathrm{erfc}\left[\sqrt{\frac{R}{2}}(1+\delta)\right]+\frac{1}{2}\,\mathrm{erfc}\left[\sqrt{\frac{R}{2}}(1-\delta)\right]
\tag{8.41}
$$

where

$$
\delta = \frac{\Delta V_\mathrm{th}+\Delta V_\mathrm{D}+\dfrac{1}{2}\,\Delta V_\mathrm{H}}{V_\mathrm{D}}.
\tag{8.41a}
$$

The signal-to-noise ratio degradation due to the decision threshold shift is plotted in Fig. 8.18 for QPSK transmission. Similar expressions can be derived for other modulation methods.

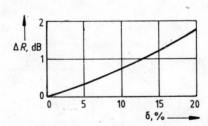

Fig. 8.18. Signal-to-noise ratio degradation as a function of the decision threshold shift for QPSK transmission.

8.7.4 Addition of degradations

In Sections 8.7.1 to 8.7.3, factors resulting in signal-to-noise ratio degradation have been discussed. A further degradation factor, the inaccuracy of bit time recovery, has been covered in Section 5.3. A simple addition of the individual degradations will result in a value somewhat different from the actually experienced overall degradation. However, it can be shown [34] that a more accurate overall signal-to-noise ratio degradation is obtained by calculating first the individual equivalent eye diagram reductions from the individual degradations:

$$\Delta E_i = 1 - 10^{-(\Delta R_i/20)} \tag{8.42}$$

next calculating the overall eye diagram reduction by simple addition of these reductions:

$$\Delta E_0 = \sum_i \Delta E_i \tag{8.43}$$

and finally finding the overall signal-to-noise ratio degradation from this overall reduction:

$$\Delta R_0 = -20 \lg (1 - \Delta E_0) \quad \text{[dB].} \tag{8.44}$$

In the above equations, the following notation has been used:

ΔE_i individual eye diagram opening reduction,
ΔR_i individual signal-to-noise ratio reduction expressed in dB,
ΔE_0 overall eye diagram opening reduction,
ΔR_0 overall signal-to-noise ratio reduction.

Table 8.1. Typical degradations in digital radio equipment

Parameter resulting in degradation	ΔR, dB		ΔE, %	
	4PSK	16QAM	4PSK	16QAM
Intersymbol interference due to band limiting	1.4		15	
Decision threshold shift	0.25		3	
Phase error of the reference signal (4PSK: 8°, 16QAM: 3°)	0.96	1	10.5	11
Inaccuracy of bit time recovery	0.3		3	
Interchannel interference	0.2		2	
Overall degradation	3.5	3.6	33.5	34

In the literature, degradation is expressed either by ΔR or by ΔE, but by using Eq. (8.42), ΔR can be converted into ΔE for addition purposes. Table 8.1 is a survey of typical degradations experienced in actual microwave digital radio equipment. The overall degradations shown in the table cannot be neglected and should be taken into account during the design procedure of the radio equipment.

8.8 Demodulation of FSK signals

Before discussion of the individual demodulator types, an expression will be derived for the noisy FSK signal at the demodulator input which will be utilized. An FM signal can be expressed as

$$V(t) = A \cos\left[\omega_0 t + \theta + \mu(t)\right] \tag{8.45}$$

where $\mu(t)$ is the instantaneous phase comprising the information to be transmitted and θ is the phase at $t=0$. The instantaneous phase is given by

$$\mu(t) = \int_{-\infty}^{t} s(x)\,dx \tag{8.46}$$

where $s(t)$ is the frequency time function corresponding to the modulation signal. For an FSK signal, $s(t)$ can be expressed as follows:

$$s(t) = \Delta\omega \sum_{n=-\infty}^{+\infty} a_n g(t-nT) \tag{8.47}$$

where $\Delta\omega$ is the peak frequency deviation; $a_n = \pm 1$ for a two-state FSK signal; $g_n(t-nT)$ is the unit gate function: it has a value of 1 in the interval $(n-1)T...nT$; and a value of 0 outside this interval.

In the following discussion, the frequency deviation related to the bit rate will be used:

$$h = \frac{\Delta\omega T}{\pi} = 2\Delta f T. \tag{8.48}$$

Figure 8.19 is a representation of an FSK radio section, showing the superposition of Gaussian noise to the received signal.

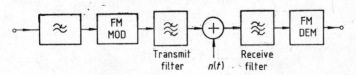

Fig. 8.19. FSK transmission path.

Let us resolve the noise into two components, one in phase with the FSK signal, and the other in phase quadrature with the FSK signal:

$$n(t) = n_c(t) \cos [\omega_0 t + \mu(t) + \theta] - n_s(t) \sin [\omega_0 t + \mu(t) + \theta]. \qquad (8.49)$$

The noisy signal can then be expressed in the following form:

$$V(t) = r(t) \cos [\omega_0 t + \varphi(t) + \theta] \qquad (8.50)$$

where

$$r(t) = [(A + n_c)^2 + n_s^2]^{1/2} \qquad (8.51)$$

$$\varphi(t) = \mu(t) + \eta(t) \qquad (8.52)$$

$$\eta(t) = \tan^{-1} \frac{n_s(t)}{A + n_c(t)}. \qquad (8.53)$$

The relative noise power is given by

$$\sigma^2 = \frac{n_c^2}{2} + \frac{n_s^2}{2} \qquad (8.54)$$

and the relative signal power is given by

$$P_s = \frac{A^2}{2} \qquad (8.55)$$

so the signal-to-noise ratio is

$$R = \frac{A^2}{2\sigma^2}. \qquad (8.56)$$

8.8.1 Demodulator with matched filter

This demodulator is shown in Fig. 8.20. The two bandpass filters are tuned to the mark and space frequencies:

$$f_1 = f_0 - \Delta f$$
$$f_2 = f_0 + \Delta f. \qquad (8.57)$$

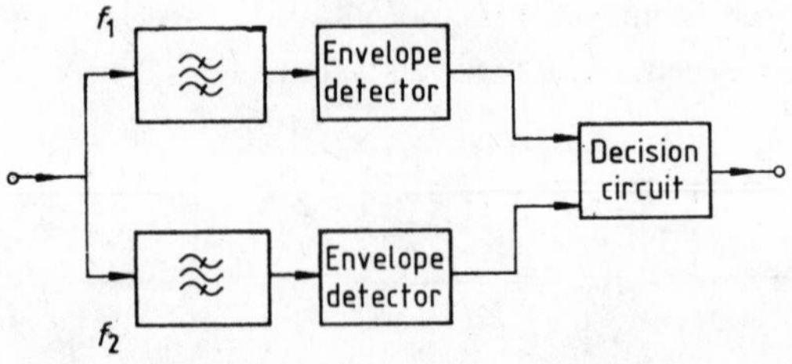

Fig. 8.20. Matched filter FSK demodulator.

In the case of orthogonal FSK, $h=1$, *i.e.* $\Delta f=1/2T$. It can be shown that this non-coherent FSK demodulator is optimal for the demodulation of an FSK signal if the two states have equal probabilities and powers, and the noise powers at the output ports of the bandpass filters are equal.

In the demodulator circuit shown in Fig. 8.20, either envelope detectors or r.m.s. detectors can be applied, as shown by the following consideration. The decision is based on the relative levels of the two signals feeding the detectors. Denoting these by r_1 and r_2, evidently

$$P(r_2 > r_1) = P(r_2^2 > r_1^2) \tag{8.58}$$

so both detector types will yield the same decision.

With the application of an optimal FSK demodulator, the error probability of FSK transmission, according to Ref. [38], is given by

$$P_E = \frac{1}{2} e^{-R/2} \tag{8.59}$$

where R is the signal-to-noise ratio at the demodulator input. An orthogonal FSK signal can also be demodulated by a coherent demodulator [38], the error probability then being given by

$$P_E = \frac{1}{2} \operatorname{erfc} \sqrt{\frac{R}{2}}. \tag{8.60}$$

However, coherent FSK demodulation is not used in practical systems because the more complicated circuit would not be justified in less expensive equipment for which FSK transmission is usually applied.

8.8.2 Limiter–discriminator type demodulator

This type of demodulator is used in analog FM equipment. In the following, its properties for FSK demodulation will be investigated, and the optimal frequency deviation and transmission bandwidth will be determined.

The limiter–discriminator type demodulator is shown in Fig. 8.21. The low-pass filter following the discriminator has the function of attenuating the high

Fig. 8.21. Limiter–discriminator type FSK demodulator.

frequency components, and has no effect on the signal waveform [39]. The expression of the modulated signal, given by Eq. (8.45), is now re-written in the following form:

$$V(t) = x(t) \cos(\omega_0 t + \theta) - y(t) \sin(\omega_0 t + \theta) \tag{8.61}$$

where

$$x(t) = A \cos \mu(t)$$
$$y(t) = A \sin \mu(t). \tag{8.62}$$

If the impulse response of the receiver bandpass filter, shown in Fig. 8.21, is given by $h(t)$ then the two components of the filter output signal can be expressed by convolution integrals as follows:

$$x_1(t) = A \int_{-\infty}^{t} h(t-\tau) \cos \mu(\tau) \, d\tau$$

$$\tag{8.63}$$

$$y_1(t) = A \int_{-\infty}^{t} h(t-\tau) \sin \mu(\tau) \, d\tau$$

and the phase of the filtered signal is given by

$$\Phi(t) = \tan^{-1} \frac{y_1(t)}{x_1(t)}. \tag{8.64}$$

The instantaneous signal frequency is given by the time derivative of the phase:

$$\dot{\Phi}(t) = \frac{x_1 \dot{y}_1 - y_1 \dot{x}_1}{x_1^2 + y_1^2}. \tag{8.65}$$

The noisy filtered signal can be expressed in the following form:

$$V(t) = x_2(t) \cos(\omega_0 t + \theta) - y_2(t) \sin(\omega_0 t + \theta) \tag{8.66}$$

where

$$x_2(t) = x_1(t) + n_c(t)$$
$$y_2(t) = y_1(t) + n_s(t). \tag{8.67}$$

Denoting the phase of the noisy signal by $\varphi(t)$, the instantaneous signal frequency will be given by

$$\dot{\varphi} = \frac{x_2 \dot{y}_2 - y_2 \dot{x}_2}{x_2^2 + y_2^2} \tag{8.68}$$

or by the more detailed expression of

$$\dot{\varphi} = \frac{(x_1 + n_c)(\dot{y}_1 + \dot{n}_s) - (y_1 + n_s)(\dot{x}_1 + \dot{n}_c)}{(x_1 + n_c)^2 + (y_1 + n_s)^2}. \tag{8.69}$$

The relative noise power at the filter output is given by

$$\sigma_0^2 = \frac{N_0}{2} \int_{-\infty}^{+\infty} |H(\omega)|^2 \, df \tag{8.70}$$

where $N_0/2$ is the two-sided spectral density of the noise and $H(\omega)$ is the filter transfer function.

We shall now investigate the probability distribution of the instantaneous frequency. For this purpose, it is sufficient to consider the numerator of the expression giving $\dot\phi(t)$ because the denominator is the sum of two squares which is always positive. Let us introduce the new variable $z = x_2 \dot y_2 - y_2 \dot x_2$. For a given x_2, y_2, the probability density function will be given, according to Ref. [39], by the following expression:

$$p(z|x_2, y_2) = \frac{1}{\sigma \sqrt{2\pi}} \exp\left[-\frac{(z - z_0)^2}{2\sigma^2}\right] \tag{8.71}$$

where

$$z_0 = x_2 E\{\dot y_2\} - y_2 E\{\dot x_2\} \tag{8.72}$$

and $E\{\ \}$ represents the expectation. σ^2 is the sum of the variances of $x_2 \dot y_2$ and $y_2 \dot x_2$:

$$\sigma^2 = (x_2^2 + y_2^2)\sigma_1^2 \tag{8.73}$$

$$\sigma_1^2 = \frac{N_0}{2} \int_{-\infty}^{+\infty} \omega^2 |H(\omega)|^2 \, df. \tag{8.74}$$

In the case of $f_0 + \Delta f$, when $\dot\Phi > 0$, a wrong decision takes place for $z < 0$ while in the case of $f_0 - \Delta f$, when $\dot\Phi < 0$, a wrong decision takes place for $z > 0$. The error probabilities in the two cases are given by

$$P_+ = \int_{-\infty}^{0} p(z) \, dz \quad (\dot\Phi > 0)$$

$$\tag{8.75}$$

$$P_- = \int_{0}^{\infty} p(z) \, dz \quad (\dot\Phi < 0).$$

A useful lower limit for the error probability is given by

$$P_e = \frac{1}{\pi} \int_{0}^{\pi/2} \exp\left[-\frac{P_0 \dot\Phi}{\sigma_1^2 \sin^2\theta + \sigma_0^2 \dot\Phi^2 \cos^2\theta}\right] d\theta \tag{8.76}$$

or expressed in another form,

$$P_e = \frac{1}{\pi} \int_{0}^{\pi/2} \exp\left[-\frac{a^2 R}{\sin^2\theta + a^2 \cos^2\theta}\right] d\theta \tag{8.77}$$

where

$P_0 = A^2/2$ is the received signal power,

$$a^2 = \frac{\sigma_0^2 \dot{\Phi}^2}{\sigma_1^2},$$

$R = A^2/2\sigma_0^2$ is the signal-to-noise ratio.

A simple and accurate expression is obtained by assuming $\dot{\Phi} = \dfrac{\sigma_1}{\sigma_0}$:

$$P_e = \frac{1}{2} e^{-R}. \tag{8.78}$$

In the case of intersymbol interference-free transmission, $\dot{\Phi} = \Delta\omega$. In the foll lowing, the conditions for intersymbol interference-free FSK transmission wil- be investigated.

It is well known that in the case of AM and PM transmission, the transmission characteristic should meet the Nyquist criteria for intersymbol interference-free transmission. However, these criteria cannot be applied in general to FSK trans- mission because FSK modulation and demodulation are nonlinear operations. In the special case of continuous phase orthogonal FSK transmission, the modu- lated signal can be expressed as the sum of a carrier component and a component having a continuous spectrum. This latter part carries the information to be transmitted and can be regarded as a double sideband AM signal with suppressed carrier [40]. This means that the Nyquist criteria also apply for orthogonal FSK transmission $(h=1)$.

For non-orthogonal FSK transmission $(h \neq 1)$, the condition for intersymbol interference-free transmission is expressed as a condition for the overall transfer function

$$H(\omega) = H_t(\omega) H_r(\omega) \tag{8.79}$$

where $H_t(\omega)$ is the transmit filter characteristic and $H_r(\omega)$ the receive filter char- acteristic. This condition, given in Ref. [41], is more stringent than the Nyquist criteria.

For any frequency deviation, transmit and receive filters with optimal char- acteristic for minimizing intersymbol interference can be determined, and the error probability *versus* signal-to-noise ratio curves are then identical within about 1 dB, independently of the frequency deviation [42]. However, filters of microwave radio equipment are, in general, not optimal for two reasons.

(1) In the case of microwave modulators, the optimal transmit filter cannot be realized in the microwave frequency range because of the small relative bandwidth required.

(2) The resultant optimal characteristic of the transmit and receive filters is realized only if there is no nonlinear element in between. FSK transmission would tolerate a nonlinear transmission path, but this advantage would be lost by meeting the above requirement.

Because of these reasons, FSK systems are designed in practice by determining the optimal deviation and transmission bandwidth for the filter characteristic applied (*e.g.*, Butterworth, Gaussian *etc.*). In this case, the transmission path nonlinearity does not affect the transmission. This design can be implemented by computer simulation.

The main data of FSK transmission depend on the baseband encoding method. The optimal frequency deviation and transmission bandwidth values are summarized in Table 8.2 while the error probability curves are presented in Fig. 8.22.

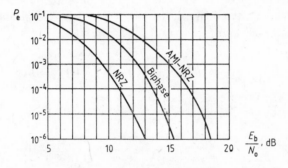

Fig. 8.22. Error probability as a function of signal-to-noise ratio for FSK transmission. B_{opt}, fourth-order Butterworth filter.

Table 8.2. Calculated optimal frequency deviation and transmission bandwidth values for FSK transmission and a fourth-order Butterworth filter

Encoding	h_{opt}	$B_{opt} T_b$
NRZ	0.7	1.0
AMI–NRZ	1.0	1.0
Biphase	1.4	1.9

8.8.3 Limiter–discriminator type demodulator with integrate-and-dump filter

This type of demodulator is similar to the limiter–discriminator type shown in Fig. 8.21, the difference being in the post detection low-pass filter which has a much lower cut-off frequency $(B \sim 0.45/T)$. By the noise reduction thus obtained, a small improvement in signal-to-noise ratio is achieved. In the analysis of this demodulator, it is possible to replace the post detection filters by an integrate-and-dump filter network [43].

The noisy filtered input signal can be written as

$$V(t) = r(t) \cos \left[\omega_0 t + \Phi(t) + \eta(t) \right] \tag{8.80}$$

where, for simplicity, the initial phase θ has been assumed to be zero. Here the voltage amplitude $r(t)$ is given by Eq. (8.51), the noise phase $\eta(t)$ by Eq. (8.53) and the phase of the filtered signal, $\Phi(t)$ by Eq. (8.64). The probability density function of $\eta(t)$ is well known:

$$p(\eta) = \int_0^\infty \frac{x}{\pi} \exp \left[-x^2 + R(t) - 2x\sqrt{R}\,(t) \cos \eta \right] dx \tag{8.81}$$

where the expression of the signal-to-noise ratio is given by

$$R(t) = \frac{r^2(t)}{2\sigma_0^2}. \tag{8.82}$$

The filtered signal amplitude $r(t)$ depends on the modulating signal and the filter bandwidth. The latter is chosen so that the steady-state value is reached only during long 1111... or 0000... bit streams but not during 1010... transitions. The limiter output signal is given by

$$V(t) = \cos \left[\omega_0 t + \Phi(t) + \eta(t) \right] \tag{8.83}$$

while the discriminator output signal is the derivative of the phase. Integration therefore yields the phase change as given by

$$\Delta\varphi = \Phi(t) - \Phi(t-T) + [\eta(t) - \eta(t-T) \bmod 2\pi + N(T)2\pi] \tag{8.84}$$

or

$$\Delta\varphi = \Delta\Phi + \Delta\eta \bmod 2\pi + N(T)2\pi \tag{8.85}$$

where $\Delta\Phi$ is the phase change due to modulation; $\Delta\eta$ is the phase change due to noise, reduced by mod 2π; $N(T)$ is the number of clicks during the bit-time T.

Decision is based on the sign of the phase change which is equal to the sign of the frequency change. An errored bit can be generated either by continuous noise or by click noise, the latter being the effect of cycle slip:

$$P_e = P_{\text{continuous}} + P_{\text{click}}. \tag{8.86}$$

The filter bandwidth has an optimal value because a large filter bandwidth results in a high click noise while a small filter bandwidth gives a high continuous noise level. The number of clicks, N, has a Poisson distribution. The following expression gives the probability of a click number $N=k$ during a time interval T:

$$P(N = k) = \frac{e^{-\overline{N}}\overline{N}^k}{k!}. \tag{8.87}$$

The following expression holds for the average value of the number of clicks during the time interval T:

$$\overline{N} \cong T\Delta f e^{-R}. \tag{8.88}$$

The optimal deviation and transmission bandwidth for this type of demodulator is calculated in several publications [44, 45, 47]. Generally, the following design procedure is applied. A given filter type is selected (*e.g.,* Gaussian or Butterworth type) and it is assumed that intersymbol interference is introduced by only a limited number of adjacent bits. For instance, only two adjacent bits are effective in the case where the transmission bandwidth B is higher than $1/T$. Next, the error probabilities for all possible bit combinations are calculated. In another publication, two preceding and two following bits are taken into account [46]. The possible bit combinations are provided by a pseudo random bit stream having a suitable sequence length.

The optimal deviation and transmission bandwidth values and the error probability curves are close to those applying for the simple limiter–discriminator type demodulator discussed in Section 8.8.2. For NRZ encoding, again $h \sim 0.7$ and $B_{\text{opt}} T_b \sim 1$ has been obtained. In publication [44], the optimal deviation is somewhat dependent on the signal-to-noise ratio, within the range $h = 0.70$ to 0.72.

It has been shown that the number of clicks is increased by the deviation. Above the optimal deviation, errors are introduced mainly by clicks while below this deviation, continuous noise is dominant [43].

According to the error probability curves given in Refs [38, 44, 45], the error probability for $h = 0.7$ does not differ substantially from the error probability for orthogonal coherent FSK transmission given by Eq. (8.60). This means that for NRZ encoding, this equation can be used for calculation of the error probability in the case of optimal deviation and transmission bandwidth.

MSK modulation can be regarded as a special type of FSK modulation having a deviation of $h = 0.5$. MSK signals are frequently demodulated by a limiter-discriminator type demodulator in which case the optimal transmission bandwidth is again given by $B_{opt} \sim 1/T$. The signal-to-noise ratio for a given error probability is 1 to 1.5 dB higher than for the optimal value of $h=0.7$.

For FSK transmission of signals with biphase encoding, $h_{opt}=1.3$ and $B_{opt} T =1.8$ [46].

8.8.4 Differential demodulator

The differential demodulator is shown in Fig. 8.23. It is seen that the incoming filtered noisy signal is delayed by a time interval T, and following a phase shift of $\pi/2$, is multiplied with the undelayed signal. The filtered noisy signal, given by Eq. (8.80), is reproduced here:

$$V(t) = r(t) \cos [\omega_0 t + \Phi(t) + \eta(t)]. \tag{8.89}$$

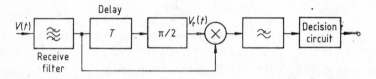

Fig. 8.23. Block diagram of differential FSK demodulator.

Following the 90° phase shift and delay, the signal is expressed by

$$V_r(t) = -r(t-T) \sin [\omega_0 t + \Phi(t-T) + \eta(t-T)]. \tag{8.90}$$

Following the multiplication, the double frequency components are attenuated by the low-pass filter so the demodulated signal after filtering is given by

$$V_d(t) = \frac{r(t) r(t-T)}{2} \sin [\omega_0 T + \Delta\varphi(T)] \tag{8.91}$$

where

$$\Delta\varphi(T) = \Delta\Phi(T) + \Delta\eta(T). \tag{8.92}$$

In the above equation, $\Delta\Phi(T)$ is the phase change due to the received signal and $\Delta\eta(T)$ is the phase change due to the noise during time interval T.

To simplify the analysis, it is normally assumed that $\omega_0 T = k2\pi$ where k is an integer. The demodulated signal is then given by

$$V_d(t) = \frac{r(t) r(t-T)}{2} \sin \Delta\varphi(T). \tag{8.93}$$

Decision is based on the sign of $V_d(t)$. The phase distribution is identical with the distribution for the limiter–discriminator type demodulator with the integrate-and-dump filter. The only difference is given by the fact that instead of the sign of $\Delta\varphi$, the sign of $\sin \Delta\varphi$ will be relevant for the decision.

It is seen from Eq. (8.93) that the highest output signal is obtained for $h=0.5$ when the phase change $\Delta\varphi$ is $\pi/2$. This signal is given by

$$V_d(t) = \frac{r(t)\,r(t-T)}{2}. \tag{8.94}$$

For $h\leq0.5$, the signal-to-noise ratio for a given error probability is the same for differential demodulation and limiter–discriminator type demodulation. For $h=0.7$, the limiter–discriminator is better by about 2 dB, and for $h=1$, this type of differential demodulator having a delay of $\tau=T$ cannot be used [48]. However, if the delay is chosen to be $\tau=T/2h$ $(h>0.5)$, the degradation is much smaller, and the demodulator can then be used even for $h=1$ [54]. The differential demodulator is widely applied in non-coherent MSK systems because for $h=0.5$, it is equivalent to the limiter–discriminator type demodulator and much simpler. The optimal bandwidth is again near $1/T$.

If the spectrum of an MSK signal has to be limited then a transmit filter with Gaussian response is applied. The output signal is then called a GMSK signal. In this case, a higher bandwidth has to be chosen for the receive filter. This has the consequence that for a given error probability, the required signal-to-noise ratio is somewhat higher for GMSK transmission than for MSK transmission.

With suitable transmit side encoding, a differential demodulator with a delay of $2T$ can also be realized. With GMSK transmission, this yields a better signal-to-noise ratio than the differential demodulator with a delay of T [49].

8.8.5 PLL demodulator

In the preceding part of Section 8.8, it has been assumed that the carrier frequency ω_0 at the input of the demodulator is constant. This is only approximately valid, but in most cases the carrier frequency shift $\Delta\omega_0$ is much smaller than the frequency deviation $\Delta\omega$. However, in some cases, for example for low speed transmission for which $\Delta\omega_0\sim\Delta\omega$, the demodulation methods previously discussed are useless, and PLL type demodulators have to be applied. A wideband input filter is then used, and the error probability is determined by the loop bandwidth of the PLL which can be smaller than the pull-in frequency range. A detailed investigation of the PLL demodulator is found in the literature [50].

8.8.6 *M*-FSK demodulator

Two basic characteristics of digital transmission are the required transmitter power and the occupied transmission bandwidth. It was shown in Section 2.1.2 that the latter can be reduced by applying multi-state modulation schemes. An *n*-fold bandwidth reduction can be achieved by utilizing *M*-ary modulation such as *M*-PSK or *M*-QAM. However, the transmitter power has to be increased exponentially to obtain the same error probability.

On the other hand, it is also possible to decrease the required transmitter power by utilizing a suitable modulation scheme. *M*-FSK modulation with a relative deviation of $h=1$, resulting in an orthogonal signal set, is suitable for this purpose. An *M*-FSK signal is generated by driving an FM modulator by *n* bit streams in which symbols comprising *n* bits are formed; $M=2^n$. The bandwidth required for the transmission of one of the modulation states is given by

$$B_1 = 1/T_s = 1/nT_b$$

so the bandwidth required for the transmission of all the $M=2^n$ states is given by

$$B = 2^n/nT_b. \tag{8.95}$$

We note that increasing the number of modulation states, will also have the effect of increasing the occupied transmission bandwidth.

For *M*-FSK transmission, the energy received during unit time is given by

$$E_s = P_M nT_b.$$

The energy required to obtain a given error probability is independent of M so the relation for the power reduction achieved by *M*-FSK with respect to 2FSK is easily derived:

$$P_M/P_2 = 1/n. \tag{8.96}$$

This expression is valid as long as the relation

$$P_M T_b/N_0 \gg -1.6 \, \text{dB}$$

holds where N_0 is the one-sided spectral density of the demodulator input noise (see Section 2.1). In practical cases, the sign $\gg$ means only a few decibels.

An orthogonal *M*-FSK signal can be demodulated by a matched filter demodulator similar to that shown in Fig. 8.20 but having M bandpass filters and M envelope detectors. The filters are tuned to the individual frequencies of the *M*-ary FSK signal and have bandwidths of $1/T_s$. The detected signals are fed to a signal processing circuit which has the function of selecting the largest signal. Other methods for demodulating *M*-FSK signals are also possible 51].

M-FSK transmission can also be applied to reduce the required transmission bandwidth by selecting a suitably reduced deviation Δf. Evidently, in this case, non-orthogonal FSK modulation is used [52]. The required bandwidth for M-FSK is given by

$$B = \frac{1}{nT_b} + (n-1)\Delta f \tag{8.97}$$

and the error probability is given by

$$P_E \cong \frac{1}{\sqrt{R}} \exp\left[-2R \sin^2\left[\frac{\pi}{2}\Delta f n T_b\right]\right] \tag{8.98}$$

where R is the signal-to-noise ratio at the demodulator input. It follows from Eq. (8.98) that in order to maintain a constant error probability while reducing the bandwidth, the transmitter power has to be increased nearly exponentially. Equation (8.98) can be used to determine the value of $\Delta f n T_b$ resulting in a given error probability P_E at a given signal-to-noise ratio R. Utilizing this value and Eq. (8.97), an optimal value for n, giving minimal transmission bandwidth, can be calculated. Denoting this optimal value by n_0, this calculation leads to the following expression:

$$\Delta f T_b n_0 = \frac{1}{n_0(\ln n_0 - 1) + 1}. \tag{8.99}$$

From Eq. (8.99), the bandwidth efficiency can be calculated:

$$\frac{f_b}{B} = 1.443\left[\ln n_0 + \frac{1}{n_0} - 1\right] \text{bit/Hz} \tag{8.100}$$

where f_b stands for the bit rate.

In practice, $M=4$ or $M=8$ is used for M-FSK, and the optimal value of n is not calculated. Instead of this, the values of $h = 2\Delta f T_s$ are calculated which yield the lowest E_b/N_0 value for a given error probability. Results of these calculations show that for $M=4$, $h=0.8$ and for $M=8$, $h=0.879$ [53].

A non-orthogonal M-FSK signal can be demodulated by the limiter–discriminator type demodulator shown in Fig. 8.21.

8.9 References

[1] Rhodes, J. D.: A low-pass prototype network for microwave linear phase filters. *IEEE Trans. Microwave Theory Techn.*, June, pp. 290–307, 1970.

[2] Reiter, G.: Stripline filters with lumped capacitances. In: *Proc. 5th Europ. Microwave Conf., Hamburg*, pp. 426–430, 1975.

[3] Atia, A. E., Williams, A. E.: Narrow bandpass waveguide filters. *IEEE Trans. Microwave Theory Techn.,* April, pp. 258–265, 1972.

[4] Barber, M. R.: Noise figure and conversion loss of the Schottky barrier diode. *IEEE Trans. Microwave Theory Techn.,* Nov., pp. 629–635, 1967.

[5] Stracca, G. B., Alpesi, F., D'Archangelo, T.: Low-noise microwave down-converter with optimum matching at idle frequencies. *IEEE Trans. Microwave Theory Techn.,* August, pp. 544–547, 1973.

[6] Reiter, G.: An improved microwave diplexer construction for mixers. In: *Proc. 4th Colloq. Microwave Commun., Budapest* 1970.

[7] Faber, M. T., Gwarek, W. K.: Nonlinear–linear analysis of microwave mixers with any number of diodes. *IEEE Trans. Microwave Theory Techn.,* Nov., pp. 1174–1181, 1980.

[8] Stracca, G. B.: Noise in frequency mixer using nonlinear resistors. *Alta Freq.,* Vol. 40, No. 6, pp. 484–505, 1971.

[9] *CCIR Report AE/9*: Choice of intermediate frequencies for high-capacity digital radio systems. 1978.

[10] Lindsey, W. C., Simon, M. K.: Carrier synchronization and detection of polyphase signals. *IEEE Trans. Commun.,* Vol. Com-20, No. 3, pp. 441–454, 1972.

[11] Arturs, E., Dym, H.: On the optimum detection of digital signals in the presence of white Gaussian noise. A geometric interpretation and a study of three basic data transmission systems. *IRE Trans. Commun. Systems,* Dec., pp. 336–372, 1962.

[12] Lindsey, W. C., Simon, M. K.: *Telecommunication Systems Engineering.* Prentice-Hall, Englewood-Cliffs 1973.

[13] Koga, K., Muratani, T., Ogawa, A.: On-board regenerative repeaters applied to digital satellite communications. *Proc. IEEE,* March, pp. 401–410, 1977.

[14] Carel, M., Lainey, G., Labasse, M.: 4-phase PSK differential demodulator using A. S. W. delay lines. *Electron Lett.,* 15th Sept., pp. 586–588, 1977.

[15] Feher, K.: *Digital Communications, Satellite Earth Station Engineering.* Prentice-Hall, Englewood-Cliffs 1983.

[16] Morais, D. H., Feher, K.: NLA-QAM: A new method for generating high-power QAM signals through nonlinear amplification. *IEEE Trans. Commun.,* Vol. Com-30, No. 3, pp. 517–522, 1982.

[17] Nannicini, M., Salerno, M., Vismara, L.: HTN-6u digital high-capacity (140 Mbit/s) microwave system. *Telettra Rev.,* 36, pp. 3–23, 1984.

[18] Dupuis, P., Joindot, M., Leclert, A., Soufflet, D.: 16-QAM modulation for high capacity digital radio system. *IEEE Trans. Commun.,* Vol. Com-27, No. 12, pp. 1771–1781, 1979.

[19] Miyauchi, K., Seiki, S., Ishio, H.: New technique for generating and detecting multilevel signal formats. *IEEE Trans. Commun.,* Vol. Com-24, No. 2, pp. 263–267, 1976.

[20] Simon, M. K., Huth, G. K., Polydoros, A.: Differentially coherent detection of QASK for frequency hopping system. Part. I, *IEEE Trans. Commun.,* Vol. Com-30, No. 1, pp. 158–164, 1982.

[21] Kurematsu, H.: The QAM 2G-10R digital radio equipment using a partial response system. *FUJITSU Sci. Tech. J.,* No. 2, pp. 27–48, 1977.

[22] Austin, M. C., Chang, M. U., Horwood, D. F., Maslov, R. A.: QPSK staggered QPSK and MSK. A comparative evaluation. *IEEE Trans. Commun.,* Vol. Com-31, No. 2, pp. 171–182, 1983.

[23] Prabhu, V. K.: MSK and offset QPSK modulation with bandlimiting filters. *IEEE Trans. Aerosp. Electron. Syst.,* Vol. AES-17, No. 1, pp. 2–8, 1981.

[24] Morais, D. H., Feher, K.: Bandwidth efficiency and probability of error performance of

MSK and offset QPSK systems. *IEEE Trans. Commun.,* Vol. Com-27, No. 12, pp. 1794–1801, 1979.

[25] Ishizuka, M., Yasuda, J.: Improved coherent detection of GMSK. *IEEE Trans. Commun.,* Vol. Com-32, No. 3, pp. 308–310, 1984.

[26] Benett, W. R., Davey, J. R.: *Data Transmission.* McGraw-Hill, New York 1965.

[27] Lucky, R. W., Salz, J., Weldon, E. J.: *Principles of Data Communication.* McGraw-Hill, New York 1968.

[28] Fredricsson, S. A.: Optimum receiver filters in digital quadrature phase-shift-keyed system with a nonlinear repeater. *IEEE Trans. Commun.,* Vol. Com-23, No. 12, pp. 1389–1399, 1975.

[29] Sunde, E. D.: Pulse transmission by AM, FM and PM in the presence of phase distortion. *Bell Syst. Tech. J.,* No. 3, pp. 353–422, 1961.

[30] Papoluis, A.: *The Fourier Integral and its Applications.* McGraw-Hill, New York 1962.

[31] Assal, F.: Approach to a near-optimum transmitter–receiver filter design for data transmission pulse shaping networks. *COMSAT Tech. Rev.,* No. 2, pp. 301–322, 1973.

[32] Borgne, M.: Comparison of high-level modulation schemes for high-capacity digital radio systems. *IEEE Trans. Commun.,* Vol. Com-33, No. 5, pp. 442–449, 1985.

[33] Clandrino, L., Crippa, G.: Digital radio relay systems with MSK modulation. *Alta Freq.,* Febr., pp. 91–97, 1976.

[34] Yamamoto, H., Morita, K., Komaki, S.: QPSK system error rate performance. *Rev. Electr. Commun. Lab.,* Vol. 25, No. 5–6, pp. 515–526, 1977.

[35] Morita, T., Hayashi, M., Ogawa, K.: 11/15 GHz band digital radio equipment. *FUJITSU Sci. Techn. J.* Vol. 12, No. 2, pp. 41–64, 1976.

[36] Jones, J.: Filter distortion and intersymbol interference effects on PSK signals. *IEEE Trans. Commun.,* Vol. Com-19, No. 2, pp. 120–132, 1971.

[37] Farell, E., Ntake, P. L.: 90 Mbit/s digital performance of Canada's 14/12 GHz ANIK C earth stations. *IEEE Trans. Commun.,* Vol. Com-29, No. 10, pp. 1502–1513, 1981.

[38] Swartz, M., Benett, W. R., Stein, S.: *Communication Systems and Techniques.* McGraw-Hill, New York 1966.

[39] Benett, W. R., Salz, J.: Binary data transmission by FM over a real channel. *Bell Syst. Tech. J.,* Sept., pp. 2387–2426, 1963.

[40] Sunde, E. D.: Ideal pulses transmitted by AM and FM. *Bell Syst. Tech. J.,* Nov., pp. 1375–1426, 1959.

[41] Cattermole, K. W.: Digital transmission by frequency shift keying with zero intersymbol interference. *Electron Lett.,* Vol. 10, No. 17, 22th Aug., pp. 349–350, 1974.

[42] Benedek, A.: Noncoherent reception of digital frequency modulation (in Hungarian). TKI Anniversary Conf., Budapest 1985.

[43] Mazo, J. E., Salz, J.: Theory of error rates for digital FM. *Bell Syst. Tech. J.,* Nov., pp. 1511–1535, 1966.

[44] Pawula, R. F.: On the theory of error rate for narrow-band digital FM. *IEEE Trans. Commun.,* Vol. Com-29, No. 11, pp. 1634–1643, 1981.

[45] Tjhung, T. T., Wittke, P. H.: Carrier transmission of binary data in restricted band. *IEEE Trans. Commun.,* Vol. Com-18, No. 4, pp. 295–304, 1970.

[46] Tjhung, T. T., Singh, H.: Performance of narrow-band Manchester coded FSK with discriminator detection. *IEEE Trans. Commun.,* Vol. Com-31, No. 5, pp. 659–667, 1983.

[47] Cartier, D. E.: Limiter–discriminator detection performance of Manchester and NRZ coded FSK. *IEEE Trans. Aerosp. Electron. Syst.,* Vol. AES-13, Jan., pp. 62–70, 1977.

[48] Simon, M. K., Wang, C.: Differential versus limiter–discriminator detection of narrow-band FM. *IEEE Trans. Commun.,* Vol. Com-31, No. 11, pp. 1227–1234, 1983.

[49] Simon, M. K., Wang, C.: Differential detection of Gaussian MSK in a mobile radio environment. *IEEE Trans. Vehicular Technol.*, Vol. VT-33, No. 4, Nov., pp. 307–320, 1984.

[50] Lindsey, W. C., Simon, M. K.: Detection of digital FSK and PSK using a first-order phase locked loop. *IEEE Trans. Commun.*, Vol. Com-25, No. 2, 200–214, 1977.

[51] Schonhoff, T. A.: Symbol error probabilities for *M*-ary CPFSK coherent and noncoherent detection. *IEEE Trans. Commun.*, Vol. Com-24, No. 6, pp. 644–652, 1976.

[52] Mazo, J. E., Rowe, H. E., Salz, J.: Rate optimization for digital frequence modulation. *Bell Syst. Techn. J.*, Nov., pp. 3021–3030, 1969.

[53] Yue, O.: Performance of frequency-hopping multiple-access multilevel FSK systems with hard-limited and linear combining. *IEEE Trans. Commun.*, Vol. Com-29, No. 11, pp. 1687–1694, 1981.

[54] Ekanayeke, N.: On differential detection of binary FM. *IEEE Trans. Commun.*, Vol. Com-32, No. 4, pp. 469–470, 1984.

CHAPTER 9

CARRIER RECOVERY CIRCUITS

It was shown in Section 2.1.2 that coherent detection is more efficient than non-coherent detection, and most of the demodulator types discussed in Chapter 8 are coherent (PSK, MSK, QAM). For coherent demodulators, a reference signal has to be generated from the received modulated signal, even though this is normally a suppressed carrier signal, and the recovered carrier should have a phase which is suitable for the demodulator applied.

The transmit side carrier which has to be recovered at the receive side is given by

$$V_c(t) = \sqrt{2} \cos [\omega_0 t + \Phi_t] \tag{9.1}$$

where ω_0 is the carrier frequency and Φ_t is the initial phase of the carrier. In general, the modulated signal comprises both amplitude modulation and phase modulation:

$$V(t) = c(t) \cos [\omega_0 t + \varphi(t) + \Phi_t] \tag{9.2}$$

where $c(t)$ is the amplitude modulation function and $\varphi(t)$ is the phase modulation function.

On the received signal, Gaussian noise is superimposed, which can be resolved into in-phase and quadrature components with respect to the carrier:

$$n(t) = n_c(t) \cos (\omega_0 t + \Phi_t) - n_s(t) \sin (\omega_0 t + \Phi_t). \tag{9.3a}$$

The carrier recovery circuit has the function of generating from the incoming modulated, noisy signal, a reference signal given by

$$V_r(t) = \sqrt{2} \cos [\omega_0 t + \Phi_R] \tag{9.3b}$$

where Φ_R is the initial phase suitable for the demodulator. For carrier recovery, a nonlinear process is applied. In the following, the carrier recovery circuits applied for differing modulation schemes will be investigated.

9.1 M-PSK transmission

9.1.1 Mth power law device

For M-PSK transmission, the phase function in Eq. (9.2) can be expressed as

$$\varphi_M(t) = \frac{2\pi}{M} n; \quad n = 0, 1, ..., M-1. \tag{9.4}$$

The amplitude being constant, the modulated signal is given by

$$V(t) = \sqrt{2}\, A \cos\left(\omega_0 t + \frac{2\pi}{M} n + \Phi_t\right). \tag{9.5}$$

Let this signal pass through a nonlinear circuit having input–output characteristic $y = x^M$ where x is the input voltage and y is the output voltage. Several frequency components will appear at the output from which a filter is used to select the component of frequency $M\omega_0$:

$$V_0(t) = \frac{A^M 2^{M/2}}{2^{M-1}} \cos(M\omega_0 t + 2\pi n + M\Phi_t). \tag{9.6}$$

This can be simplified by omitting the phase term $2\pi n$ because of the periodicity of the cosine function:

$$V_0(t) = \frac{A^M 2^{M/2}}{2^{M-1}} \cos(M\omega_0 t + M\Phi_t). \tag{9.7}$$

It is thus seen that the phase modulation of the signal of frequency $M\omega_0$ has been canceled by the Mth power law device. The reference signal of frequency ω_0 is obtained by driving a frequency divider by M with this signal.

Figure 9.1 shows a few variants of this type of carrier recovery circuit. The principle of operation is shown in Fig. 9.1a; the slicer inserted between the band-pass filter and the frequency divider has the function of driving the frequency divider by a square-wave signal. PLL-type filtering is used extensively, as illustrated by the variants of Figs 9.1b and 9.1c. In Figure 9.1b, the VCO is operated at frequency $M\omega_0$, and is followed by the frequency divider while in Fig. 9.1c, the VCO is operated directly at the carrier frequency ω_0, and the VCO signal is multiplied by a second multiplier for frequency comparison.

The recovered carrier may have M different phases, differing from each other by $2\pi/M$:

$$V_r(t) = \sqrt{2} \cos\left(\omega_0 t + \Phi_t + \frac{2\pi}{M} k\right). \tag{9.8}$$

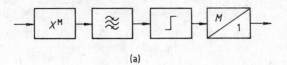

(a)

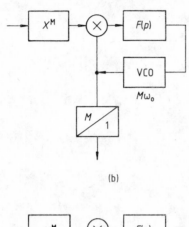

(b)

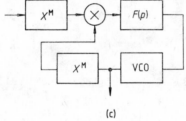

(c)

Fig. 9.1 (a), (b), (c). Variants of *M*th power law carrier recovery circuit.

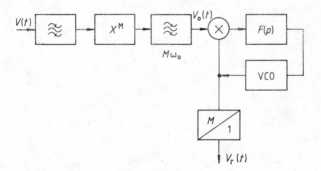

Fig. 9.2. Model of *M*th power law carrier recovery circuit.

This phase ambiguity of the reference signal can be eliminated by differential encoding at the transmit side and differential decoding at the receive side.

The preceding investigations apply for a noise-free input signal. Figure 9.2 shows a model to be used for investigations assuming a noisy input signal which can be expressed by applying Eqs (9.2) and (9.3):

$$V(t) = \sqrt{2} \, A \cos(\omega_0 t + \varphi_M(t) + \Phi_t) + n_c \cos(\omega_0 t + \Phi_t) - n_s \sin(\omega_0 t + \Phi_t). \quad (9.9)$$

This can be re-written as

$$V(t) = r(t) \cos\left[\omega_0 t + \frac{2\pi}{M} n + \Phi_t + \varphi_n(t)\right] \quad (9.10)$$

where

$$r(t) = \{[\sqrt{2} \, A + n_c(t)]^2 + n_s^2(t)\}^{1/2} \quad (9.11)$$

and

$$\varphi_n(t) = \tan^{-1} \frac{n_s(t)}{\sqrt{2} \, A + n_c(t)}. \quad (9.12)$$

At the output of the Mth power law device, the component of frequency $M\omega_0$ which is utilized is given by

$$V_0(t) = \frac{r^M(t)}{2^{M-1}} \cos[M\omega_0 t + M\Phi_t + M\varphi_n]. \quad (9.13)$$

The signal-to-noise ratio can be expressed in a normalized form for a signal of frequency ω_0:

$$R_0 = \frac{M R^M}{(M-1)! \sum_{i=0}^{M-1} \binom{M}{i} \dfrac{R^i}{i!}} \quad (9.14)$$

where

$$R = \frac{A^2}{BN_0}. \quad (9.15)$$

Here B denotes the noise bandwidth of the filter preceding the demodulator, and N_0 denotes the one-sided power density of the noise. From a knowledge of the input signal-to-noise ratio R, the signal-to-noise ratio at frequency $M\omega_0$ can be calculated through the expression

$$R_M = \frac{R_0}{M^2}. \quad (9.16)$$

By using linear approximation, and assuming unity signal power, the variance of the phase error which is characteristic for the PLL loop phase noise is given by

$$\sigma_\varphi^2 = \frac{1}{\varrho_L} \quad (9.17)$$

where ϱ_L is the signal-to-noise ratio in the PLL loop. The subscript φ refers to $\Phi_t - \Phi_R$, the phase difference between the input and reference signals. The VCO is operated at frequency $M\omega_0$ so the signal-to-noise ratio in the PLL loop can be calculated by applying Eq. (9.16):

$$\varrho_{LM} = \frac{B}{B_L} R_M = \frac{B}{B_L} \frac{R_0}{M^2} \tag{9.18}$$

where B_L denotes the one-sided noise bandwidth of the PLL loop. It thus follows that the variance of the phase error at frequency $M\omega_0$ is given by

$$\sigma_{M\varphi}^2 = \frac{B_L}{B} \frac{M^2}{R_0}. \tag{9.19}$$

The operation of the PLL loop, together with the relevant relations, has been discussed in Chapter 3. The basic PLL equation is reproduced here:

$$M\dot{\varphi} = M\Delta\omega - KF(p)[2^{(1-M)/2}A^M \sin M\varphi + N(\varphi, t)] \tag{9.20}$$

where $M\Delta\omega$ is the difference between the multiplied signal frequency and the VCO free running frequency, $M\varphi$ is the phase of the VCO signal, $F(p)$ is the transfer function of the loop filter, and K is the overall loop gain.

The probability density function of the phase distribution is given by

$$p(M\varphi) = \frac{\exp(\varrho \cos M\varphi)}{2\pi I_0(\varrho)}; \quad -\pi < M\varphi < \pi \tag{9.21}$$

where I_0 is the modified Bessel function of zero order. In this expression, $\varrho = \varrho_{LM}$ so the phase distribution and the cycle slip will be determined by R_M. According to Eq. (9.16), this is less than R_0, the difference being 6 dB for frequency doubling and 12 dB for frequency quadrupling.

The VCO control voltage as a function of $M\varphi$ has a periodicity of 2π, and as a function of φ, a periodicity of $2\pi/M$. The variance of the reference signal phase error, at frequency ω_0, is given by

$$\sigma_{\varphi}^2 = \frac{B_L}{B} \frac{(M-1)!}{MR^M} \sum_{i=0}^{M-1} \binom{M}{i} \frac{R^i}{i!}. \tag{9.22}$$

For 2PSK transmission $M=2$, the variance is given by

$$\sigma_{\varphi}^2 = \frac{B_L}{B} \frac{1}{R} \left[1 + \frac{1}{2R}\right] \tag{9.23}$$

and for 4PSK transmission $M=4$, by

$$\sigma_{\varphi}^2 = \frac{B_L}{B} \frac{1}{R} \left[1 + \frac{9}{2} \frac{1}{R} + \frac{6}{R^2} + \frac{3}{2R^3}\right]. \tag{9.24}$$

It is seen from the above expressions that through the processes of frequency multiplication and division, the variance is increased by the bracketed factor. The degradation can also be expressed by

$$\sigma_\varphi^2 = \frac{1}{\varrho S_L} = \frac{B_L}{B} \frac{1}{R S_L} \tag{9.25}$$

where S_L is called "squaring loss", and is given by the reciprocal value of the bracketed expressions in Eqs (9.23) and (9.24). The degradation is due to the intermodulation products of the signal and noise components generated during the process of frequency multiplication.

If the model shown in Fig. 9.2 had a limiter following the frequency multiplier, this would have the function of presenting a constant input signal level to the PLL and would result in a constant pull-in range, noise bandwidth and phase error which depend on the signal amplitude. Another function of the limiter is the elimination of excess driving voltages and thus achieving a constant operation point of the multiplier acting as the PLL phase detector. On the other hand, a so-called small signal suppression is caused by the limiter at low signal-to-noise ratios, resulting in a degradation in the signal-to-noise ratio. The signal-to-noise ratio can also be improved at high signal-to-noise ratios with a carrier recovery circuit having an x^M characteristic.

9.1.2 Taking the absolute value

Figure 9.3 shows a carrier recovery circuit for 2PSK transmission by taking the absolute value: the input signal is sliced by a limiter, and the sliced signal is multiplied by the original signal [11]. The waveform of the sliced signal is given by

$$V_0(t) = \frac{4}{\pi} \sum_{k=0}^{\infty} \frac{(-1)^k}{2k+1} \cos\{(2k+1)[\omega_0 t + \varphi(t) + \varphi_n(t)]\}. \tag{9.26}$$

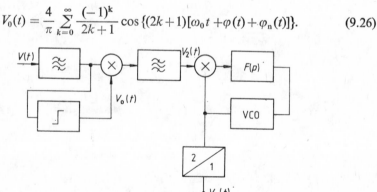

Fig. 9.3. Carrier recovery by taking the absolute value.

It is seen that the sliced signal is made up of only odd harmonics. Multiplying the sliced signal by the original one and filtering out the frequency component of $2\omega_0$ yields the following output signal:

$$V_2(t) = \frac{4}{3\pi} r(t) \cos\left[2\omega_0 t + 2\varphi_M(t) + 2\varphi_n(t)\right]. \tag{9.27}$$

From this signal, the reference signal is derived by bandpass filtering and frequency division. The small signal suppression of the limiter is again present, resulting in a higher phase noise at small signal-to-noise ratios compared with the square-law device. This type of carrier recovery is also applicable to 4PSK transmission because the sliced signal also comprises a third harmonic component.

9.1.3 Costas loop

Let us first investigate a Costas loop realized by analog circuits, and suitable for 2PSK transmission (see Fig. 9.4). The noisy input signal can be expressed in the following form:

$$V(t) = \sqrt{2}\, Aa(t) \sin\left(\omega_0 t + \Phi_t\right) + \tag{9.28}$$
$$+ \sqrt{2}\, n_c(t) \cos\left(\omega_0 t + \Phi_t\right) + \sqrt{2}\, n_s(t) \sin\left(\omega_0 t + \Phi_t\right)$$

where $a(t) = \pm 1$. Equation (9.3b), expressing the reference signal, is reproduced here:

$$V_r(t) = \sqrt{2} \cos\left(\omega_0 t + \Phi_R\right). \tag{9.29}$$

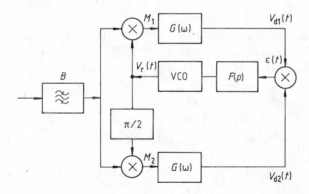

Fig. 9.4. Costas loop.

By introducing the notation $\varphi = \Phi_t - \Phi_R$, the two components of the demodulated 2PSK signal are given by

$$V_{d1}(t) = [Aa(t) + n_s(t)] \sin \varphi + n_c(t) \cos \varphi$$
$$V_{d2}(t) = [Aa(t) + n_s(t)] \cos \varphi - n_c(t) \sin \varphi. \tag{9.30}$$

The error voltage appearing at the output of multiplier M_3 in Fig. 9.4 is expressed by

$$\varepsilon(t) = V_{d1}(t)V_{d2}(t) = \frac{1}{2} \{[Aa(t) + n_s(t)]^2 - n_c^2(t)\} \sin 2\varphi +$$
$$+ [Aa(t) + n_s(t)] n_c(t) \cos 2\varphi. \tag{9.31}$$

Performing the operations, we have

$$\varepsilon(t) = \frac{1}{2} A^2 a^2 \sin 2\varphi + [n_s^2 + n_s Aa - n_c^2] \sin 2\varphi + [n_c Aa + n_c n_s] \cos 2\varphi. \tag{9.32}$$

The first term represents the desired signal, and the other two terms stand for the two components of the noise. The well known fundamental equation of the PLL loop is the following:

$$2\dot{\varphi} = 2\Delta\omega - KF(p)[A^2 \sin 2\varphi + N_1(t, \varphi)] \tag{9.33}$$

where $\Delta\omega$ is the difference between the incoming signal frequency ω_0 and the free running VCO frequency ω_r, K is the loop gain, and $F(p)$ is the loop transfer function. In the bracketed expression, the signal and noise are represented by the first and second terms, respectively.

It can be shown that the substitution of $M = 2$ will transform Eq. (9.20) into Eq. (9.33) [4]. The variance of the phase error is given by

$$\sigma_\varphi^2 = \frac{B_L}{2G} \frac{1}{R} \left[1 + \frac{1}{2R}\right] \tag{9.34}$$

where G is the noise bandwidth of the low-pass filter, in the case when $B \gg 2G$. If the bandwidth is determined by B, i.e. $2G \gg B$, then B has to be replaced instead of $2G$. In this case, Eq. (9.34) is even formally identical to Eq. (9.23). The control voltage is proportional to $\sin 2\varphi$ so the control voltage/phase function has a periodicity of π.

A practical problem of the Costas loop is the drift in the multipliers and the DC amplifier. A solution to this problem is the use of a limiter in the branch V_{d2}, resulting in $V'_{d2} = \text{sign}(V_{d2})$, and the application of a switching type multiplier to multiply V_{d1} by V'_{d2}. This results in improved stability in the case of a shifting operation point [13]. At high signal-to-noise ratios, the Costas loop with limiter and the normal Costas loop according to Fig. 9.4 give identical results. At low signal-to-noise ratios, there will be a degradation of 0.8 to 1 dB due to the small signal suppression.

In a Costas loop, false lock may be generated by a distorted signal waveform. Investigations have shown that false lock, *i.e.* a condition in which $f_{in} \neq f_{VCO}$, may be generated whenever $f_{in} - f_{VCO} = n/2T = (f_b/2)n$, *i.e.* the difference between the input signal frequency and the VCO frequency is a multiple of the halved bit frequency. Difficulties can arise when the initial frequency difference is higher than this value. While the VCO frequency is nearing the input frequency during the pull-in procedure, a false lock may take place at the first of these frequencies [14].

There are several ways of indicating the false lock. One method is to apply a DC comparator since the control voltage at false lock is smaller. Another method is to indicate the AC component which appears at the multiplier output in the case of false lock. The false lock state can be eliminated by superimposing on the control voltage a pulse which is generated by the false lock indication signal.

The Costas loop for carrier recovery can also be applied for *M*-PSK transmission [17].

In the following, a four-phase Costas loop, realized by digital circuits and shown in Fig. 9.5, will be investigated for a noise-free input signal. The incoming 4PSK signal is given by

$$V(t) = \sqrt{2} \cos\left(\omega_0 t + n\frac{2\pi}{4} + \Phi_t\right) \tag{9.35}$$

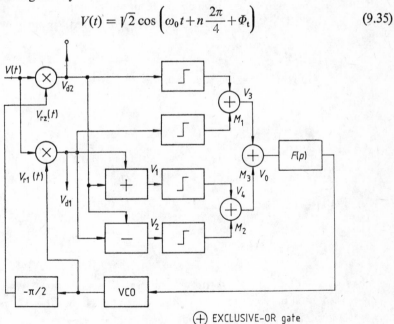

Fig. 9.5. Baseband carrier recovery utilizing digital circuits.

where $n = 0, 1, 2, 3$. The two quadrature components of the reference signal to be generated are given by

$$V_{r1}(t) = \sqrt{2} \cos (\omega_0 t + \Phi_R)$$
$$V_{r2}(t) = -\sqrt{2} \sin (\omega_0 t + \Phi_R). \tag{9.36}$$

Let us introduce the following notation:

$$\varphi = \Phi_t - \Phi_R \tag{9.37}$$

$$\Phi = n \frac{2\pi}{4} + \varphi. \tag{9.38}$$

The two components of the demodulated signal are given by

$$V_{d1}(t) = \cos \Phi$$
$$V_{d2}(t) = \sin \Phi. \tag{9.39}$$

The signals appearing at the combining and subtracting circuits in Fig. 9.5 are expressed by

$$V_1 = \cos \Phi + \sin \Phi$$
$$V_2 = \cos \Phi - \sin \Phi. \tag{9.40}$$

The above four signals are used to drive limiters, and the sliced signals are input to the EXCLUSIVE-OR gates M_1 and M_2 which perform the multiplication of the signals:

$$V_3 = \text{sign } V_{d1} \oplus \text{sign } V_{d2} = \text{sign } (V_{d1} V_{d2}). \tag{9.41}$$

Simple trigonometrical relations lead to the following expression:

$$V_3 = \text{sign } (\sin 2\Phi) \tag{9.42}$$

and similarly,

$$V_4 = \text{sign } (V_1 V_2) = \text{sign } (\cos 2\Phi). \tag{9.43}$$

The VCO control voltage at the input of the low-pass filter is given by

$$V_0 = \text{sign } (\sin 4\Phi). \tag{9.44}$$

By taking into account the ralation

$$4\Phi = n \cdot 2\pi + 4\varphi = 4\varphi \tag{9.45}$$

we have

$$V_0 = \text{sign } (\sin 4\varphi). \tag{9.46}$$

This control voltage is passed through the low-pass filter for driving the VCO which is used to supply the recovered carrier. It is seen that the control voltage does not comprise the modulation components but only the phase of the carrier as required.

Equation (9.46) shows that the control voltage/phase function has a square-wave form, with a periodicity $\pi/2$. The delay within the PLL loop may give rise to oscillations which can be eliminated if the comparators are realized by suitable hysteresis.

It can be shown [26] that the condition for loop stability is given by

$$\delta > \frac{2A_d K_{v0} \tau_2 T_0}{\pi \tau_1} \qquad (9.47)$$

where δ is the hysteresis, expressed in phase values, K_{v0} is the control slope of the VCO, T_0 is the delay in the PLL loop, τ_1, τ_2 are the time constants of the second-order loop filter, A_d is the signal amplitude at the output of the exclusive-or gate M_3.

9.1.4 Decision feedback loop

In the following the two-phase and four-phase loops will be investigated. A general treatment of the *M*-ary loop is also possible [4, 20, 21].

9.1.4.1 Two-phase loop

The designation is explained by the decision circuit applied in the PLL loop which has the function of generating the sampling signal and carrying out the sampling process. The two-phase loop shown in Fig. 9.6 is similar to the Costas loop. The decision circuit is in the lower (in-phase) branch, and the signal thus obtained is multiplied by the demodulated signal obtained in the quadrature branch. The decision time of one bit is taken into account by the T_b delay circuit

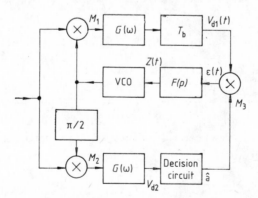

Fig. 9.6. Two-phase decision feedback loop.

in the other branch. The signals denoted by V_{d1} and V_{d2} are those obtained by the Costas loop and given in Eq. (9.30). The output signal of the decision circuit is given by

$$V'_{d2} = \hat{a} \tag{9.48}$$

where $\hat{a}$ is the output voltage of the decision circuit. The output signal of the other branch is given by

$$V_{d1} = (Aa + n_s) \sin \varphi + n_c \cos \varphi \tag{9.49}$$

where A is the signal amplitude, and n_c is the amplitude of the cosine noise component. The control voltage at the output of multiplier M_3 is then

$$\varepsilon(t) = \hat{a} V_{d1}. \tag{9.50}$$

Disregarding first the delay T_b, this can be expressed as

$$\varepsilon(t) = Aa\hat{a} \sin \varphi + \hat{a}n_s \sin \varphi + \hat{a}n_c \cos \varphi. \tag{9.51}$$

The loop bandwidth is much lower than the bit frequency allowing $a\hat{a}$ to be substituted by its expectation value:

$$E\{a\hat{a}\} = P\{a = \hat{a}\} - P\{a \neq \hat{a}\} = 1 - 2P_E. \tag{9.52}$$

Substituting this into (9.51) and taking into account the delay and the loop filter response $F(p)$, the VCO control voltage is obtained:

$$Z(t) = F(p) \exp(-pT_b)\{A(1 - 2P_E) \sin \varphi + \hat{a}N(\varphi, t)\} \tag{9.53}$$

where

$$N(\varphi, t) = n_c(t) \cos \varphi + n_s(t) \sin \varphi. \tag{9.53a}$$

Taking into account the considerations regarding the Costas loop, the fundamental equation of the PLL loop can be expressed as

$$\dot{\varphi} = \Delta\omega - KF(p)\{A(1 - 2P_E) \sin \varphi + \hat{a}N(\varphi, t)\}. \tag{9.54}$$

It can be shown that neither the spectral density nor the distribution of $N(\varphi, t)$ will be affected by the factor $\hat{a}$. This means that the above equation is identical with Eq. (9.20) which applies to a VCO at frequency $M\omega_0$, after substituting $M = 1$. The only difference is the fact that because of the finite error probability, the amplitude of the VCO control is decreased in proportion to $(1 - 2P_E)$.

Assuming linear approximation, the variance of the phase error can be determined:

$$\sigma_\varphi^2 = \frac{N_0 B_L}{A^2(1 - 2P_E)^2} = \frac{B_L}{B} \frac{1}{R} \frac{1}{(1 - 2P_E)^2}. \tag{9.55}$$

Comparing this expression with (9.23) derived for the second power law device or the Costas loop, it is seen that the phase noise of the decision feedback system is lower.

The control voltage/phase characteristic can be calculated by solving the stochastic differential equation (9.54). The characteristic has a periodicity of 180° which gives a phase ambiguity of the same value [4]. The noise bandwidth of a first-order PLL loop is given by

$$W_L = \frac{AK \, \text{erf} \, \sqrt{R}}{2} \tag{9.56}$$

where, by definition,

$$\text{erf} \, x = \frac{2}{\sqrt{\pi}} \int_0^x e^{-t^2} dt. \tag{9.56a}$$

It is seen that the noise bandwidth depends on the signal-to-noise ratio R. This is explained by the fact that the control voltage amplitude depends on the error probability and thus on the signal-to-noise ratio.

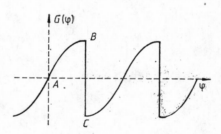

Fig. 9.7. Phase characteristic of two-phase decision feedback loop.

The control voltage/phase characteristic is shown in Fig. 9.7. Point A corresponds to the condition in which the reference signal phase is equal to the received signal phase. The control voltage is increased by increasing phase difference, having a value highest at $\varphi=90°$ (point B). At this point, the control voltage changes sign, and the control voltage jumps to point C which is the starting point of another sine wave. A further increase of φ will move the control voltage along this sine wave. This figure also shows the control voltage/phase periodicity of π, the stable operation points being at 0, π, 2π, etc.

9.1.4.2 Four-phase loop

It has been shown in Section 8.2.3 that the demodulation of a 4PSK signal yields two demodulated signals (channels "*a*" and "*b*"). In this case, the decision has to be performed in both branches as shown in Fig. 9.8. The received noisy signal at the input of the carrier recovery circuit is given by

$$V(t) = Aa(t)\cos(\omega_0 t + \Phi_t) - Ab(t)\sin(\omega_0 t + \Phi_t) +$$
$$+ n_c(t)\cos(\omega_0 t + \Phi_t) - n_s(t)\sin(\omega_0 t + \Phi_t) \tag{9.57}$$

where

$$A = \sqrt{P}, \quad a(t) = \pm 1, \quad b(t) = \pm 1,$$

P denoting the received signal power at the input of the carrier recovery circuit. The two reference signals to be generated are given by

$$V_{r1}(t) = 2\cos(\omega_0 t + \Phi_R) \tag{9.58}$$

$$V_{r2}(t) = -2\sin(\omega_0 t + \Phi_R). \tag{9.59}$$

The two demodulated signals are expressed by

$$V_{d1}(t) = [Aa(t) + n_c(t)]\cos\varphi - [Ab(t) + n_s(t)]\sin\varphi \tag{9.60}$$

$$V_{d2}(t) = [Ab(t) + n_s(t)]\cos\varphi + [Aa(t) + n_c(t)]\sin\varphi. \tag{9.61}$$

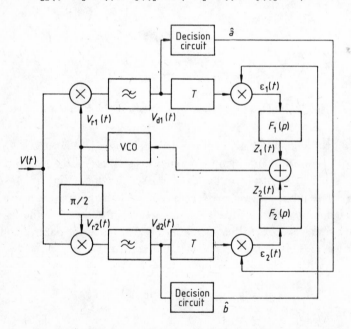

Fig. 9.8. Four-phase decision feedback loop.

Expressing the decision results by $\hat{a}$ and $\hat{b}$, the control voltages at the input ports of the low-pass filters are given by

$$\varepsilon_1(t) = V_{d1}\hat{b} = Aa\hat{b} \cos \varphi - Ab\hat{b} \sin \varphi + n_c \hat{b} \cos \varphi - n_s \hat{b} \sin \varphi \qquad (9.62)$$

$$\varepsilon_2(t) = V_{d2}\hat{a} = Ab\hat{a} \cos \varphi + Aa\hat{a} \sin \varphi + n_s \hat{a} \cos \varphi + n_c \hat{a} \sin \varphi. \qquad (9.63)$$

By taking into account the delays and the loop filters, the two input voltages of the summing network will be given by

$$Z_1(t) = F_1(p) \exp(-pT) \times$$
$$\times [Aa\hat{b} \cos \varphi - Ab\hat{b} \sin \varphi + n_c \hat{b} \cos \varphi - n_s \hat{b} \sin \varphi] \qquad (9.64)$$

$$Z_2(t) = F_2(p) \exp(-pT) \times$$
$$\times [Ab\hat{a} \cos \varphi + Aa\hat{a} \sin \varphi + n_c \hat{a} \sin \varphi + n_s \hat{a} \cos \varphi]. \qquad (9.65)$$

In practical cases, the PLL bandwidth is much less than the symbol frequency, allowing substitution of the expectation values instead of the products $a\hat{b}$ and $b\hat{b}$:

$$E\{a\hat{a}\} = 1 - 2P_{E2}, \quad E\{b\hat{b}\} = 1 - 2P_{E1}, \quad E\{\hat{a}b\} = 0, \quad E\{\hat{b}a\} = 0. \quad (9.66)$$

The fundamental PLL equation, similar to the above, is the following:

$$\dot{\varphi} = \Delta\omega - K_1 F_1(p)[A(1 - 2P_{E1}) \sin \varphi + N_1(\varphi, t)] -$$
$$- K_2 F_2(p)[A(1 - 2P_{E2}) \sin \varphi + N_2(\varphi, t)]. \qquad (9.67)$$

The value of expressions $N_1(\varphi, t)$ and $N_2(\varphi, t)$ can be calculated from Eqs (9.64) and (9.65). K_1 and K_2 denote the loop gains. In practical cases, the two loops are identical so that $K_1 = K_2 = K$, resulting in the following equation:

$$\dot{\varphi} = \Delta\omega - K2F(p)[A(1 - 2P_E) \sin \varphi + N(\varphi, t)]. \qquad (9.68)$$

The spectral density and the distribution of $N(\varphi, t)$ is obtained from Eq. (9.21) with the substitution of $M = 1$. For linear approximation, the variance of the phase error is given by

$$\sigma_\varphi^2 = \frac{N_0 B_L}{A^2(1 - 2P_E)^2} = \frac{B_L}{B} \frac{1}{R} \frac{1}{(1 - 2P_E)^2}. \qquad (9.69)$$

It was shown in Section 2.1.3 that for a given signal-to-noise ratio, the error probability is higher for 4PSK than for 2PSK. The phase noise of the four-phase decision feedback loop, as compared with the two-phase loop, is higher in the same proportion. A shortcoming of the decision feedback system is the delay required in the PLL loop, which has the effect of decreasing the pull-in range.

9.1.5 Inverse modulator and remodulator

In digital microwave radio equipment, the reference signal is frequently obtained by the inverse modulation or remodulation method. In principle, the two methods are identical, differing only in the practical realization. In both systems, the demodulator is followed by a modulator. According to the inverse modulation method shown in Fig. 9.9a, the incoming signal is demodulated, sliced, and the sliced signals are used to modulate the incoming signal so as to cancel the phase modulation. The phase discriminator M is used to compare this unmodulated signal with the VCO signal, and supplies the control voltage to the VCO. According to the remodulation method shown in Fig. 9.9b, the recovered reference signal, supplied by the VCO, is modulated by the demodulated signal, and the VCO control voltage is obtained from the comparison of the modulated signal thus obtained, and the incoming signal [23, 24, 25, 26]. In the following, only the four-phase inverse modulator will be investigated in detail (Fig. 9.10).

The modulation and slicing of the signal introduce a delay of T_1. The modulator must therefore be preceded by a suitable delay T_2 in order to ensure that the modulation takes place at a suitable time instant. Similarly, the VCO signal

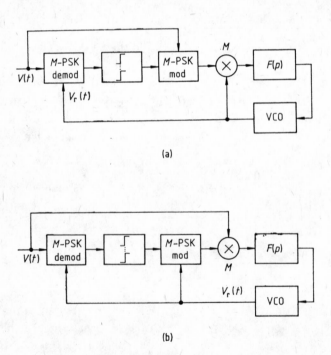

(a)

(b)

Fig. 9.9. (a) Inverse modulator; (b) remodulator.

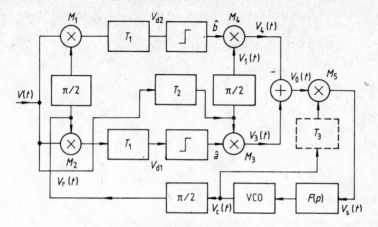

Fig. 9.10. Four-phase inverse modulator.

must also be delayed by a suitable time interval T_3 to obtain a correct phase comparison. The condition for correct operation is given by $T_1=T_2=T_3$.

The input signal is expressed in the conventional manner:

$$V(t) = [Aa(t)+n_c(t)] \cos \omega_0 t + [Ab(t)+n_s(t)] \sin \omega_0 t. \tag{9.70}$$

The reference signal to be generated is given by

$$V_r(t) = 2 \cos (\omega_0 t + \Phi_R) \tag{9.71}$$

where $\Phi_R=0$ for a correct reference signal phase.

The two demodulated signals are

$$V_{d1}(t) = Aa(t)+n_c(t) \tag{9.72}$$

$$V_{d2}(t) = Ab(t)+n_s(t) \tag{9.73}$$

and the two comparator output signals are

$$\hat{a} = \text{sign} [Aa(t)+n_c(t)]$$

$$\hat{b} = \text{sign} [Ab(t)+n_s(t)]. \tag{9.74}$$

The input signal drives the multiplier M_3 directly and the multiplier M_4 *via* a delay of $\pi/2$:

$$V_1(t) = [Aa(t)+n_c(t)] \sin \omega_0 - [Ab(t)+n_s(t)] \cos \omega_0 t. \tag{9.74a}$$

The multiplier output signals are given by

$$V_3(t) = [Aa(t)\hat{a}+\hat{a}n_c(t)] \cos \omega_0 t + [Ab(t)\hat{a}+\hat{a}n_s(t)] \sin \omega_0 t$$

$$V_4(t) = [Aa(t)\hat{b}+\hat{b}n_c(t)] \sin \omega_0 t - [Ab(t)\hat{b}+\hat{b}n_s(t)] \cos \omega_0 t. \tag{9.74b}$$

The difference between these two signals is generated by the circuit in Fig. 9.10:

$$V_0(t) = A[a\hat{a} + b\hat{b}] \cos \omega_0 t + A[\hat{a}b - b\hat{a}] \sin \omega_0 t +$$
$$+ [\hat{a}n_c + \hat{b}n_s] \cos \omega_0 t + [\hat{a}n_s - \hat{b}n_c] \sin \omega_0 t. \tag{9.74c}$$

The first term gives the desired signal, the second term gives a quadrature signal, while the third and fourth terms represent thermal noise. In practical cases, the PLL bandwidth is much less than the symbol frequency so instead of the products $a\hat{a}$, etc., their expected values can be used for the calculation:

$$E\{a\hat{a}\} = 1 - 2P_E, \quad E\{b\hat{b}\} = 1 - 2P_E, \quad E\{a\hat{b}\} = 0, \quad E\{b\hat{a}\} = 0. \tag{9.74d}$$

Substituting these into Eq. (9.74c), we have

$$V_0(t) = 2A(1 - 2P_E) \cos \omega_0 t + N(t, \varphi) \tag{9.75}$$

where $N(\varphi, t)$ can be calculated from Eq. (9.74c).

Let us express the variance of the second term:

$$\sigma_x^2 = E\{x^2\} - E^2\{x\}. \tag{9.76}$$

Substituting the expressions given in (9.74d), we have the variance due to the so-called pattern jitter:

$$\sigma_p^2 \cong 4A^2 P_E. \tag{9.77}$$

The relative signal power is given by $P_s = A^2(1 - 2P_E)^2$; the relative thermal noise power $P_n = (n_c^2 + n_s^2)$; and the relative noise power due to modulation by $P_{sn} = A^2 4P_E$.

In order to write down the fundamental equation of the PLL loop, the expression of the VCO signal is required:

$$V_c(t) = -\sin(\omega_0 t + \Phi_R). \tag{9.78}$$

The input signal to the multiplier M_5, in the noise-free case, is given by

$$V_0(t) = 2A(1 - 2P_E) \cos \omega_0 t \tag{9.79}$$

and the phase discriminator output signal is given by

$$V_s(t) = A(1 - 2P_E) \sin \varphi \tag{9.80}$$

where $\varphi = -\Phi_R$ because of the assumed value of $\Phi_t = 0$. The fundamental PLL loop equation, expressed in the conventional manner, is the following:

$$\dot{\varphi} = \Delta\omega - F(p)K[A(1 - 2P_E) \sin \varphi + N(t, \varphi)]. \tag{9.81}$$

In order to determine the control voltage/phase characteristic $G(\varphi)$, we note that when φ reaches the value of $\pi/4$, the control voltage changes sign and jumps

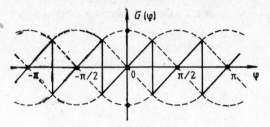

Fig. 9.11. Phase characteristic of the inverse modulator.

to another sine wave. As shown in Fig. 9.11, the characteristic has a saw-tooth waveform and a periodicity of $\pi/2$.

In Ref. [19], the PLL loop stability has been investigated, and found to be conditionally stable. The stability range is given by $0.5 < \xi < 0.9$ where

$$\xi = \frac{1}{2} \tau_2 \sqrt{\frac{AK}{\tau_1}} \tag{9.82}$$

is the attenuation factor, K is the overall loop gain, and τ_2, τ_1 are the time constants of the second-order filter. The calculation of the PLL loop bandwidth has resulted in the expression

$$B_L = \frac{AK}{4} F_0(1 + F_0 T_3 AK) \tag{9.83}$$

where $F_0 = \tau_2/\tau_1$. With linear approximation, the variance due to thermal noise is

$$\sigma_\varphi^2 = \frac{B_L}{B} \frac{1}{R(1-2P_E)^2} \tag{9.84}$$

where B is the noise bandwidth of the receive filter. Note that this expression is identical with expression (9.55) applicable to the decision feedback case. However, P_E is now higher because the decision is replaced by a comparison, without sampling. If a decision circuit is used instead of the comparator then the inverse modulator arrangement will be transformed into a decision feedback system, the delay T_1 being equal to the symbol time. This solution is applied in some high speed transmission systems.

The second term of Eq. (9.74c) also represents phase noise, distributed evenly in the frequency range of B. The variance due to the pattern jitter can then be expressed by

$$\sigma_{p\varphi}^2 = \frac{B_L}{B} \frac{4P_E}{(1-2P_E)^2} . \tag{9.85}$$

The overall variance is the sum of the variances due to thermal noise and pattern jitter.

9.1.6 Effect of band limiting

In the preceding sections, it has been assumed that the signal is transmitted by the receive bandpass filter and the post-demodulation low-pass filter without distortion. Filter properties have been taken into account only as far as the signal-to-noise ratio in the expression of σ_φ^2 is governed by the filter bandwidth. However, the filters also have band limiting effect resulting in the distortion of the transmitted signal. This effect will be investigated in the following.

The transmitted symbol stream can be expressed as the sum of elementary symbols:

$$V(t) = \sum_k b(t-knT) \cos\left[\omega_0 t + \Phi_k(t)\right] \tag{9.86}$$

where $b(t-knT)$ is a square-wave gate function. The complex envelope of a signal transmitted by a filter will be given by

$$V(t) = \sum_k e^{j\Phi_k} \int_{-\infty}^{+\infty} h(\tau) b(t - nkT - \tau)\, d\tau \tag{9.87}$$

where $h(t)$ is the impulse response of the filter. A PSK signal has nearly constant amplitude as the transition from one phase to another takes place in a short time interval. However, filtering has the effect of changing the signal amplitude at the time instants of phase jumps. Figure 9.12 shows the amplitude and phase changes of a QPSK signal for 180° and 90° phase changes. In the recovered carrier, pattern jitter will be introduced by these amplitude changes as these take place at the modulation rate.

Another harmful effect of band limiting is a reduction in signal energy. In the following, this will be calculated for a carrier recovery circuit employing an Mth

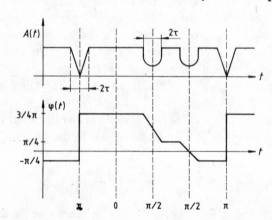

Fig. 9.12. Effect of band limiting on the modulated signal.

power law device. Let us introduce the attenuation factor according to Refs [4, 11]:

$$D = \int\limits_{-\infty}^{+\infty} S(\omega)|H(\omega)|^2 \, df \qquad (9.88)$$

where $S(\omega)$ is the power density spectrum of the modulated signal, and $H(\omega)$ is the filter characteristic.

The Mth law device has the effect of reducing the power level of the $M\omega_0$ frequency component by D^M [8]:

$$S_1(M\omega_0) = D^M S_0(M\omega_0) \qquad (9.89)$$

where $S_0(M\omega_0)$ is the power level without band limiting.

A further effect of band limiting is the increase in squaring loss:

$$S_L = S_{L0} D \qquad (9.90)$$

which has the following form for the 2PSK case:

$$S_L = \frac{D}{1 + \dfrac{1}{2R}}. \qquad (9.91)$$

It follows from this expression that the bandwidth of the receive filter has an optimal value which is explained as follows. Increasing B will result in increased D values but in this case, the thermal noise level reaching the carrier recovery circuit will also increase, thus decreasing the signal-to-noise ratio R. On the other hand, decreasing B will result in increased R values but in this case, D will be reduced. This squaring loss function is shown in Fig. 9.13, showing an optimal value in the vicinity of $BT=1$. This is close to the bandwidth giving the lowest error probability.

The phase noise due to the filter amplitude and phase characteristic can be determined by computer simulation [23, 26].

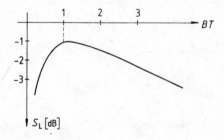

Fig. 9.13. Squaring loss as a function of the relative bandwidth.

In the following, the pattern jitter introduced by the residual amplitude modulation of the recovered carrier due to the filter band limiting will be investigated. The following simple expressions can be derived from the amplitude and phase functions shown in Fig. 9.12. The spectrum of the distortion products due to band limiting has the relative value of

$$S_c(\omega_0) \cong \frac{T_s}{64} \frac{1}{(BT_s)^2} \tag{9.92}$$

for the case of a fourth power law carrier recovery circuit, and the relative value of

$$S_c(\omega_0) \cong \frac{T_s}{400} \frac{1}{(BT_s)^2} \tag{9.93}$$

for an inverse modulator type carrier recovery circuit.

From these equations, the phase noise variance is calculated to be

$$\sigma_{p\varphi}^2 = W_L S_c(\omega_0). \tag{9.94}$$

The overall phase error variance is given by the sum of variances due to thermal noise and pattern jitter:

$$\sigma_\varphi^2 = \sigma_{t\varphi}^2 + \sigma_{p\varphi}^2. \tag{9.95}$$

It is known that the phase error and the pull-in range of a PLL loop is a function of input signal amplitude. The phase error is given by

$$\Delta\Phi = \frac{\Delta\omega}{AK} \tag{9.96}$$

where A is the signal amplitude, K is the overall loop gain and $\Delta\omega$ is the difference between the free running VCO frequency and the incoming signal frequency. The pull-in range is given by

$$\omega_{B0} = AK\sqrt{2F_0} \tag{9.97}$$

where $F_0 = \tau_2/\tau_1$, and τ_1, τ_2 are the time constants of the second-order loop filter. Retaining our previous notation, we have

$$A^2 \sim S_1(M\omega_0). \tag{9.98}$$

By utilizing Eq. (9.89), the phase error increment due to band limiting is calculated to be

$$\Delta\Phi = \frac{\Delta\Phi_0}{D^{M/2}} \tag{9.99}$$

and the pull-in range reduction due to band limiting is calculated to be

$$\omega_B = \omega_{B0}D^{M/2}. \tag{9.100}$$

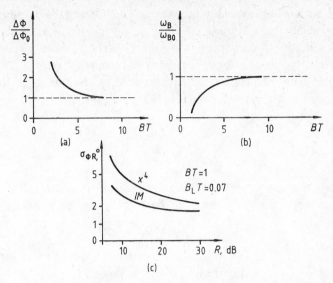

Fig. 9.14. Parameters of fourth power type and inverse modulator type carrier recovery circuits: (a) phase error; (b) pull-in range; (c) phase noise.

The above parameters are plotted in Figs 9.14a and 9.14b as a function of relative bandwidth. Figure 9.14c shows the phase noise as a function of the signal-to-noise ratio [27]. The thermal noise is dominant at low signal-to-noise ratios while the pattern jitter has a decisive effect at high signal-to-noise ratios.

It should be noted that the Costas loop is equivalent to the *M*th power law device so the above results obtained for the *M*th power law device are also applicable to the Costas loop.

The expressions for the control characteristic and phase noise of the inverse modulator are identical to those obtained for the decision feedback systems. This means that there is a similarity between these two carrier recovery methods which resembles the similarity between the Costas loop and the *M*th power law device. Consequently, the band limiting effect on the decision feedback system can be well approximated by the application of the inverse modulator expressions.

9.1.7 Comparison of different types

Several comparison methods can be applied to evaluate the carrier recovery methods utilized for the coherent demodulation of *M*-PSK signals [19, 23, 6].

All carrier recovery systems of continuous mode transmission equipment utilize a PLL loop. Usually, it is required that the phase error of the PLL should be less than $5°$ within the frequency range of $f_0 + \Delta f$ where f_0 is the nominal value of the intermediate frequency and Δf is the change in intermediate frequency due to the finite stability of oscillators. Also, the PLL should have a suitable pull-in range to accommodate the above instability frequency range.

The intermediate frequency signal is generated by the mixing of the transmit oscillator signal and the receive local oscillator signal:

$$f_0 = f_t - f_L$$

or   (9.101)

$$f_0 = f_L - f_t$$

where f_t are the transmit and f_L the receive local oscillator frequencies. Let the relative frequency stability of the microwave oscillators be given by

$$k = \Delta F / f_t$$

where ΔF is the change in microwave frequency due to the finite oscillator stability. The change in intermediate frequency is then given by

$$\Delta f = \Delta F_t + \Delta F_L \cong 2 k f_t \qquad (9.102)$$

and the relative frequency stability of the intermediate frequency is

$$\frac{\Delta f}{f_0} \sim 2k \frac{f_t}{f_0}. \qquad (9.103)$$

By noting that the usual value of f_t / f_0 is around 100, it is seen that the relative frequency instability of the intermediate frequency is high. The realization of a suitable pull-in range and an effective noise filtering is a difficult task, especially in low and medium speed transmission equipment. The delay in the PLL loop has the effect of severely reducing the pull-in range; the decision feedback system, comprising a delay of one symbol time, can only be applied in high speed microwave digital equipment.

The Mth power law devices can be easily realized for 2PSK and 4PSK transmission. In the case of 8PSK transmission, the processing of the recovered carrier of frequency $8f_0 = 560$ MHz (with an intermediate frequency of $f_0 = 70$ MHz) may present difficulties. At this frequency, lumped parameter circuits are difficult to apply while distributed parameter circuits have large dimensions. Therefore, Costas loops with digital circuits [28] or inverse modulators are commonly applied for 8PSK transmission. Similar difficulties are experienced also in QPSK transmission if the intermediate frequency is high.

In the inverse modulator and remodulator arrangements, a complete i.f. modulator has to be realized in the carrier recovery circuit, but these arrangements have the advantage of requiring the processing of only f_0 frequency and baseband frequency signals. This allows the use of identical type amplifiers and phase detectors in the modulators and demodulators.

From the two variants of the Costas loop, the digital variant is more widely applied because in the analog variant, the drift of the cascaded multipliers and amplifiers may be troublesome. Several solutions have been developed for the indication and elimination of the false lock phenomenon [14, 15, 16, 17].

If the noise of the reference signal is taken as a quality criterion, then the decision feedback system is the best carrier recovery arrangement, and this is followed by the inverse modulator, the *M*th power device, the Costas loop and other baseband procedures.

9.1.8 Fast acting types

In TDMA transmission systems, the receiver has to process signals arriving in rapid succession from different transmitters. The carrier frequencies of these signals differ by values governed by the frequency stability of the transmitters, and their phases may show a random variation in the range of 0 to 2π. Under these circumstances, the carrier recovery circuit has to lock onto the subsequent carriers within short time intervals. Conventional PLL circuits are unsuitable for this application because of their slow response. The response time may also be substantially increased by the so-called "hang-up" phenomenon, according to which the slope of the control voltage time function is decreased in the vicinity of unstable operation points. For QPSK transmission, these unstable operation points are at phase differences of $\pi/4$, $3\pi/4$, *etc.*, while stable operation is obtained at 0, $\pi/2$, $3\pi/2$ *etc.*

Instead of a PLL which can be regarded as an active filter, a passive bandpass filter and limiter can be applied to obtain rapid synchronization [29]. The PLL phase noise expressions are applicable also to passive filters if B_L, the one-sided loop bandwidth of the PLL is substituted by $B_N/2$ where B_N is the noise bandwidth of the passive filter. It has been shown that the "hang-up" phenomenon may appear even in inverse modulators in which passive filters are applied. This is explained by the fact that inverse modulator arrangements comprise feedback even without a PLL [27]. However, experiments have shown that inverse modulators with passive filters are applicable for fast carrier recovery in TDMA systems [30]. In most cases, a single resonant circuit is used for the bandpass filter because its phase/frequency slope is lowest.

The noise bandwidth of a single tuned bandpass filter is given by

$$B_N = \frac{\pi}{2} \frac{f_0}{Q} \tag{9.104}$$

where f_0 is the resonant frequency and Q is the Q-factor. The filter time constant is given by

$$\tau = \frac{Q}{\pi}\frac{1}{f_0} = \frac{1}{2B_N}. \tag{9.105}$$

Owing to a signal burst arriving during TDMA transmission, the envelope of the carrier appearing at the output of the filter is given by

$$V(t) = V_0 \exp(-t/\tau) \tag{9.106}$$

at the end of the burst, and by

$$V(t) = V_0[1-\exp(-t/\tau)] \tag{9.107}$$

at the beginning of the burst. Between bursts, the two carriers are simultaneously present, and this overlapping appears as interference noise in the recovered carrier, increasing the error probability.

Rapid synchronization and good noise filtering are conflicting requirements because the time constant of the filter is inversely proportional to the noise bandwidth. Taking into account both requirements, the time constant is usually chosen to be in the range $16T<\tau<32T$ where T is the bit time.

Figure 9.15 shows two variants of fast acting carrier recovery. In Fig. 9.15a, a fourth power law device carrier recovery is illustrated while Fig. 9.15b is an example of an inverse modulator type carrier recovery.

The phase error of the reference signal has several components as shown by the following equation:

$$\Delta\Phi = \Delta\Phi_t + \Delta\Phi_f + \Delta\Phi_1 + \varphi_h. \tag{9.108}$$

Here $\Delta\Phi_t$ is the phase error due to the detuning of the filter, $\Delta\Phi_f$ is the phase error resulting from the input signal frequency change, $\Delta\Phi_1$ is the phase error caused by the AM-to-PM conversion of the applied limiter, changing with the input signal level, and φ_h is the phase error during the transition time.

In the case of TDMA transmission over a satellite, the input signal frequency change has two components:

$$\Delta f = \Delta f_1 + \Delta f_2. \tag{9.109}$$

Here Δf_1 is due to the satellite oscillator instability and, in case of a non-stationary satellite, to the Doppler frequency variation, and thus represents a slow change which is present equally in all time slots. Δf_2 is the instability of the individual earth station oscillators, rapidly changing in each successive time slot. The earth station frequency stability (10^{-7} to 10^{-6}) is much better than the satellite stability (about 10^{-5}). This means that the frequency change is made up of a slowly changing larger part and a rapidly changing smaller part. The slow

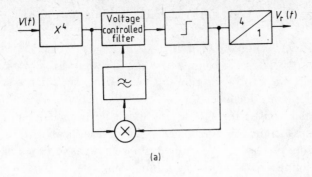

(a)

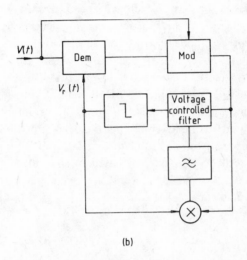

(b)

Fig. 9.15. Fast acting carrier recovery systems: (a) fourth power type; (b) inverse modulator type.

contribution can be canceled by the application of an automatic phase control loop (APC): a voltage controlled filter is applied in the carrier recovery circuit which is tuned automatically to the input signal frequency by the adjustment of lowest phase error. This control loop has a large time constant and thus will not affect the ability to follow rapid frequency changes.

A PLL loop can also be applied in a fast acting carrier recovery circuit by utilizing a special arrangement to affect fast operation [31]. Another method is to have a large filter bandwidth during the pull-in period which is subsequently switched to a smaller value in the steady-state condition.

9.2 O-QPSK and MSK transmission

It was shown in Section 7.5 that the O-QPSK signal is a modified variant of the QPSK signal, the difference being in the half unit time delay between the modulating signals in the two quadrature channels. However, the four phases of the modulated signal are identical for the two modulation methods so all carrier recovery methods given for QPSK transmission are equally applicable for O-QPSK transmission too: the fourth power law device, the inverse modulation and the Costas loop. Evidently, the required phase shift has to be taken into account when a decision feedback arrangement is applied.

In an MSK signal, a continuous phase change takes place and there are no specific phase values of the signal which, with the exception of the phase transition intervals, prevail. Therefore, the fourth power law device and the corresponding Costas loop cannot be applied. From the carrier recovery methods discussed earlier, the remodulation type is applicable for MSK transmission.

In the following, a special carrier recovery method developed specifically for MSK transmission will be outlined. MSK is really frequency modulation with small frequency deviation, having a continuous spectrum without discrete spectrum lines. By applying frequency doubling, the deviation will be doubled and the spectrum will comprise two spectrum lines at $2f_1$ and $2f_2$.

Several carrier recovery circuits are based on this principle [34, 35], one of which is shown in Fig. 9.16.

Let the two extreme values of the original MSK signal frequencies be given by

$$f_1 = f_0 - \Delta f$$
$$f_2 = f_0 + \Delta f. \tag{9.110}$$

Following frequency doubling, the extreme frequencies will be $2f_1$ and $2f_2$ which are filtered out by two PLL's. A multiplier is driven by the two filtered signals, and the sum and difference frequency components, appearing at the

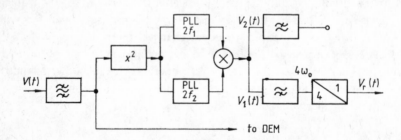

Fig. 9.16. Carrier recovery of an MSK system.

multiplier output, are separated by low-pass and high-pass filters. At the output of the high-pass filter, we have

$$V_1(t) = \cos 4\omega_0 t. \tag{9.111}$$

This signal is presented to a frequency divider to obtain the reference signal of frequency ω_0. The phase ambiguity is resolved by normal means.

At the output of the low-pass filter, we have

$$V_2(t) = \cos (2\pi 4\Delta f t). \tag{9.112}$$

For MSK transmission, $2\Delta fT=1/2$ where T is the bit time. Substituting this into Eq. (9.112) yields

$$V_2(t) = \cos \frac{2\pi}{T} t. \tag{9.113}$$

We note that the output signal of the low-pass filter is the recovered clock signal.

Another way to resolve the phase ambiguity is to apply a frequency demodulator. The quality of the signal thus obtained is worse than that obtained by phase demodulation, but it is still suitable for determining the correct reference signal phase.

9.3 *M*-QAM transmission

A QAM signal is the sum of two modulated signals which are in quadrature, and can be expressed, during symbol time interval, by

$$V(t) = \sqrt{2}\, p(t)[a_i \cos (\omega_0 t + \Phi_t) - b_j \sin (\omega_0 t + \Phi_t)] \tag{9.114}$$

where $p(t)=1$ in the interval $(k-1)T<t<kT$ and $p(t)=0$ outside this interval. Φ_t is the phase of the transmit carrier, and ω_0 is the carrier frequency. For an

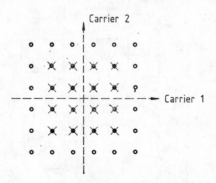

Fig. 9.17. Signal space diagram of *M*-QAM transmission, also showing the necessary reference signal phases. The points of the 16QAM diagram are denoted by ✕ .

M-QAM signal, $M=L^2$ where L can be any even number. For 16QAM, $L=4$ and for 64QAM, $L=8$. a_i, $b_j=\pm1$, ±3, ±5, ..., $\pm(L-1)$ are the values of the modulating signals in the interval k. The signal space diagram of M-QAM transmission is illustrated in Fig. 9.17.

9.3.1 Baseband carrier recovery

Figure 9.18 shows the block diagram of an M-QAM carrier recovery circuit [36, 37, 41]. The additive Gaussian noise of the input signal is expressed by

$$n(t) = \sqrt{2}\,[n_c(t)\cos(\omega_0 t + \Phi_t) - n_s(t)\sin(\omega_0 t + \Phi_t)] \tag{9.115}$$

and the two components of the reference signal are given by

$$V_{r1}(t) = \sqrt{2}\cos(\omega_0 t + \Phi_R)$$
$$V_{r2}(t) = -\sqrt{2}\sin(\omega_0 t + \Phi_R). \tag{9.116}$$

Introducing the notation $\varphi = \Phi_t - \Phi_R$, the demodulated signal components, generated by the two multipliers M_1 and M_2, are given by

$$V_{d1}(t) = a_i\cos\varphi - b_j\sin\varphi + n_c\cos\varphi - n_s\sin\varphi$$
$$V_{d2}(t) = b_j\cos\varphi + a_i\sin\varphi + n_s\cos\varphi + n_c\sin\varphi. \tag{9.117}$$

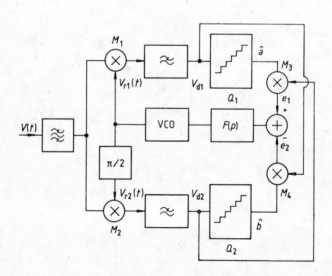

Fig. 9.18. M-QAM carrier recovery circuit.

The two low-pass filters following the two multipliers serve to suppress the sum frequency components. The two quantizers $Q1$ and $Q2$ have the function of generating the quantized signals $\hat{a}$ and $\hat{b}$ having $L=\sqrt{M}$ levels.

The VCO control signal $e(t)$ is generated by taking the difference between signals $e_1(t)$ and $e_2(t)$ supplied by multipliers $M3$ and $M4$. Denoting the delay introduced by the low-pass filters and quantizers by τ, we have the following expression for the control voltage:

$$e(t) = \exp(-p\tau)\times$$
$$\times\{[a_i\hat{a}_i + b_j\hat{b}_j]\sin\varphi + [a_i\hat{b}_j - b_j\hat{a}]\cos\varphi + \hat{b}N_c(\varphi) - \hat{a}N_s(\varphi)\} \tag{9.118}$$

where

$$N_c(\varphi) = n_c(t)\cos\varphi - n_s(t)\sin\varphi$$
$$N_s(\varphi) = n_s\cos\varphi + n_c\sin\varphi. \tag{9.119}$$

Taking these into account, the fundamental PLL equation is expressed as

$$\dot{\phi} = \Delta\omega - KF(p)\exp(-p\tau)\times$$
$$\times[(a\hat{a} + b\hat{b})\sin\varphi + (a\hat{b} - b\hat{a})\cos\varphi + \hat{b}N_c(\varphi) - \hat{a}N_s(\varphi)] \tag{9.120}$$

where $\Delta\omega$ is the difference between the incoming and VCO frequencies, K is the overall loop gain, $F(p)$ is the response of the loop filter, $\exp(-p\tau)$ represents the delay.

The calculation is usually based on the following assumption [36, 40]:

$$W_L \ll \frac{1}{T_s} \ll B \ll f_0 \tag{9.121}$$

where W_L is the two-sided noise bandwidth of the loop, T_s is the symbol time, B is the 3 dB bandwidth of the input filter, f_0 is the carrier frequency.

The right-hand side inequality, allowing the narrow-band approximation, has been taken into account in formulating Eq. (9.115). The centre inequality means that the signal is assumed to have no distortion, *i.e.* the amplitude and phase do not change within symbol time interval. This condition is not met in practice but is assumed for simplicity. The inequality $B \gg W_L$ means that $N(\varphi, t)$ can be approximately regarded as expressing white noise. The left-hand side inequality means that the phase function $\varphi(t)$ changes much more slowly than the signal and noise functions, so we can take these into account by their expectation values.

Neglecting the delay term in Eq. (9.120) will not alter the phase characteristic, but only reduce the PLL pull-in range. We then have

$$\dot{\phi} = \Delta\omega - KF(p)\times$$
$$\times[E\{(a\hat{a} + b\hat{b})\}\sin\varphi + E\{(a\hat{b} - b\hat{a})\}\cos\varphi + (E\{\hat{a}^2/\varphi\} + E\{\hat{b}^2/\varphi\})^{1/2}N_1(t)] \tag{9.122}$$

where $E\{\ \}$ denotes the statistical average, and $N_1(t)$ is the baseband noise process.

The evaluation of Eq. (9.122) is rather complicated, and is performed for the case when the quantizers are preceded by integrators [36]. In this case, the control voltage characteristic is expressed by the following normalized function:

$$g(\Phi) = \frac{4\delta^2}{MG'(0)} \sum_{ij} (i \sin \Phi + j \cos \Phi) T_{ij}(\Phi) \tag{9.123}$$

where

$$T_{ij}(\Phi) = \sum_l Q[\Delta(l - i \cos \Phi + j \sin \Phi)]; \quad l = 0, \pm 2, \pm 4, ..., \pm(L-2) \tag{9.124}$$

and

$$Q = \frac{1}{2\sqrt{2\pi}} \int_x^\infty \exp(-y^2/2)\, dy \tag{9.125}$$

$$\Delta = \sqrt{\frac{R}{s}}, \quad R = \frac{P_s}{\sigma^2}, \quad s = \frac{M-1}{3}. \tag{9.126}$$

The demodulator is followed by a low-pass filter which is intended only to suppress the sum frequency terms. The noise is then determined by the input filter, and the variance of the input noise is given by

$$\sigma^2 = N_0 B. \tag{9.127}$$

Here N_0 denotes the one-sided noise power density. The slope of the phase characteristic at $\Phi = 0$ is

$$G'(0) = \frac{4\delta^2}{L} \sum_i i \left\{ \sum_l Q[\Delta(l-i)] - \frac{M-1}{3\sqrt{2\pi}} \Delta \sum_l \exp\left[-(l-i)^2 \frac{\Delta^2}{2}\right] \right\}. \tag{9.128}$$

The two extreme values of the function $g(\Phi)$ are given by

$$\lim_{\Delta \to 0} g(\Phi) = \frac{1}{4} \sin 4\Phi \tag{9.129}$$

$$\lim_{\Delta \to \infty} g(\Phi) = \frac{4\delta^2}{MG'(0)} \sum_{ij} (i \sin \Phi + j \cos \Phi) \times$$

$$\times \sum_l \left[\frac{1}{2} - \frac{1}{2} \operatorname{sign}(l - i \cos \Phi + j \sin \Phi) \right] \tag{9.130}$$

$$\lim_{\Delta \to \infty} G'(0) = \frac{4\delta^2}{L} \sum_i \left\{ i \sum_l \left[\frac{1}{2} - \operatorname{sign}(l - i) \right] \right\}. \tag{9.131}$$

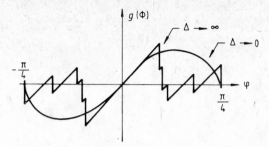

Fig. 9.19. Phase characteristic of 16QAM carrier recovery.

For 16QAM transmission, the straight line section covers $\pm16°$, and the phase characteristic has a periodicity of $\pi/2$ (see Fig. 9.19).

In Ref. [38], a carrier recovery circuit suitable for the demodulator shown in Fig. 8.10 is investigated. This has also a piecewise linear phase characteristic which is tangent to the centre portion of the characteristic $g(\Phi) = \sin 4\Phi$.

In Refs [39, 40], a baseband carrier recovery circuit comprising digital circuit elements is discussed. Here the phase characteristic has a square-wave form in the noise-free case, which is rolled-off by the noise. The width of the usable portion is approximately $\pm15°$.

For the reception of QAM signals, a decision feedback carrier recovery circuit has also been developed [44].

9.3.2 Fourth power law device

The QAM signal given by Eq. (9.114) can also be expressed as

$$V(t) = \sqrt{2}\, c(t) \cos [\omega_0 t + \Phi(t) + \Phi_t] \qquad (9.132)$$

where

$$c(t) = [a^2(t) + b^2(t)]^{1/2} \qquad (9.133)$$

$$\Phi(t) = \tan^{-1} \frac{b(t)}{a(t)}. \qquad (9.134)$$

It is seen that the signal has not only amplitude modulation but also phase modulation, allowing the use of carrier recovery circuits applicable to PSK transmission. Driving a fourth power law frequency multiplier by the modulated

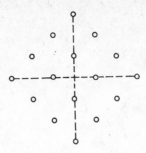

Fig. 9.20. Signal space diagram of 16QAM transmission, related to the vector diagram of 4PSK transmission.

signal, the $4\omega_0$ component output signal will be given by

$$V_0(t) = \left[\frac{a^4 + b^4}{2} - \frac{3}{2}\, a^2 b^2\right] \cos 4\,\omega_0 t + ab(a^2 - b^2) \sin 4\omega_0 t. \qquad (9.135)$$

Here again, the expectation value of the signal can be taken into account. The second term is canceled because $E\{a^2\} = E\{b^2\}$.

This method is usually applied for 16QAM transmission. Investigation of the signal space diagram shows (see Fig. 9.20) that 8 of the 16 signal vectors have phases coinciding with those of the 4PSK signal. The remaining 8 vectors are pairwise symmetrically located with respect to these four phases. Their effect can be canceled by scrambling [6], as this process yields vector pairs with equal probability of occurrence.

9.4 QPRS transmission

The QPRS transmission method investigated in Section 7.4 has 8 phases, the ninth vector being in the centre. The reference signal can thus be generated by an eighth power law device applied for 8PSK transmission [28]. Figure 9.21a shows a quadrature demodulator for demodulating the QPRS signal, together with an eighth power law carrier recovery circuit. The reference signal may have 2 phases, the correct phase being shown in Fig. 9.21b, and the incorrect phase in Fig. 9.21c. In the case of an incorrect phase lock, a comparator is used to initiate a switch-over, inserting a phase shift of $\pi/4$ into the reference signal path. This results in a change-over into the correct phase lock position.

The QPRS signal can be regarded as a QAM signal so the carrier recovery methods given in Section 9.3 for QAM signals are applicable also for QPRS signals.

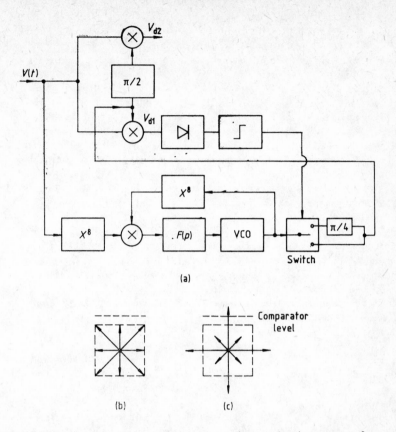

(a)

(b)   (c)

Fig. 9.21. (a) QPRS carrier recovery circuit; (b) vector diagram showing correct reference signal phase; (c) vector diagram showing incorrect reference signal phase.

9.5 References

[1] Costas, J. P.: Synchronous communications. *Proc. IRE 44*, No. 12, pp. 1713–1718, 1956.

[2] Lindsey, J. W. C.: *Synchronous Systems in Communication and Control*. Prentice-Hall, Englewood-Cliffs 1972.

[3] Viterbi, A.: *Principles of Coherent Communication*. McGraw-Hill, New York 1966.

[4] Lindsey, J. W. C., Simon, M. K.: *Telecommunication System Engineering*. Prentice-Hall, Englewood-Cliffs 1973.

[5] Stiffler, I. I.: *Theory of Synchronous Communications*. Prentice-Hall, Englewood-Cliffs 1971.

[6] Franks, L. E.: Carrier and bit synchronization in data communication. A tutorial review. *IEEE Trans. Commun.*, Vol. Com-28, No. 8, pp. 1107–1121, 1980.

[7] Lindsey, J. W. C., Simon, M. K.: Carrier synchronization and detection of polyphase signals. *IEEE Trans. Commun.*, Vol. Com-20, No. 3, pp. 442–454, 1972.

[8] Butman, S. A., Lesh, J. R.: The effect of bandpass limiters on n-phase tracking systems. *IEEE Trans. Commun.,* Vol. Com-25, No. 6, pp. 569–576, 1977.

[9] Sasase, I., Mori, S.: The effect of bandpass limiters on Mth phase tracking systems in the presence of sinusoidal interference and noise. *IEEE Trans. Commun.,* Vol. Com-30, No. 8, pp. 1842–1847, 1982.

[10] Springett, I. C., Simon, M. K.: An analysis of the phase coherent–incoherent output of bandpass limiter. *IEEE Trans. Commun.,* Vol. Com-19, No. 1, pp. 42–49, 1971.

[11] Lindsey, W. C., Woo, K. T.: Analysis of squaring circuit mechanizations in Costas loops. *IEEE Trans. Aerosp. Electron. Syst.,* Vol. AES-14, No. 5, Sept., pp. 756–763, 1978.

[12] Lindsey, W. C., Simon, M. K.: Optimum design of Costas receivers containing soft bandpass limiters. *IEEE Trans. Commun.,* Vol. Com-25, No. 8, pp. 822–831, 1978.

[13] Simon, M. K.: Tracking performance of Costas loop with hard-limited in-phase channel. *IEEE Trans. Commun.,* Vol. Com-26, No. 4, pp. 420–331, 1978.

[14] Hedin, G. L., Holmes, J. K., Lindsey, W. C., Woo, K. T.: Theory of false lock in Costas loops. *IEEE Trans. Commun.,* Vol. Com-26, No. 1, pp. 1–11, 1978.

[15] Shimamura, T.: On false-lock phenomena in carrier tracking loops. *IEEE Trans. Commun.,* Vol. Com-28, No. 8, pp. 1326–1334, 1980.

[16] Ohlson, M. L.: False lock detection in Costas demodulator. *IEEE Trans. Aerosp. Electron. Syst.,* Vol. AES-11, No. 2, March, pp. 180–182, 1975.

[17] Chan, C. R.: Improving frequency acquisition of a Costas loop. *IEEE Trans. Commun.,* Vol. Com-25, No. 12, pp. 1453–1459, 1977.

[18] Osborne, H. C.: A generalized "polarity-type" Costas loop for tracking MPSK signals. *IEEE Trans. Commun.,* Vol. Com-30, No. 10, pp. 2289–2296, 1982.

[19] Tamashita, T., Sakata, T., Iguchi, K.: Synchronous phase demodulators for high speed quadrature PSK transmission systems. *FUJITSU Sci. Tech., J.,* Dec., pp. 57–80, 1975.

[20] Lindsey, W. C., Simon, M. K.: Data-aided carrier tracking loop. *IEEE Trans. Commun.,* Vol. Com-19, No. 2, pp. 157–168, 1971.

[21] Natali, F. D., Walbesser, W. J.: Phase-locked loop detection of binary PSK signals utilizing decision feedback. *IEEE Trans. Aerosp. Electron. Syst.,* Vol. AES-5, No. 1, pp. 83–90, 1969.

[22] Daikoku, K.: Analysis of phase-locked loop acquisition with delay time. *Electron. Commun. Jpn,* Vol. 57-A, No. 12, pp. 25–32, 1972.

[23] Dodo, J., Iwai, S., Kawai, K.: Computer simulation of carrier recovery circuit in microwave PCM–PSK system. *FUJITSU Sci. Tech. J.,* Sept., pp. 59–81, 1972.

[24] Weber, C. L., Alem, W. K.: Demod–remod coherent tracking receiver for QPSK and SQPSK. *IEEE Trans. Commun.,* Vol. Com-28, No. 12, pp. 1945–1954, 1980.

[25] Weber, C. L., Alem, W. K.: Performance analysis of demod–remod coherent receiver for QPSK and SQPSK input. *IEEE Trans. Commun.,* Vol. Com-28, No. 12, pp. 1954–1968, 1980.

[26] Yamamoto, H., Hirade, K., Watanabe, Y.: Carrier synchronizer for coherent detection of high-speed four-phase-shift-keyed signals. *IEEE Trans. Commun.,* Vol. Com-20, No. 4, pp. 803–808, 1972.

[27] Gardner, F. M.: Comparison of QPSK carrier regenerator for TDMA application. In: *Proc. ICC Conf. Minneapolis,* pp. 43.B.1–43.B.5, 1974.

[28] Hogge, C. R.: Carrier and clock recovery for 8PSK synchronous demodulation. *IEEE Trans. Commun.,* Vol. Com-26, No. 5, pp. 528–533, 1978.

[29] Gardner, F. M.: Rapid synchronization carrier and clock recovery for high-speed digital communication. *Microwave Syst. News,* Febr–March, pp. 57–63, 1976.

[30] Nakajima, S., Sioda, H.: PSK modem for domestic TDMA system. *Rev. Electr. Commun. Lab.*, No. 1–2, pp. 108–117, 1974.

[31] Taylor, D. P., Tang, S. K., Mariuz, S.: The limit-switched loop. A phase-locked-loop for burst mode operation. *IEEE Trans. Commun.*, Vol. Com-30, No. 2, pp. 396–407, 1982.

[32] Yamada, K., Tsuji, Y., Fukui, M., Katoh, T.: Satellite communication. *FUJITSU Sci. Tech. J.*, July, pp. 259–284, 1985.

[33] Nakajima, S., Furuya, N.: Gaussian filtered amplitude limited MSK. *Trans. IECE Jpn*, Nov., pp. 716–723, 1981.

[34] De Buda, R.: Coherent demodulation of frequency-shift keying with low deviation ratio. *IEEE Trans. Commun.*, Vol. Com-20, No. 3, pp. 429–435, 1972.

[35] Benedetto, S., Castellani V., Pent, M.: Modified PSK modulation schemes for space communications. *Alta Freq.*, Vol. 47, No. 1, pp. 26–39, 1978.

[36] Simon, M. K., Smith, J. G.: Carrier synchronization and detection of QASK signal sets. *IEEE Trans. Commun.*, Vol. Com-22, No. 2, pp. 98–106, 1974.

[37] Moss, J. F., Beyer, F. B.: Universal carrier recovery for APK signals. *Proc. IEEE*, Vol. 71, No. 7, pp. 905–907, 1983.

[38] Washio, M., Shimamura, T., Komiyama, N., Takimoto, I.: 1.6 Gb/s 16-level superposed APSK modem with baseband signal-processing coherent demodulator. *IEEE Trans. Microwave Theory Techn.*, Vol. MTT-26, No. 12, pp. 945–954, 1978.

[39] Verdot, G., Leclert, A.: Universal high bit rate modem for digital radio systems. In: *Proc. ICC Conf., Philadelphia*, pp. 3.B.1–3.B.4, 1982.

[40] Leclert, A., Vandamme, P.: Universal carrier recovery loop for QASK and PSK signal sets. *IEEE Trans. Commun.*, Vol. Com-31, No. 1, pp. 130–136, 1983.

[41] Simon, K.: Further results on optimum receiver structures for digital phase and amplitude modulated signals. In: *Proc. ICC Conf., Toronto*, pp. 42.1.1.–42.1.7, 1978.

[42] Miyauchi, K., Seiki, S., Ishio, A.: New technique for generating and detecting multilevel signal formats. *IEEE Trans. Commun.*, Vol. Com-24, No. 2, pp. 263–267, 1976.

[43] Horikawa, I., Saito, Y.: 16-QAM carrier recovery with selective gated PLL. *Trans. IECE Jpn*, Vol. E-63, No. 7, Abstracts, pp. 548–549, 1980.

[44] Moridi, S., Sari, H.: Analysis of four decision-feedback carrier recovery loops in the presence of intersymbol interference. *IEEE Trans. Commun.*, Vol. Com-33, No. 6, pp. 543–550, 1985.

CHAPTER 10

BASEBAND SIGNAL PROCESSING

In this chapter, all operations required to obtain optimal relations between the source and the digital modulator, and also between the digital demodulator and the sink, will be considered as forming parts of the baseband signal processing. In several variants of microwave digital transmission, a variety of baseband signal processing schemes are applied. A few basic building blocks of baseband signal processing which are applied in all cases are shown in Fig. 10.1; in some cases, the realization of these blocks presents severe difficulties.

The operations of the transmit side signal processing between the source and the modulator, and the receive side signal processing between the demodulator and the sink, are partially inversely related; however, at the receive side, the regeneration and the clock recovery needed for regeneration, characteristic operations in digital signal transmission, are always present.

There may be interface units at both the transmit and receive sides, according to the source and sink specifications. It will be assumed that at the processing side of these interfaces, the information is present in the form of NRZ (Non-Return-to-Zero) binary bit streams. The randomness of the transmit side bit stream may be enhanced by the transmit side scrambler which requires a receive side descrambler to perform the inverse operation. In many cases, multi-state digital modulation is applied requiring the series-to-parallel conversion of the serial bit stream into dibit, tribit, *etc.* signal streams, and the inverse operation by the parallel-to-series converter. The main part of the matching to the digital modulation scheme is performed by the encoding circuits. The error multiplication due to the memory of the digital encoding system may be reduced (*e.g.*, Gray encoding). The elimination of the inherent phase ambiguity in the modulation system may also be compensated (*e.g.*, differential encoding). Finally, the encoding circuits may also be utilized for the transmission of additional information (*e.g.*, by intentional and periodical violation of the encoding rule). The decoding circuits perform the inverse operations while the function of the regenerator is to refresh the (normally amplitude limited) bit stream received from the digital demodulator, by sampling at optimal time instants.

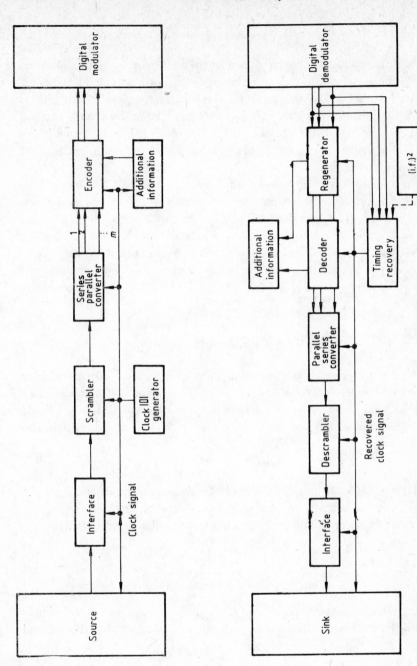

Fig. 10.1. Building blocks of typical baseband signal processing.

The timing of the baseband signal processing circuits is governed by the clock signal. The clock signal can be derived from the source or an external clock signal can be used to supply the source (as shown by the arrows of the clock signal line in Fig. 10.1); in the latter case, the transmit side signal processing circuit is provided by an internal clock generator. At the receive side, the clock information is extracted from the transmitted information bit stream, either by a serial method (solid line) or a parallel method (dashed line). The recovered clock signal is utilized to drive the regenerator circuit and the other signal processing circuits, and is also available to the sink.

The main topics of this chapter are the following: survey of a few more significant baseband digital codes, a few application problems (spectral properties, application of a scrambler, Gray and differential encoding), and finally, application examples will be presented for biphase ASK, 2PSK, 4PSK, 8PSK, O-PSK, MSK, QAM and QPRS systems. In the discussion of the scrambler, the pseudo random generators and their application for error probability measurement will be outlined. The regeneration and clock recovery process is dealt with separately in Chapter 3.

10.1 Baseband code systems

In this section, the properties of serial digital signal streams, transmitted over either a digital channel or over a channel normally used for the transmission of analog signals, will be investigated. Digital signals can be regarded as amplitude modulated pulse trains, and can thus be described in the time domain by the following convolution:

$$S(t) = S_0(t) * \sum a_n \delta(t - nT), \tag{10.1}$$

where $s_0(t)$ represents the time function of the elementary pulse and a_n is the discrete range of the actual pulse train element. The Fourier transform of $s(t)$ yields the following expression for the spectrum of the digital pulse train:

$$S(w) = S_0(w) \sum a_n e^{-j\omega nT}. \tag{10.2}$$

From the above expression, the following conclusions can be drawn.

(i) The elementary signal shape of time function $s_0(t)$ is equivalent to a simple filtering function in the frequency domain, independently of the spectral properties of the applied code system.

(ii) It is seen that

$$\frac{S}{S_0}\left(w + k\frac{2\pi}{T}\right) = \frac{S}{S_0}(w), \tag{10.3}$$

where k is an arbitrary integer, and

$$\frac{S}{S_0}\left(\frac{\pi}{T} + w\right) = \frac{S^*}{S_0}\left(\frac{\pi}{T} - w\right). \tag{10.4}$$

It can be shown by using (10.3) and (10.4) that the spectrum of an arbitrary pulse train can be uniquely characterized by the spectrum part falling into the frequency range $0 < \omega < \pi/T$ as expressed also by the first Nyquist criterion [1]. Assume that the transmission channel is characterized by an ideal low-pass filter having a cut-off frequency $F_{max} = 1/2T$; this allows a transmission free of intersymbol interference. However, this criterion cannot be exactly met in practice, leaving one of the two following choices: an antimetric cut-off with a smaller slope (higher bandwidth), or on the contrary, application of partial response transmission (smaller bandwidth), and thus exceeding the bit rate limit following from the first Nyquist criterion [5, 6].

The average spectral density function resulting from a specified code system is given by the Fourier transform of the encoded signal autocorrelation function [22, 32, 33]. In several applications, the zero or almost zero value of the DC spectrum component has a special significance. Code systems not having DC components are characterized by the digital sum variation DSV which is the difference between the highest and lowest digital sum values of an arbitrary number of elementary symbols:

$$DSV = \left[\sum_{n=M}^{N} a_n\right]_{max} - \left[\sum_{n=M}^{N} a_n\right]_{min}, \tag{10.5}$$

where N and M denote arbitrary limits of the investigation. Evidently, the minimization of the DSV value is advantageous.

There are several methods of classifying baseband code systems. In the following, two-level and multilevel code systems will be distinguished. Multilevel code systems allow the generation of both linear and nonlinear codes. The more significant linear codes are the partial response code, the AMI code, the duobinary code, *etc.*

10.1.1 Two-level codes

In digital transmission, the binary signal type, having a signal set of zeros and ones, is most frequently applied. In most cases, the symbol values correspond to voltage levels. The binary signal streams have two variants. In the NRZ (Non-Return-to-Zero) variant, the voltage levels characterizing the symbols are maintained throughout the symbol time or bit time while in the RZ (Return-to-Zero) variant, the voltage level pertaining to the symbol one is maintained only in a part of the bit time. Figure 10.2 shows the generation method and the waveform of the NRZ signal and the RZ signal with 50 per cent duty ratio. The RZ signal can be derived from the NRZ signal by utilizing an AND combination of the NRZ signal and the clock signal:

$$B_{RZ} = B_{NRZ} ck. \tag{10.6}$$

A high DC component of binary signals is frequently unwanted. For this reason and others, biphase encoding is frequently applied [3, 11], according to which the original NRZ information signal and the clock signal are summed in a mod 2 sense. Two variants of the biphase encoding are applied. According to the WAL_1 encoding, the clock signal is in phase with the NRZ signal while according to the WAL_2 encoding, there is a phase difference of 90° between the two signals*, as shown in Fig. 10.2:

$$B_{WAL_1} = B_{NRZ} + ck \bmod 2 \tag{10.7}$$

$$B_{WAL_2} = B_{NRZ} + ck_{90} \bmod 2. \tag{10.8}$$

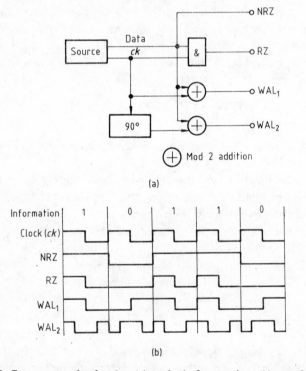

(a)

(b)

Fig. 10.2. Common two-level codes: (a) method of generation; (b) waveforms.

* The WAL notation refers to the similarity with the Walsh functions.

The bit rate of the pulse train generated by WAL encoding is formally higher than the bit rate of the original NRZ signal: twice higher for WAL_1 and four times higher for WAL_2 as shown in Fig. 10.2b, but with favourable code restriction. In the case of WAL_1 encoding, the highest square-wave frequency $1/T$ is obtained for a continuous run of zeros or ones, while a train of ...1010... results in the lowest square-wave frequency $1/2T$. In the case of the WAL_2 encoding, a continuous run of zeros or ones generates a pulse train of medium frequency $1/T$ while a binary train of ...1010... results in an NRZ train of ...00101101... which is a pattern of period $T/2$, comprising both the highest frequency component $2/T$ and the lowest frequency component $1/2T$. Applying biphase encoding, all resulting patterns of the original NRZ signal will have frequencies between the above limits which is extremely advantageous for obtaining lowest intersymbol interference and pattern independence of the clock frequency component.

The symbol sets and the spectral power densities of the preceding codes are shown in Fig. 10.3. Investigation of the symbol sets shows that WAL encoding can be realized in mode A and mode B, illustrating the phase ambiguity involved: during the transmission of continuous zeros or ones, there is no reference at the receive side to allow us to obtain the phase information. This phase ambiguity is

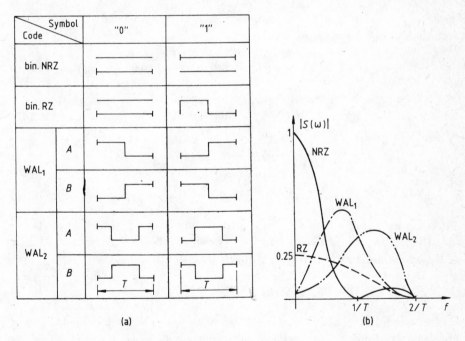

(a)   (b)

Fig. 10.3. Characteristics of two-level codes in the case of random binary information: (a) symbol set; (b) spectral power density; T is the bit time.

in this case also advantageously resolved by the application of differential encoding. Figure 10.3b shows that because of the more frequent transitions, the RZ signal spectrum has a more extended character compared with the NRZ signal spectrum. The spectra of WAL encoded signals have no DC component, the WAL$_1$ spectrum being more concentrated than the WAL$_2$ spectrum. All spectra shown in Fig. 10.3b are valid for random digital signals. The DSV value referring to the absence of the DC component can be calculated from Eq. (10.5). DSV$=\infty$ for binary signals, DSV$=1$ for WAL$_1$ signals and DSV$=3$ for WAL$_2$ signals.

10.1.2 Linear multilevel codes

A multilevel code system comprising binary information is said to be linear if the digital values of the code are given by the linear combination of K consecutive binary signal elements:

$$a_n = \sum_{k=0}^{k} b_{n-k}\alpha_k, \tag{10.9}$$

where a_n is the amplitude of the multilevel code, α_k is the weight characterizing the code generation, and b_{n-k} is the value of the binary signal at a given time instant. The time function of the above multilevel digital signal stream is given, according to Eq. (10.1), as follows:

$$s(t) = \sum_{m=-\infty}^{\infty} b_m\delta(t-mT) \sum_{k=0}^{k} \alpha_k\delta(t-kT) * s_0(t). \tag{10.10}$$

It can be seen that a linearly multilevel coded signal with binary information content can also be characterized by the time function of a binary code which has a signal element characterized by the function

$$s_0'(t) = s_1(t) * s_0(t), \tag{10.11}$$

where

$$s_1(t) = \sum_{k=0}^{k} \alpha_k\delta(t-kT). \tag{10.12}$$

The binary signal is thus converted into a multilevel signal by a linear transformation corresponding to a filtering operation. The frequency characteristic of the operation is expressed by the Fourier transform of $s_1(t)$:

$$S_1(w) = \sum_{k=0}^{k} \alpha_k e^{j\omega kT}. \tag{10.13}$$

From all multilevel codes, the three-level codes, to be discussed in the following, have greatest significance. Special considerations will be given to the partial response codes which are frequently higher than three-level codes.

10.1.3 Partial response transmission codes

The essential feature of partial response binary transmission and encoding is the generation of a controlled intersymbol interference by not fully meeting the Nyquist criteria and by suitably selecting the transmission channel parameters, thus realizing a decision model for a multilevel digital signal [6, 18, 19, 20, 29]. The block diagram of partial response transmission is shown in Fig. 10.4. The task is the transmission of the binary signal generated by the source. According to the model shown in Fig. 10.4a, the input of the band limited channel is driven directly by the signal of the source. At the output of the channel, the regeneration is carried out according to the multilevel digital signal $\{c\}$, generated by the controlled distortion of the channel. A suitable decoding results in the binary signal $\{a'\}$ which is, in principle, equal to the binary signal $\{a\}$. In the model shown in Fig. 10.4b, the same principle is applied but the multilevel signal $\{c\}$ is already generated at the transmit side, i.e. it is not generated by the band limiting property of the channel. The variant shown in Fig. 10.4b is utilized if the primary goal is to limit the radiated signal spectrum width.

In the models shown in Fig. 10.4, the digits c_n of the multilevel signal $\{c\}$ can be expressed as the weighted sum of the binary bits $\{a\} = a_1, \ldots; a_n$:

$$C_n = k_1 a_n + k_2 a_{n-1} + \ldots + k_n a_1. \tag{10.14}$$

The band limiting in the frequency range $0 < f < F = 1/T$ can be given by the following characteristic:

$$H(w) = \int_{-\infty}^{\infty} \left[k_1 \delta(t) + k_2 \delta\left(t - \frac{1}{2F}\right) + \ldots + k_n \delta\left(t - \frac{n-1}{2F}\right) \right] e^{-j\omega t} \, dt. \tag{10.15}$$

This expression shows that owing to the controlled intersymbol interference, a common weighted evaluation of n binary symbols, according to weights k_i, takes place at the receive side.

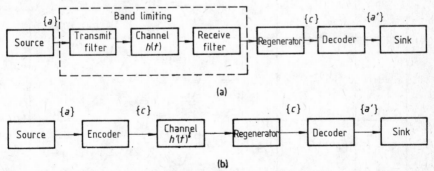

(a)

(b)

Fig. 10.4. Block diagram of partial response binary transmission: (a) encoding by band limiting; (b) digital multilevel encoding.

10.1.3.1 Classification

There is an infinite number of partial response transmission methods, and it is usual to classify them into classes and orders [6, 19]. The class number refers to the weight values and symmetries, while the order refers to the size of the system memory. The more significant partial response systems are summarized in Fig. 10.5,

Class (C)	Order (n)	Weights $k_1\ k_2\ k_3 \ldots k_n$	Number of levels	Weight function $h(t)$	Spectrum $0 < f < F$
0	1	1	2		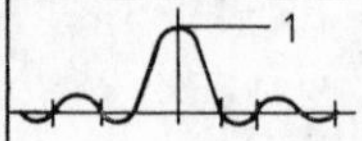 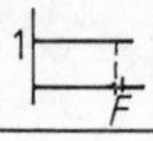 1
1	2	1 1	3		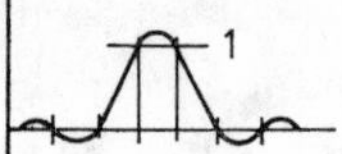 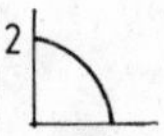 $2\cos\left(\dfrac{\pi}{2}\dfrac{f}{F}\right)$
	3	1 1 1	4		 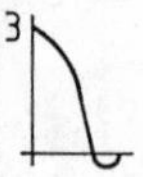 $\dfrac{\sin\left(\dfrac{3\pi}{2}\dfrac{f}{F}\right)}{\sin\left(\dfrac{\pi}{2}\dfrac{f}{F}\right)}$
	n	1 1 1 1	$n+1$		$\dfrac{\sin\left(\dfrac{n\pi}{2}\dfrac{f}{F}\right)}{\sin\left(\dfrac{\pi}{2}\dfrac{f}{F}\right)}$
2	3	1 2 1	5		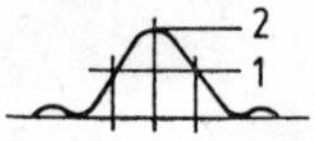 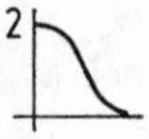 $1+\cos\left(\pi\dfrac{f}{F}\right)$
	n	1 2 3 . . . 3 2 1	$\dfrac{(n+1)^2}{4}+1$		$\dfrac{2}{n+1}\ \dfrac{1-\cos\left(\dfrac{n+1}{2}\pi\dfrac{f}{F}\right)}{1-\cos\left(\pi\dfrac{f}{F}\right)}$
3	3	2 1 −1	5		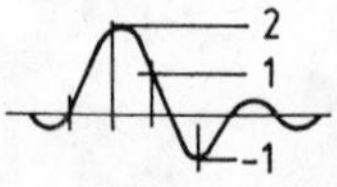 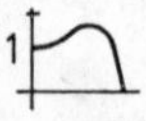 $2+\cos\left(\pi\dfrac{f}{F}\right)-$ $-\cos 2\pi\left(\dfrac{f}{F}\right)+$ $+j\left[\sin\left(\pi\dfrac{f}{F}\right)-\sin\left(2\pi\dfrac{f}{F}\right)\right]$
4	7	1 2 1 0 −1 −2 −1	3		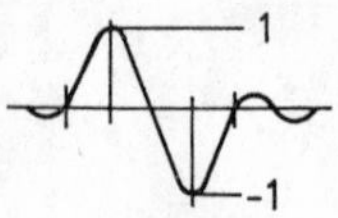 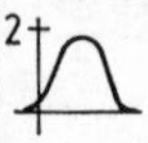 $2\sin\left(\pi\dfrac{f}{F}\right)$
5	5	−1 0 2 0 −1	5		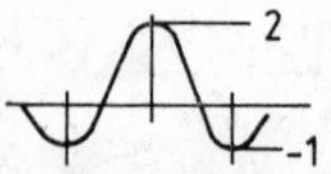 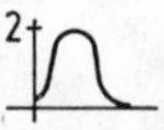 $1-\cos\left(2\pi\dfrac{f}{F}\right)$

Fig. 10.5. Characteristics of frequently used partial response transmission modes.

presenting the weight numbers characterizing the system, the levels of the transmitted partial response digital signal, the characteristic weight function, and finally, the spectrum part within the Nyquist bandwidth. The first row of the figure corresponds to binary transmission: a channel characterized by an ideal low-pass filter $0 < f < F$ generates the well known weight function $\sin x/x$. However, if the ideal band limiting is substituted by the limiting given by the frequency characteristic $2\cos\left[\dfrac{\pi}{2}\dfrac{f}{F}\right]$ in the Nyquist range, then the weighting function will be broadened. In this case, for the transmission of a binary "1", most of the response energy will be spread into twice the bit time interval, instead of a single bit interval. This transmission mode is summarized in the second row of Fig. 10.5 ($C=1$, $n=2$).

The operation of the system $C=1$, $n=2$ is explained in Fig. 10.6. Figure 10.6a shows the variant in which the multilevel signal is generated by the channel band limiting. In this case, the transmission of the binary signal 001100 results in a receive side response which is the response sum $1' + 1''$ for an input level drive $1'$ and $1''$. The regenerator will then produce the multilevel signal 001210... Figure 10.6b shows the variant in which the multilevel signal is generated by digital means at the transmit side. In this case, the channel should meet the requirement shown in the figure, evidently with efficient band utilization.

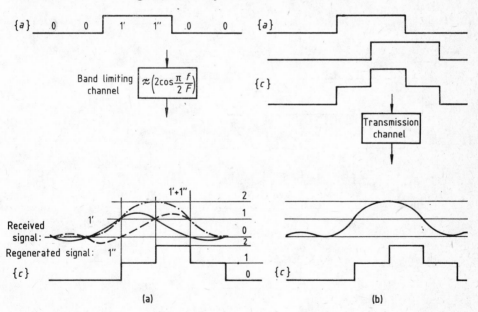

Fig. 10.6. Duobinary partial response transmission, $C=1$, $n=2$: (a) encoding by band limiting; (b) digital three-level encoding.

From the variants shown in Fig. 10.5, the class $C=1$ is frequently applied because of its simplicity. An example for this class, the QPRS system, will be investigated in this chapter. The variant $C=4$, $n=3$ also has high significance because of the zero DC component of its spectrum, and is widely used for data transmission.

10.1.3.2 Decoding

It follows from the above investigations that a regenerator suitable for n-level decision is needed for the realization of partial response binary transmission. The multilevel pulse train which appears as a result of the decision has to be decoded according to the block diagrams shown in Fig. 10.4. A simple decoder is realized by the so-called McColl circuit shown in Fig. 10.7, comprising a recursive digital filter and a comparator [19]. The multilevel pulse train $\{c\}$ is connected to one comparator input while the other input is driven by the output signal of the shift register. The latter signal can be expressed as the following sum expression:

$$S_n = \sum_{i=1}^{n-1} k_{i+1} a_{n-i}. \tag{10.16}$$

The values of the weights k_{i+1} are given in Fig. 10.5. and Table 10.2. It is evident from the code generation rule expressed by Eq. (10.14) that

$$a_n = \begin{cases} 0 & (C_n - S_n = 0) \\ 1 & (C_n - S_n = k_1) \end{cases} \tag{10.17}$$

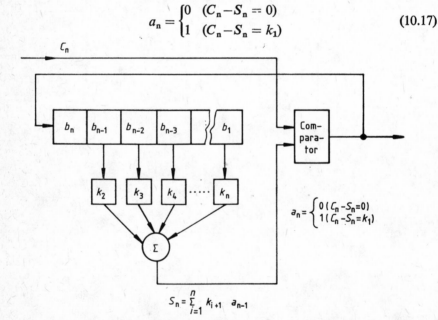

Fig. 10.7. McColl circuit [19] for the decoding of partial response multilevel signals.

and the realization of the binary transmission can be comprehended. If suitable precoding is applied then the decoding operation can be carried out by simple even–odd detection, as will be explained in Section 10.1.4.

Performance

The performance of partial response transmission will be characterized by two fundamental parameters.

(i) How far should the signal-to-noise ratio be increased with respect to the binary transmission?

(ii) To what extent should the Nyquist rate be exceeded to obtain the first closure of the eye diagram?

For the transmission modes shown in Fig. 10.5, these data are presented in Table 10.1.

Table 10.1. Performance of a few partial response systems (with respect to binary transmission)

Partial response system		Signal-to-noise ratio degradation, dB	Bit rate tolerance above Nyquist rate*, %
Class C	Order n		
0**	1	0	0
1	2	2.1	43
	4	3.5	
	5	6.1	
2	3	6.0	40
	7	12.0	
3	3	1.2	38
4	3	2.1	15
5	5	6.0	8

$$* \; 1 + \text{tolerance} = \frac{\text{Bit rate at first closure of the eye diagram}}{\text{Nyquist rate}}.$$

** binary transmission.

One drawback of partial response systems is the error multiplication *i.e.*, the generation of a long burst of errors, instead of a single error, created by a false decision of the regenerator. This drawback can be eliminated by the application of precoding.

10.1.4 Precoding

Because of the memory present in partial response systems, the error multiplication phenomenon, mentioned earlier, may arise. For instance, the decoding
rule can be constructed from Eq. (10.17) by following the operation algorithm of
the simple system characterized by $C=1$, $n=2$: a binary one corresponds to
level two, a binary zero corresponds to level zero, and the negated version of the
preceding binary signal corresponds to level one. Let us investigate the error
multiplication by the following example:

$\{a\}$ ... 0 1 1 0 1 0 1 0 0 1 ...   input binary signal

$\{c\}$ ... 0 1 2 1 0 1 1 1 0 1 ...   faulty signal: 0

$\{a'\}$ ... 0 1 1 0 0 1 0 1 0 1 ...   erroneously decoded bits: 0 1 0 1.

In this example, one decision error resulted in four bit errors. In the case of
higher classes and orders, a further drawback of partial response systems is the
complicated structure of the decoder unit shown in Fig. 10.7.

The error multiplication and the complicated decoding structure can be avoided
by precoding, allowing parity detection. The partial response system with precoding is shown in Fig. 10.8: the bit sequence $\{b\}$, generated by the precoder from the
bit sequence $\{a\}$ to be transmitted, has elements for which the following expression holds:

$$a_n = \sum_{i=1}^{n} k_i b_{n+1-i} = k_1 b_n + k_2 b_{n-1} + ... + k_n b_1 \bmod 2, \qquad (10.18)$$

where k_i denote the weights setting the parameters of the partial response transmission, summarized in Table 10.2. It can be shown that for even k_i values,
$k_i b_{n+1-i} = 0 \pmod 2$, and for odd k_i values, $k_i b_{n+1-i} = 1 \pmod 2$. On the receive
side, the multilevel bit sequence at the regenerator output is expressed by

$$C_n = \sum_{i=1}^{n} k_i b_{n+1-i} = k_1 b_n + k_2 b_{n-1} + ... + k_n b_1. \qquad (10.19)$$

Evaluation of Eqs (10.18) and (10.19) allows the following conclusions to be
drawn for the decoding:

(i) if $a_n = 1$, this means odd parity according to (10.18), so C_n is odd;
(ii) if $a_n = 0$, this means even parity according to (10.18), so C_n is even.

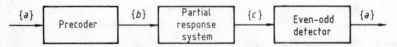

Fig. 10.8. Principle of precoded partial response transmission.

Table 10.2. Weights of a few partial response systems

Partial response system		Weights								
Class C	Order n	k_1	k_2	k_3	k_4	k_5	k_6	k_7	k_8	k_9
1	2	1	1							
	3	1	1	1						
	4	1	1	1	1					
2	3	1	2	1						
	7	1	2	3	4	3	2	1		
3	3	2	1	-1						
	7	4	3	-3	3	-2	1	-1		
4	3	1	0	-1						
	7	1	2	1	0	-1	-2	-1		
5	5	-1	0	2	0	-1				
	9	1	0	-2	0	3	0	-2	0	1

Table 10.3. Precoding rules of partial response systems

Class C	Order n	$b_n = a_n \oplus k_2 b_{n-1} \oplus \ldots \oplus k_n b_1$
1	2	$b_n = a_n \oplus b_{n-1}$
	3	$b_n = a_n \oplus b_{n-1} \oplus b_{n-2}$
	4	$b_n = a_n \oplus b_{n-1} \oplus b_{n-2} \oplus b_{n-3}$
2	3	$b_n = a_n \oplus b_{n-2}$
	7	$b_n = a_n \oplus b_{n-2} \oplus b_{n-4} \oplus b_{n-8}$
3	3	$b_{n-1} = a_n \oplus b_{n-2}$
	7	$b_{n-1} = a_n \oplus b_{n-2} \oplus b_{n-6}$
4	3	$b_n = a_n \oplus b_{n-2}$
	7	$b_n = a_n \oplus b_{n-2} \oplus b_{n-4} \oplus b_{n-6}$
5	5	$b_n = a_n \oplus b_{n-4}$
	9	$b_n = a_n \oplus b_{n-4} \oplus b_{n-8}$

$\oplus$ = addition in the modulo 2 sense.

The investigation above shows that by the use of precoding, the decoding process can be implemented by a simple even/odd detection operation, and there is no error multiplication. Precoding rules have been summarized in Table 10.3.

Design example

A 9-level partial transmission code with $C=4$, $n=7$, has to be designed. The precoding rule is given by

$$b_n = a_n + b_{n-2} + b_{n-4} + b_{n-6} \bmod 2. \qquad (10.20)$$

From Table 10.2, the values of the weights k_i are given by 1, 2, 1, 0, -1, -2, -1. The digital realization of the encoder results in the recursive digital filter shown in Fig. 10.9. Equations (10.19) and (10.20) allow numerical evaluation of the encoder as follows:

signal to be transmitted	a	0 1 1 0 1 0 0 0 0 1 1 1 0 0 1 0 1 0
precoded signal	c	0 1 1 1 0 0 1 0 0 0 0 1 1 1 0 0 0 0
partial response signal	c	0 1 3 4 3 0 2 2 2 1 1 1 2 4 3 0 3 4
parity of partial response signal		E D D E D E E E E D D D E E D E D E
received signal	a'	0 1 1 0 1 0 0 0 0 1 1 1 0 0 1 0 1 0
(E = even, D = odd)		

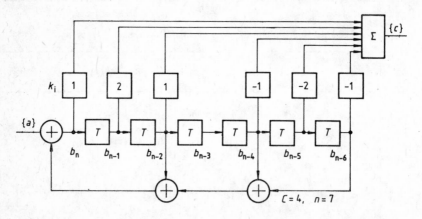

Fig. 10.9. Generation of precoded partial response signal stream with $C=4$, $n=7$.

10.1.5 Three-level linear codes

Usable three-level codes are either partial response codes or codes of other types. Let us now investigate two of these latter.

The duobinary three-level code is generated according to the following rule:

$$C_n = a_n - a_{n-1}. \qquad (10.21)$$

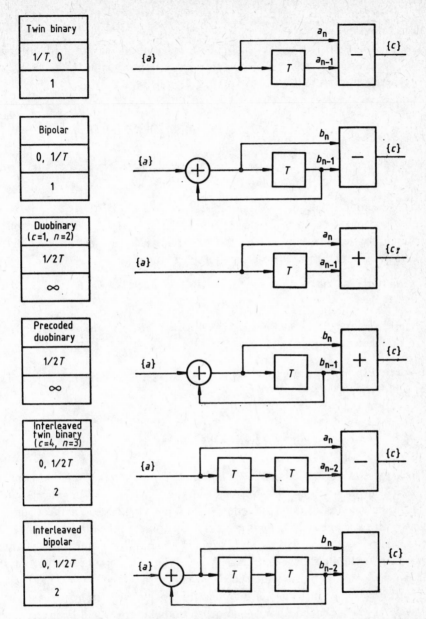

Fig. 10.10. Main characteristics and generation methods of frequently used three-level linear codes. Below the code designation at the left, the upper data are the zero frequencies of the spectrum while the lower number is the DSV value.

The spectrum of this code has no DC component, and its digital sum variation DSV is unity.

The precoded version of the duobinary code is the alternate mark inversion (AMI) code which is a bipolar code frequently applied in PCM transmission. It is generated according to the following rule:

$$\left.\begin{aligned} C_n &= b_n - b_{n-1}, \\ b_n &= a_n + b_{n-1} \bmod 2 \end{aligned}\right\}. \tag{10.22}$$

According to this code, the binary signal $\{a\}$ is transformed into the bipolar signal $\{c\}$ according to the following:

— zeros become zeros,
— ones become ones which, however, have alternating signs, independent of the number of zeros. Errors in the bipolar signal have the effect of violating the bipolar rule, and there is no error multiplication.

The partial response code has the following three-level variants:

— duobinary code with $C=1$, $n=2$, and its precoded variant,
— interleaved twinbinary code with $C=4$, $n=3$ and its precoded variant, the interleaved bipolar code.

Figure 10.10 shows the derivation of these six codes, together with the zero-spectrum frequency and DSV values.

10.1.6 Three-level nonlinear codes

There exists a variety of nonlinear codes having binary information content, designated in the literature as "alphabetical" and "high density" nonlinear codes [4, 5]. In these codes, 3^n different code words can be generated in the course of n consecutive symbols. The code is generated by grouping the binary stream to be encoded into m-bit binary words, and assigning 2^m set of code words to the ternary symbol set. It can be shown that $2^m \leq 3^n$ which leads to $m \leq 1.58n$. This means that in principle, the transmission speed can be increased by 58 per cent. There is no speed increase with the use of high density bipolar type ternary codes. In the following, a few alphabetical and high density bipolar codes will be investigated.

PST (Pair Selected Ternary) code. This is the most widely used alphabetical ternary code, generally used for primary level PCM transmission. The binary signal to be transmitted is grouped into dibits, and the following correspondence is realized for the $3^2 = 9$ ternary signal set:

Dibit	(+) mode PST	(−) mode PST
00	− +	− +
01	Ø +	Ø −
10	+ Ø	− Ø
11	+ −	+ −

In the above table, symbols −, +, Ø represent the levels of the ternary code. Only the 4/9 part of the PSK code set is actually utilized, and this redundancy allows continuous error monitoring. Clock recovery is easily realized, making possible binary transmission almost without code restriction. On the other hand, DSV=3 and compared to the bipolar AMI code, a signal-to-noise ratio about 1.5 dB higher is required for a given error probability.

3B–2T code. According to this code, two-symbol ternary groups are assigned to three-bit binary words ($m=3$, $n=2$). The speed increase is 50 per cent. This code has a DC component.

4B–3T code. This is a more widely used code as it has no DC component. It has parameters $m=4$, $n=3$, and a unique correspondence realized between code words allows the elimination of the DC component. From the possible $3^3=27$ symbol sets, ØØØ is excluded, and a unique correspondence to the following symbols with zero digital sum is realized: $+Ø−$, $+−Ø$, $−Ø+$, $Ø−+$, $−+Ø$. The further 20 elements of the symbol set is grouped into the following complementary pairs: $+++$ and $−−−$; $−ØØ$ and $+ØØ$, *etc*. To each of the latter 10 complementary pairs, specific binary words are assigned which alternate during the encoding. It can be shown that for this code, DSV=7.

A desirable property of the alphabetical codes is the possibility of detection without error multiplication.

High density bipolar codes. These have gained importance with the increasing number of PCM transmission systems. These codes eliminate arbitrarily long runs of zeros which is the unfavourable property of the bipolar AMI code. High density bipolar codes are thus suitable for transmission without code restriction. All high density encoding systems include a monitoring of the binary–bipolar transcoding result, and if the number of consecutive zeros exceeds a specified limit, these zeros are replaced by a distinguishable symbol. This symbol is recognized at the receive side, and is re-substituted by the suitable number of zeros. Several types of high density bipolar codes are used, depending on the property of whether or not the same symbol is used for all long zero runs. The recognition of these symbols at the receive side is based on the defined violation of the bipolar encoding rule.

The bipolar rule means that the binary zero always converts to a ternary zero, *i.e.* $0 \rightarrow Ø$, while the binary 1 alternately becomes a ternary $+1$ or -1. This operation can be denoted by $1 \rightarrow B$ by taking into account the alternating operation.

B6ZS (Bipolar-with-six-Zero-Substitution) code. This is a line code used in the American PCM hierarchy. Its encoding rule corresponds to the bipolar encoding rule as long as the number of zeros ($\emptyset$) in the ternary signal exceeds five. In this case, a special symbol $B\emptyset VB\emptyset V$ is inserted in the place of the sixth ternary zero where V denotes the violation of the above bipolar coding rule.* For the B6ZS code, DSV=3.

HDB$_n$ (High Density Bipolar) and CHDB$_n$ (Compatible High Density Bipolar) codes. Compared with the preceding codes, these codes have better spectral properties. Subscript n denotes the highest allowed number of consecutive $\emptyset$ symbols. If the number of zeros in the binary signal to be transmitted is less than or equal to n, then the encoding corresponds to the known bipolar rule. However, if the bipolar rule resulted in a character made up of $n+1$ $\emptyset$ symbols, then these would be substituted by one of the following characters:
— in the case of an HDB$_n$ code, by

$$B\emptyset\emptyset, ..., \emptyset V, \tag{10.23}$$

or

$$\emptyset\emptyset\emptyset, ..., \emptyset V \tag{10.24}$$

and in the case of a CHDB$_n$ code, by

$$\emptyset, \emptyset, ..., B\emptyset V \tag{10.25}$$

or

$$\emptyset, \emptyset, ..., \emptyset\emptyset V. \tag{10.26}$$

The characters should always be chosen to leave an even number of B symbols between two V symbols. The substituting characters (10.23) and (10.25) are applied if the preceding V of the ternary stream is followed by an even number of B's while the substituting characters (10.24) and (10.26) are applied if the preceding V is followed by an odd number of B's.

It can be shown for these codes that DSV=$n-1$. The frequency spectra of the two codes are practically identical, but the CHDB$_n$ code requires less hardware; also, a CHDB$_n$ decoder can be applied for any number of n. In spite of this, the HDB$_3$ code is widely used internationally, so in the following, we summarize briefly the encoding and decoding rules of this code.

* The bipolar encoding rule is violated if the consecutive ternary $+1$ or -1 symbols do not alternate.

HDB₃ encoding rules

(i) A binary 0 becomes a ternary Ø, but if the character is made up of more than three zeros then rule *(iii)* applies.

(ii) A binary 1 becomes a ternary symbol B according to the bipolar rule, but if the character is made up of more than three zeros then the bipolar rule has to be violated.

(iii) The character 0000 of the binary signal is encoded as follows:

— the first binary 0 becomes a ternary Ø if the preceding B and V of the ternary signal are of opposite polarity;

— the preceding binary 0 becomes a ternary B not violating the bipolar rule if the preceding B and V of the ternary signal are of identical polarity;

— the second and third binary 0 becomes a ternary Ø;

— the fourth binary 0 becomes a ternary V violating the bipolar rule.

HDB₃ decoding rules

(i) A ternary Ø always becomes a binary 0.

(ii) A ternary B becomes either a binary 1 if the B is not followed by a ØØV character, or a binary zero if the B is followed by a ØØV character.

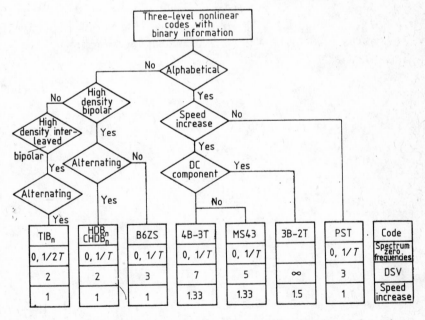

Fig. 10.11. Characteristics of three-level nonlinear codes with binary information.

(iii) A ternary V becomes either a binary 0 if the V is preceded by an $MB\emptyset\emptyset$ or $M\emptyset\emptyset\emptyset$ character (where M means either B or V), or a binary one if it is not preceded by one of these characters.

The second zero frequency of the HDB_n and $CHDB_n$ codes is equal to $1/T$ which is a shortcoming from the aspect of regeneration.

On the other hand, the second zero frequency of the TIB_n (Transparent Interleaved Bipolar) code is equal to $1/2T$ thus improving the conditions of bit timing recovery. This is not a bipolar code but an interleaved bipolar code treated in Sections 10.1.4 and 10.1.5 and shown in Fig. 10.10 [5]. The main parameters of the ternary codes investigated in the preceding are summarized in Fig. 10.11.

10.1.7 More than three-level nonlinear codes

In digital microwave transmission, binary signals are usually not transmitted by applying more than three-level nonlinear codes. However, these codes may appear in certain phases of the baseband signal processing, for example, in realizing 8PSK modulation and demodulation.

The binary signal $\{a\}$ is transformed by a series–parallel converter into dibit, tribit, tetrabit, *etc.* parallel signals $\{b\}$. An analog multiplexer M is applied to transfer to the output Y a signal c_i which corresponds to the control of the parallel signal vector $\{b\}$ (see Fig. 10.12). This transcoding can also be realized by a weighting network. The symbol rate of the m-level symbol appearing at the multiplexer output is given by

$$S_m = \frac{S}{\log_2 m},$$
(10.27)

where s denotes the bit rate of the binary signal $\{a\}$.

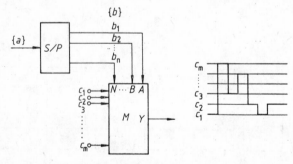

Fig. 10.12. Generation of multilevel nonlinear codes by a digitally driven analog multiplexer

10.2 Spectral behaviour of encoded signals

The spectral properties of various code types have already been discussed in Section 10.1. The frequency spectrum of a given code type pertaining to a given signal is not in itself meaningful because in addition, the effect of statistical and deterministical signal changes on the spectrum must also be known.

In radio transmission, binary codes have great significance. However, the spectrum character of these codes depends essentially on the information content of the digital signal, which is a drawback. A way out of this problem is the scrambling of the binary signal as discussed in Section 10.3.

The spectra of three or more level codes are less dependent on the information to be transmitted, because the redundancy due to the binary information content has the effect of reducing the spectrum variation [32, 33]. Still more favourable are the biphase WAL_1 and WAL_2 codes which have a structure resulting in spectra which are practically independent of the transmitted information.

The independence of the spectrum from the transmitted information and thus the constant clock frequency component of the spectrum is of importance for straightforward clock recovery. On the other hand, the performance of the digital modulator and demodulator depends also on the spectral behaviour of the transmitted digital signal. The clock recovery process requires frequent signal transitions while the lack of transitions is often preferred by the carrier recovery process. Strong line components in the spectrum, a result of periodic patterns of relatively short duration, have adverse effects on both the clock recovery and the carrier recovery processes. A relatively good compromise is provided by a digital signal which is random and has a spectrum which is only slightly dependent on the information content.

In digital microwave transmission, most of the signal energy should be concentrated within a given bandwidth. In Table 10.4, the energy relations of a few well known code types, discussed in Section 10.1, are summarized by assuming the transmission of a random binary signal. It is seen from this table that for the precoded duobinary code ($C=1$, $n=2$), more than 90 per cent of the total signal energy is concentrated within the Nyquist bandwidth [11, 22].

Some applications require that the spectrum of the transmitted signal should not contain a DC component. A more stringent requirement is set by the DSV quantity defined by Eq. (10.5) which refers to the short term signal average of the code and thus allows the estimation of the near-zero frequency components of the spectrum. Unity DSV, a property of only bipolar (AMI) and duobinary codes, is most favourable. DSV values of a few discussed code types are summarized in Table 10.5.

Table 10.4. Percentage energy concentration of a few codes discussed in Section 10.1, by assuming binary signal transmission. Total signal energy $= 100$ per cent, $T =$ bit time

Code	Bandwidth			
	$\dfrac{1}{2T}$	$\dfrac{1}{T}$	$\dfrac{2}{T}$	$\dfrac{3}{T}$
Binary NRZ	77.3	90.2	95.0	96.5
Binary RZ (50%)	46.9	77.5	90.3	93.1
Biphase (WAL$_1$)	16.2	64.4	85.6	89.6
Bipolar NRZ	64.4	85.5	92.5	94.9
Bipolar RZ (50%)	44.8	79.6	90.1	93.2
Precoded duobinary	90.2	94.9	96.5	96.6
PST	68.3	86.9	93.2	95.4

Table 10.5. DSV values of a few codes discussed in Section 10.1, characterizing the DC and near-zero frequency spectral components

Code	DSV
Binary	∞
WAL$_1$	2
WAL$_2$	3
Bipolar	1
Twin binary	1
Interleaved binary	2
Interleaved twin binary	2
TIB$_n$	$n-1$
CHDB$_n$	$n-1$
HDB$_n$	$n-1$
PST	3
B6ZS	3
MS43	5
4B—3T	7
3B—2T	∞

10.3 "Whitening" of the binary information (scrambler)

We have seen that in many cases, it is an advantage if the digital baseband signal has a random behaviour, independently of the information to be transmitted. This is why the "scrambler–descrambler" principle has great significance in digital radio transmission [26, 27, 28] even though the nonlinear encoding methods discussed in Section 10.1.6 (*e.g.*, HDB$_n$, B6ZS *etc.*) are also intended for spectrum improvement.

The scrambler as an encryption device has been known for many years [2]. The evolution of digital integrated circuit technology has resulted in scramblers which can be realized economically and simply, resulting in many applications*.

According to the scrambler principle, a random digital bit stream having a prescribed sequence is added to the digital signal to be transmitted. On the receive side, knowledge of this sequence allows the subtraction of this bit stream and thus the recovery of the transmitted signal. The scrambler principle can be realized by several code systems but in practice, so-called pseudo random bit streams are frequently applied for this purpose. In the following, the methods of generating pseudo random signals will be surveyed.

10.3.1 Generation of pseudo random bit streams

Pseudo random bit streams are generated by a shift register and a suitable feedback network as shown in Fig. 10.13** [23, 24, 25, 28].

The blocks $a_1, ..., a_m$ form a shift register which has a sequence length m. Each block is responsible for the delay of one symbol, corresponding to the period of the clock signal drive. For a symbol set with p elements, the feedback network represents a mod p summing network. The range of weights C_i is 0 and 1; the former value means an open loop, and the latter a closed loop. In the following, only the binary case represented by $p=2$ will be investigated, but it can be proved that our results apply to other values of p as well [15].

Pseudo random generators allow the generation of binary bit streams which can have arbitrarily long sequence lengths by increasing the number of shift register elements, and which have the following properties.

---

* Scramblers are now widely used for data transmission, *e.g.* in CCITT Rec. V. 27, a scrambler is specified for 9600 bit/s data transmission.

** A digital bit stream is designated random if the symbol values are independent of each other, and the appearance of any symbol value has the same probability.

(i) Balance of symbols: the difference between the number of ones and zeros within a sequence is at most unity;

(ii) Correlation property: comparison of a group of bits with another group shifted by an arbitrary number of bits shows that the number of invariant bits differs by not more than one from the number of bits which have changed value.

(iii) Number of NRZ sequences: the length of sequences without transitions within a period is one bit for the half of the sequences, 2 bits for the quarter of the sequences, 3 bits for the eighth of the sequences, *etc*. The statistical distribution of the bit stream is thus uniquely defined, and is not altered by the code-to-code relative change of the NRZ sections in question.

(iv) General property: the addition in mod 2 sense of a pseudo random sequence to the same sequence shifted by an arbitrary number of bits results in the original sequence. However, the phase of this new sequence differs from the phases of both terms.

At any junction of the shift register shown in Fig. 10.13, a pseudo random sequence having a sequence length of $2^m - 1$ will appear if the feedback network meets certain requirements, and the shift register will go through all 2^m states according to a specified order, with the exception of the 0 state.

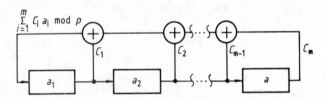

Fig. 10.13. Pseudo random generator made up from a feedback shift register. Weights $C_1...C_m$ should be suitably chosen.

The shift register state at the time of the nth bit is represented by the vector

$$\mathbf{A_n} = \begin{bmatrix} a_1 \\ \vdots \\ a_m \end{bmatrix}, \tag{10.28}$$

where a_i can be zero or one. The relation between this vector and that representing the state at the next, $(n+1)$th bit time is given by the matrix $\mathbf{T}$:

$$\mathbf{A_{n+1}} = \mathbf{T}\mathbf{A_n} \bmod 2, \tag{10.29}$$

where

$$\mathbf{T} = \mathbf{E} + \mathbf{U_m}\mathbf{C^*} \bmod 2. \tag{10.30}$$

Matrix $\mathbf{E}$ represents the shift register without the feedback network while the row vector $\mathbf{C}^*$ represents the feedback network. U_m is the unity vector. Thus

$$\mathbf{T} = \begin{bmatrix} C_1 & C_2 & \cdots & C_{m-1} & C_m \\ 1 & 0 & \cdots & 0 & 0 \\ 0 & 1 & \cdots & 0 & 0 \\ & & \vdots & & \\ 0 & 0 & \cdots & 1 & 0 \end{bmatrix} = \begin{bmatrix} 0 & 0 & \cdots & 0 & 0 \\ 1 & 0 & & 0 & 0 \\ 0 & 1 & & 0 & 0 \\ & & \vdots & & \\ 0 & 0 & & 1 & 0 \end{bmatrix} + \begin{bmatrix} 1 \\ 0 \\ 0 \\ \vdots \\ 0 \end{bmatrix} [C_1 \ldots C_m] \bmod 2. \quad (10.31)$$

By starting from the shift register initial state $\mathbf{A}_0$, corresponding to the zeroth bit time, it can be shown that the shift register state at the nth bit time will be given by

$$\mathbf{A}_n = \mathbf{T}^n \mathbf{A}_0 \bmod 2. \quad (10.32)$$

The feedback shift register states $\mathbf{A}_i$ are determined by matrix $\mathbf{T}$ and its powers, so the operation of the pseudo random generator shown in Fig. 10.13 can be understood from the properties of the matrices $\mathbf{T}^i$. If

$$\mathbf{T}^N = \mathbf{I} \bmod 2, \quad (10.33)$$

where $\mathbf{I}$ is unit matrix, then the matrix $\mathbf{T}$ is periodic according to N, i.e. the period of matrix $\mathbf{T}$ is N. On the other hand, the periodicity N will have maximal length of $N = 2^m - 1$ if the characteristic polynomial $h(x)$ of matrix $\mathbf{T}$ is primitive in the mod 2 sense [23, 24, 26], meaning that

$$h(x) = x^m - C_1 x^{m-1} - \ldots - C_m \bmod 2. \quad (10.34)$$

Polynomial $h(x)$ is primitive in the mod 2 sense if it cannot be factored, i.e. there is no other polynomial (with the exception of itself and 1) by which it is divisible in mod 2 sense. Further, the polynomial

$$(x^p - 1)/h(x) \quad (10.34a)$$

exists if $p = 2^m - 1$, but no finite polynomial exists if $p < 2^m - 1$ [2, 26, 34].

How should the row vector $\mathbf{C}^x$ defining the feedback be chosen to obtain a maximal sequence length? If, for instance, $m = 4$, then the conditions are met even by the polynomial $h(x) = x^4 - x - 1$ which cannot be factored; however, in a mod 2 sense,

$$x^{15} - 1 = (x^4 - x - 1)(x^{11} + x^8 + x^7 + x^3 + x^2 + x + 1) \quad (10.34b)$$

and furthermore, the polynomial fraction $(x^p - 1)/(x^4 - x - 1)$ has a residuum if $p < 2^4 - 1 = 15$.

The primitive variants, in the mod 2 sense, of characteristic polynomial (10.34) are discussed in Ref. [2] and treated in detail in Ref. [34]. A few simple polynomials are summarized in Table 10.6 which should be interpreted as shown by the following example. By using Eq. (10.34) and Fig. 10.13, the polynomial can be expressed as $h(x)=x^5-x^3-1$. In other words, the shift register network has a length of $m=5$, and in this case, the existing feedback loops are assigned by $C_3=C_5=1$. Further, $C_1=C_2=C_4=0$ denote non-realized feedback loops.

Table 10.6. A few possible series of coefficients in the polynomial (10.34) characterizing pseudo random generators

	C_1	C_2	C_3	C_4	C_5	C_6	
$m=2$	1	1	1				
$m=3$	1	0	1	1			
	1	1	0	1			
$m=4$	1	0	0	1	1		
	1	1	0	0	1		
$m=5$	1	0	0	1	0	1	
	1	0	1	0	0	1	
	1	0	1	1	1	1	
	1	1	1	0	1	1	
	1	1	0	1	1	1	
	1	1	1	1	0	1	
$m=6$	1	0	0	0	0	1	1
	1	0	1	0	1	1	1
	1	1	0	0	0	0	1
	1	1	0	0	1	1	1
	1	1	0	1	1	0	1
	1	1	1	0	0	1	1

Meaning of the underlined series: $h(x)=x^5-x^{-3}-1$ mod 2.

It is seen from Table 10.6 that several maximal length primitive polynomials pertain to a specific value m. This means that for example a 5 bit length shift register network is capable of generating 6 distinguishable pseudo random bit streams which all have the same statistical properties. The number of these distinguishable bit streams increases rapidly but not monotonously with m according to the relation

$$L(m) = \frac{\Phi(2^m-1)}{m}, \tag{10.35}$$

where $\Phi(x)$ is the Euler function defined as

$$\Phi(x) = \begin{cases} 1, & \text{if } x = 1 \\ \prod_{i=1}^{k} p_i^{\alpha_i - 1}(p_i - 1), & \text{if } x > 1, \end{cases} \qquad (10.36)$$

(see. Ref. [2]) where p_i denotes the prime factors of the number x.*

Table 10.7 is useful for the design of pseudo random bit generators, showing realization possibilities for m-bit generators up to $m=40$. We also give the number of possible variants $L(m)$, up to $m=24$.

Table 10.7. A few variants of the m-bit pseudo random generators shown in Fig. 10.13, and the total number of possible variants $L(m)$

m	$2^m - 1$	$L(m)$	Feedback points $C_i \neq 0$	m	$2^m - 1$	$L(m)$	Feedback points $C_i \neq 0$
2	3	1	$i = 1, 2$	22	4 194 303	120 032	$i = 21, 22$
3	7	2	2, 3	23	8 388 607	356 906	18, 23
4	15	2	3, 4	24	16 777 215	276 480	20, 21, 23, 24
5	31	6	3, 5	25			22, 25
6	63	6	5, 6	26			18, 19, 25, 26
7	127	18	6, 7	27			19, 20, 26, 27
8	255	16	2, 3, 4, 8	28			25, 28
9	611	48	5, 9	29			27, 29
10	1 023	60	7, 10	30			14, 15, 29, 30
11	2 047	176	8, 11	31			28, 31
12	4 095	144	2, 10, 11, 12	32			1, 27, 28, 32
13	8 191	630	1, 11, 12, 13	33			13, 33
14	16 383	756	2, 12, 13, 14	34			1, 14, 15, 34
15	32 767	1 800	14, 15	35			33, 35
16	65 535	2 048	11, 13, 14, 16	36			25, 36
17	131 071	7 710	14, 17	37			2, 10, 12, 37
18	262 143	8 064	11, 18	38			32, 33, 37, 38
19	524 287	27 594	14, 17, 18, 19	39			35, 39
20	1 048 575	24 000	17, 20	40			2, 19, 21, 40
21	2 097 151	84 672	19, 21				

* The interpretation of function Φ is illustrated by the following example. Calculate the number of distinguishable pseudo random bit streams which can be generated by an $m=12$ stage shift register. $2^{12}-1=4095=3^2 \times 5 \times 7 \times 13$, so according to Eq. (10.36), $\Phi(4095)= =3 \times 2 \times 4 \times 6 \times 12 = 1728$, and so $L(12) = \Phi(4095)/12 = 144$.

In all pseudo random generators, the single forbidden register state $\mathbf{A_0} \equiv 0$ (which could be generated following switch-on, or in the case of an operational failure) has to be eliminated. This task can be solved by either a logic network or a dynamic network counting the number of consecutive zeros.

10.3.2 Reset scrambler

A digital bit stream can be "whitened" by the application of the reset scrambler shown in Fig. 10.14. According to the transmits ide of this arrangement, a symbol-by-symbol addition of the pseudo random signal $\{g\}$ to the signal to be transmitted $\{D\}$, in a mod p sense, is performed:

$$\{D_s\} = \{D\} + \{g\} \bmod p. \tag{10.37}$$

Instead of the original signal $\{D\}$, the signal $\{D_s\}$ is transmitted. At the receive side, we have a pseudo random generator which is identical to the transmit side generator, and its signal $\{g\}$ is subtracted, in mod p sense, from the transmitted signal D_s:

$$\{D\} = \{D_s\} - \{g\} \bmod p, \tag{10.38}$$

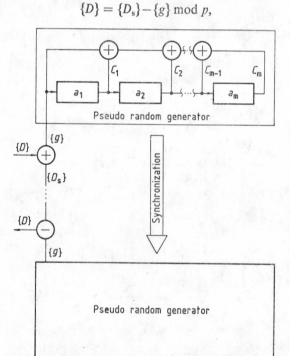

Fig. 10.14. Reset scrambler and descrambler.

thus recovering the original signal $\{D\}$. For a binary signal $(p=2)$, Eq. (10.38) becomes

$$\{D\} = \{D\}+\{g\}\, \text{mod}\, 2. \tag{10.39}$$

Obviously, the transmit and receive side pseudo random generators have to be synchronized. This is easily realized by utilizing the specified frame structure of hierarchical bit streams according to CCITT recommendations. By excluding from the scrambling operation at the transmit side, the synchronizing words intended for receive side frame recognition, these words are recognized at the receive side and can be applied for synchronization of the pseudo random generator. If the descrambler is placed near the PCM demultiplexer equipment, then the frame synchronizing word, generated by the latter, can also be utilized for the synchronization of the receive side pseudo random generator.

A disadvantage of the reset scrambler is the need for synchronization. However, it has the advantage that the channel is memoryless, meaning that the errors in the transmitted signal $\{D_s\}$ are not multiplied.

10.3.3 Self-synchronizing scrambler

The simplest realization of the scrambling principle is the application of the basic self-synchronizing scrambler [26, 27] shown in Fig. 10.15.

According to this arrangement, the previously applied pseudo random generator principle is again applied. It can also be seen that the scrambler shown in the upper part of the figure is a recursive digital filter. The lower part shows the descrambler which is essentially a non-recursive digital filter. The scrambler transformation

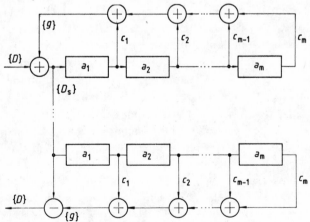

Fig. 10.15. Self-synchronizing scrambler and descrambler.

$\{D\} \rightarrow \{D_s\}$ and the descrambler transformation $\{D_s\} \rightarrow \{D\}$ are inverse operations and this also explains the spectrum forming effect involved.

Let us now investigate the scrambler operation. At the nth time instant, the nth element of the internal signal stream $\{g\}$ is expressed by the following scalar product:

$$g_n = [C_1 \ldots C_m] \begin{bmatrix} (D_s)_{n-1} \\ \vdots \\ (D_s)_{n-m} \end{bmatrix}. \tag{10.40}$$

The mod p summing at the scrambler input, according to Eq. (10.37), has the effect of generating the signal $\{D_s\}$ which is transmitted over the channel and reaches the input of the descrambler. It can be seen that there is a symbol-to-symbol correspondence between the receive side sum $\{g\}$ and the transmit side sum, with the exception of a phase difference. Consequently, the subtraction in the mod p sense results in the expected original signal $\{D\}$, as shown by Eq. (10.38) or, for the binary case, by Eq. (10.39). The self-synchronizing property is obvious — a time interval corresponding to m symbols is needed for the synchronizing process.

The self-synchronizing scrambler has two disadvantages. The worse is the error multiplication property which follows from the channel memory because of the scrambler. Investigation shows that a single errored symbol in the transmitted signal $\{D_s\}$ has the effect of

$$H \leqq 1 + \sum_{i=1}^{m} C_i \tag{10.41}$$

errors in the descrambled signal $\{D\}$. The $<$ sign refers to the possibility of canceling symbol errors in the case of high error probabilities.

The second error source of the self-synchronizing scrambler can be explained by the first theorem of Savage [26]: if the scrambler is driven by a bit stream having a periodicity of s, then the scrambled output bit stream will have a periodicity of either s or the least common multiple of $(s, p^m - 1)$. The periodicity of the output signal depends on the initial condition of the shift register. There is only a single initial condition resulting in an output signal periodicity s, and all other initial conditions result in a more favourable higher periodicity.

The main purpose of scrambling is the elimination of short periodic sequences, but this goal will not be reached if the initial condition is unfavourable. This results in worse transmission performance without resulting transmission errors.*

---

* Decreased transmission performance is the result of an unfavourable initial condition occurring with a probability of $(2^m - 1)^{-1}$ and coinciding with the beginning of a short, periodically recurring sequence, and this degradation lasts until the end of this sequence. As a result of a long run of zeros (having a periodicity of unity), the scrambling function may be degenerated, and an all-zero output signal may appear.

10.3.4 Improved self-synchronizing scrambler

Irregularities in the periodicity of the scrambled response, as explained in Section 10.3.3, can be eliminated by using the arrangement shown in Fig. 10.16. The register chain of the scrambler is extended (delay), thus making possible the control of periodicities $(s_1, s_2, \ldots s_i, \ldots s_N)$ in the signal stream $\{D_s\}$. Control is carried out by comparison of the signal $\{D_s\}$ with a replica of this signal delayed by $m+s_i$ symbols $(1 \leq i \leq N)$. Detection of a periodicity generates a correction signal which has the effect of changing the sum $\sum\limits_{i=1}^{m} C_i a_i$ into the sum

$$\left(\sum_{i=2}^{m} C_i a_1 \right) = 1 + \sum_{i=1}^{m} C_i a_i \bmod 2. \tag{10.42}$$

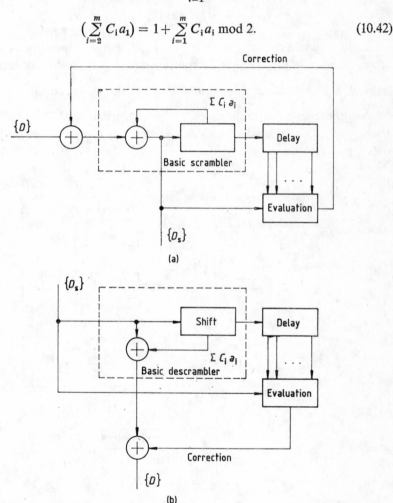

Fig. 10.16. Improved self-synchronizing scrambler and descrambler.

At the receive side, the inverse operations are applied.

The above evaluation and decision for correction is realized by the circuit shown in Fig. 10.17. The OR gates are driven by logical zeros or logical ones depending on whether the mod 2 addition results in identity or not. In the case of long term identity, *i.e.* periodicity, the counter corresponding to the periodicity overflows, according to the truth table shown in the figure, provided the threshold level of t_{si} is correctly chosen. This overflow results in the correction according to Eq. (10.42) and in the reset of all counters. The threshold level t_{si} has to be chosen according to the following inequality:

$$t_{si} \geqq (m-1) + \max_{i \neq j} [s_i] \quad (1 \leqq j \leqq N). \tag{10.43}$$

Finally, it should be noted that there is the possibility of applying the scrambling principle shown in Fig. 10.17 at the transmit side only. This is possible because the transmit side correction will occur extremely seldom, introducing $1 + \sum\limits_{i=1}^{m} C_i a_i$ errors at the receive side in the absence of inverse operations, *i.e.* by using a simple descrambler. This is usually tolerable and has only a slight effect on the overall system performance.

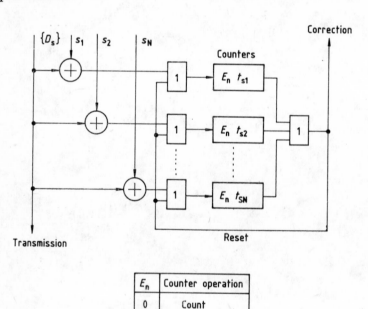

Fig. 10.17. Logic evaluation of the periodicity for obtaining a correction signal in the MCS (Multi Counter Scrambler) according to Ref. [26].

10.3.5 Spectrum improvement due to scrambling

Assume that an input bit stream having a periodicity of s results in a scrambled output bit stream which has a periodicity equal to the least common multiple of $(s, 2^m - 1)$. It can then be shown that the number of transitions Tr within one period is limited according to the following inequality:

$$s(2^{m-1} - 1) \leq Tr \leq s2^{m-1}. \tag{10.44}$$

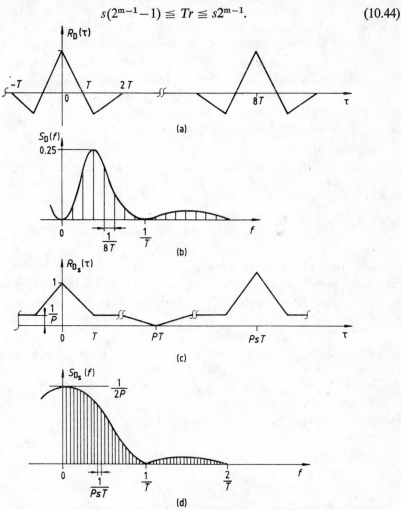

Fig. 10.18. Characteristics of a periodically recurring 10110010 NRZ binary signal. $P = 2^m - 1$, m = scrambler sequence length, T = bit time, s = periodicity of the digital signal transmitted: (a) autocorrelation function of the above signal; (b) spectrum of the above signal; (c) autocorrelation function of the above signal after scrambling; (d) spectrum of the above signal after scrambling.

The above relation shows that independently of the transmitted information, nearly 50% of all possible signal transitions will be present. The relatively tight limits given in inequality (10.44) show that the spectrum of the transmitted signal (D_s) is nearly independent of the transmitted information.

As an example, Fig. 10.18 shows the spectra of a signal $\{D\}$ made up of bit words $\{\ldots 10110010 \ldots\}$. The autocorrelation function and the spectrum of signal $\{D\}$ are shown in Figs 10.18a and 10.18b, respectively. It is seen that the spacing of the line spectrum is $1/8T$ where $1/T$ is the bit frequency, and the spectrum has a pronounced inhomogeneous distribution. The periodicity of the above bit words is $s=8$. The autocorrelation function and the spectrum of the scrambled signal $\{D_s\}$ are shown in Figs 10.18c and 10.18d, respectively. The spacing between the spectrum lines is now decreased to $1/PsT$, where $P=2^m-1$, and the spectrum envelope peak value is decreased according to the ratio $1/2P$. The spectrum distribution is thus much more even, and is hardly changed by the information content. For a sequence length of $m=12$, we have $P=4095$. This means that owing to scrambling, the spacing between spectrum lines has been decreased by a factor of 4095, and the peak value of the spectrum envelope has been decreased by a factor of 8190.

10.3.6 Simple error detection facility

Investigation of Fig. 10.15 reveals a simple error detection possibility. By driving the scrambler input with a signal $\{D\}=\{0\}$ and taking into account Eq. (10.37) it is seen that $\{D_s\}=\{g\}$, *i.e.* the scrambler operates as a pseudo random generator, and a pseudo random signal will be transmitted. Error-free transmission results in a descrambler output signal of $\{D'\}=\{0\}$, and those appearing in the output signal will be the measure of transmission errors. However, the error multiplication defined by Eq. (10.41) has to be taken into account so the actual number of errors and thus the error probability can be determined by dividing the total number D_i of errored bits by H (see Eq. (10.41)). By monitoring N symbols, the error probability will be given by the expression

$$P_E \lesssim \frac{1}{N} \frac{\sum\limits_{i=1}^{N} D_i}{1 + \sum\limits_{i=1}^{m} C_i a_i} . \tag{10.45}$$

The measured error probability will thus always be higher than the actual value but the deviation is negligible up to about 5×10^{-2}. The relation between actual and measured error probabilities is shown in Fig. 10.19.

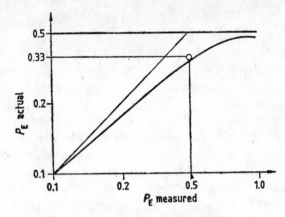

Fig. 10.19. Deviation of the measured error probability from the actual value at high P_E values, for measurement with simple scrambler–descrambler arrangement.

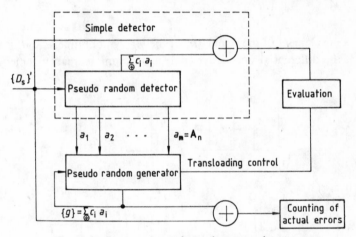

Fig. 10.20. Improved pseudo random error detector.

The measurement uncertainty due to the statistical property of error multiplication can be eliminated by the application of an improved pseudo random error detector shown in Fig. 10.20. According to this figure, not only the pseudo random detector (descrambler) but also a pseudo random generator or identical structure is applied at the receive side. The pseudo random detector, with the aid of an evaluating circuit, determines the time instants at which the state vector A_n is error free (with high probability). At these time instants, the synchronization of the pseudo random generator is affected by transloading the vector A_n into the pseudo random generator. Error measurement is performed by a symbol-to-symbol comparison of the incoming signal $\{D_s\}' = \{g\}'$ and the local signal $\{g\}$ of the receive side pseudo random generator.

10.4 Differential encoding

In digital transmission, a phase ambiguity of the received digital signal frequently appears. For instance, the phases of the recovered carrier and clock signals depend on initial switch-on conditions or other kinds of random processes [7, 8, 35].

Differential encoding and decoding are based on the transmission of the symbol-to-symbol changes of the digital signal, instead of the digital signal itself. This method for eliminating the phase ambiguity at the receive side is frequently applied, even although it has the effect of increasing the memory of the channel.

Figure 10.21a shows the general principle of a differential codec for the transmission of a p-level digital signal. It can be seen from Fig. 10.21 that simple recursive and non-recursive digital filters are applied. The encoding algorithm is given by

$$d_n = a_n + d_{n-1} \bmod p \qquad (10.46)$$

where a_n is the element of the signal $\{a\}$ at the nth symbol time, and d_n is the same element of the differential encoded signal $\{d\}$. A simple rearrangement of Eq.

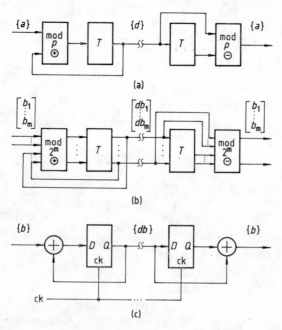

Fig. 10.21. Differential encoding: (a) conventional model; (b) model utilizing block generation from the series bit stream; (c) differential encoding and decoding of the binary signal.

(10.46) yields the result of decoding:

$$a_n = d_n - d_{n-1} \bmod p. \tag{10.47}$$

For instance, assume that a three-level signal has to be transmitted (symbol set: 0, 1, 2), and owing to some phase ambiguity, the following shift of symbols takes place: $0 \to 1$, $1 \to 2$, $2 \to 0$. Transmission is then realized as follows:

signal to be transmitted	... 01212022 ...
result of differential encoding	... 1010021 ...
result of phase ambiguity	... 2121102 ...
result of decoding	... 212110 ...
transmitted signal	... 212022 ...

In many cases, the differential codec is designed to handle the dibit, tribit, *etc.*, parallel bit streams investigated earlier. According to Eqs (10.46) and (10.47), the encoding and decoding algorithm is then given by

$$\begin{bmatrix} db_1 \\ \vdots \\ db_m \end{bmatrix} = \begin{bmatrix} b_1 \\ \vdots \\ b_m \end{bmatrix} + \begin{bmatrix} db_1 \\ \vdots \\ db_m \end{bmatrix}_{n-1} \bmod 2^m \tag{10.48}$$

and

$$\begin{bmatrix} b_1 \\ \vdots \\ b_m \end{bmatrix} = \begin{bmatrix} db_1 \\ \vdots \\ db_m \end{bmatrix}_n - \begin{bmatrix} db_1 \\ \vdots \\ db_m \end{bmatrix}_{n-1} \bmod 2^m. \tag{10.49}$$

These algorithms are realized by the arrangement shown in Fig. 10.21b, and have the advantage that they can be implemented by conventional digital integrated circuits. Figure 10.21c shows the degenerated case given by $m=1$, illustrating the differential encoding and decoding of a binary signal. It is seen that the addition and subtraction in the mod 2 sense are both realizable by the same EXCLUSIVE-OR gate circuit. Modulo 2^m adder networks are applied to implement the addition in the mod 2^m sense while mod 2^m sense subtraction is realized by the following equation:

$$A - B = A + B^k \bmod 2^m \tag{10.50}$$

where B^k is the mod 2^m complement of B. This can also be generalized in the mod p sense. The $m=2$ case has special significance; the mod 4 addition is explained as follows.

Assume that

$$\begin{bmatrix} a_1 \\ a_2 \end{bmatrix} + \begin{bmatrix} b_1 \\ b_2 \end{bmatrix} = \begin{bmatrix} \Sigma 1 \\ \Sigma 2 \end{bmatrix} \bmod 4, \tag{10.51}$$

where $\Sigma 1$ refers to the lower significance and $\Sigma 2$ refers to the higher significance. We then have

$$\Sigma 1 = a_1 + b_1 \mod 2 \qquad (10.52)$$

and

$$\Sigma 2 = a_2 + b_2 + a_1 b_1 \mod 2. \qquad (10.53)$$

One feasible solution for a mod 4 binary adder network is shown in Fig. 10.22.

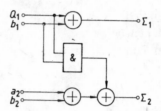

Fig. 10.22. Realization of modulo 4 binary addition.

The generation of complements needed for mod 2^m subtraction can be implemented as follows:

$$\begin{bmatrix} a_1 \\ \vdots \\ a_m \end{bmatrix}^k = \begin{bmatrix} \bar{a}_1 \\ \vdots \\ \bar{a}_2 \end{bmatrix} + \begin{bmatrix} 1 \\ \vdots \\ 0 \end{bmatrix} \mod 2^m. \qquad (10.54)$$

This means that element-by-element vector negation is carried out, and unity is added in the mod 2^m sense. It follows from expression (10.54) that in the case $m=2$, the building of complements can be realized as given by

$$\begin{bmatrix} a_1 \\ a_2 \end{bmatrix} = \begin{bmatrix} a_1 \\ a_1 \oplus a_2 \end{bmatrix}, \qquad (10.55)$$

i.e., the bit of lower significance is transmitted unchanged, while the higher significance bit of the complement vector will be the mod 2 sum of a_1 and a_2.

10.5 Error multiplication effect of differential encoding, Gray encoding

In the transmission of differentially encoded signals, error multiplication is experienced, and this error multiplication is even increased by the method of encoding the dibit, tribit, *etc.*, blocks. Assume that the following 2^m level symbol sequence has to be transmitted:

$$... \, defgh \, ...$$

Assume further that owing to noise in the transmission channel, the receive side decision for symbol f will be in error, but only by one, *i.e.* $f'-f=1$ in the mod 2^m sense.* This means that according to Eq. (10.47), this single erroneous decision will result in two faulty symbols after differential decoding: $f'-e$, and $g-f'$ (mod 2^m). The faulty decoded symbols differ from the correct ones by one value, in the mod 2^m sense. The symbol value often equals the decimal value of the binary "word" formed by the bit group belonging to the symbol itself. In this case, it can then be shown that as a result of a single erroneous decision affected by transmission channel noise, the average number of faulty bits in the receive side bit stream will be as follows: 3 bits for dibit transmission ($m=2$), 3.32 bits for tribit transmission ($m=3$), and 4.158 bits for tetrabit transmission ($m=4$). In Table 10.8, the probabilities of generating 2, 3 or 4 faulty bits as a result of a single

Table 10.8. Probability of bit errors due to a single symbol error with differential encoding

m	Number of levels 2^m	Number of errors							Average number of bit errors
		2	3	4	5	6	7	8	
		Probability of bit errors due to a single symbol error							
2	4	0.25	0.5	0.25					3.00
3	8	0.166	0.25	0.33	0.166	0.083			3.32
4	16	0.129	0.193	0.225	0.258	0.129	0.032	0.032	4.16

faulty decision are summarized. The above error multiplication can be limited by a more suitable assignment of values.

Gray encoding has the effect of limiting the error multiplication to the value of 2, independently of m. The Gray code is a member of the cyclic code family. It has a symmetrical structure and is therefore also called mirrored binary code. The Gray encoding rule is given by

$$G = g_n(2^n-1) - g_{n-1}(2^{n-1}-1) + g_{n-2}(2^{n-2}-1) \pm \ldots, \qquad (10.56)$$

where the symbol value G is given by the binary word $g_n, \ldots, g_0$, beginning with the most significant bit. The binary correspondence can be expressed by

$$B = \sum_{i=0}^{n} b_i 2^i. \qquad (10.57)$$

* It will be assumed that in the course of regeneration, the symbol in question will be in error by only one value. However, under extremely bad signal-to-noise ratio conditions, the value of the change may be higher than 1, but this case will not be investigated.

Based on the interpretation equations (10.56) and (10.57), Table 10.9 presents a comparison of the structures of Gray codes and binary codes. The important property of Gray codes can easily be recognized: only a single bit of the code is changed for any symbol value change. This means that the Gray encoding has the effect of introducing at most one bit error per transmitted bit sequence block for each wrong symbol. Accordingly, the error multiplication as a result of differential encoding will be 2, independently of the value of 2^m.

Table 10.9. Structure of binary codes and Gray codes

Symbol value	Increase of place value: ←	
	Gray code	Binary code
0	0000	0000
1	0001	0001
2	0011	0010
3	0010	0011
4	0110	0100
5	0111	0101
6	0101	0110
7	0100	0111
8	1100	1000
9	1101	1001
10	1111	1010
11	1110	1011
12	1010	1100
13	1011	1101
14	1001	1110
15	1000	1111

10.6 Realization of Gray encoding and decoding

According to Fig. 10.23, the Gray encoding of a binary signal has to begin at the least significant binary bit, and by forming interleaved dibits, the bit to be encoded has to be compared with the next more significant bit. If the values of the ith and $(i-1)$th bits are the same, then the ith bit of the Gray code will be zero; if these bits are different, the Gray code will be one:

$$g_i = \begin{cases} 0, & \text{if} \quad b_i = b_{i-1} \\ 1, & \text{if} \quad b_i \neq b_{i-1}. \end{cases} \qquad (10.58)$$

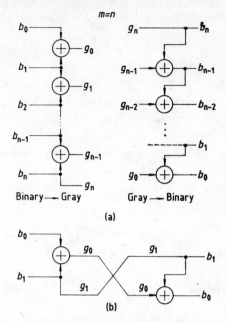

Fig. 10.23. Binary–Gray and Gray–binary transcoding: (a) general case, $m=n$, (b) $m=2$.

This means that the algorithm of Gray encoding is given by

$$g_i = b_i + b_{i-1} \bmod 2. \tag{10.59}$$

If the length n of the binary code word is even, then we have the relation $g_n = b_n$ by taking the value of the "more than most significant" bit as zero.

According to Fig. 10.23, the Gray decoding process should begin at the most significant Gray bit which is decoded without change: $b_n = g_n$. The decoding rule of less significant Gray bits into binary bits is the following:

$$b_i = \begin{cases} b_{i+1} & \text{if} \quad g_i = 0 \\ \overline{b_{i+1}} & \text{if} \quad g_i = 1. \end{cases} \tag{10.60}$$

This means that the Gray decoding algorithm is given by

$$b_i = g_i + b_{i+1} \bmod 2. \tag{10.61}$$

Figure 10.23 shows separately the case of the frequently used dibit transmission ($m=2$). In this case, it follows from Eq. (10.59) that the Gray encoding algorithm is given by

$$\begin{bmatrix} g_0 \\ g_1 \end{bmatrix} = \begin{bmatrix} b_0 + b_1 \\ b_1 \end{bmatrix} \bmod 2 \tag{10.62}$$

and it follows from Eq. (10.61) that the Gray decoding algorithm is given by

$$\begin{bmatrix} b_0 \\ b_1 \end{bmatrix} = \begin{bmatrix} g_0 + g_1 \\ g_1 \end{bmatrix} \bmod 2. \tag{10.63}$$

It is seen that in the case of $m=2$, the Gray encoding and decoding can be realized by the same circuit comprising only an EXCLUSIVE-OR gate.

Application of the Gray encoder and decoder. In the radio transmission systems discussed, the Gray codec has two possible applications. The first is intended to decrease the error multiplication, and in this case, the Gray and differential encoding and decoding blocks are interconnected as shown in Fig. 10.24a.* Another possible application is to provide a matching function between the digital modulator and demodulator.

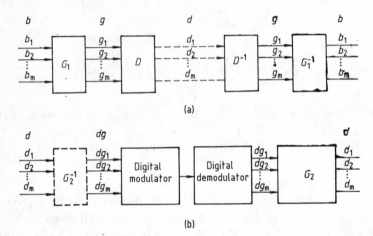

(a)

(b)

Fig. 10.24. Location of Gray encoding: (a) in conjunction with the differential codec; (b) in conjunction with the digital demodulator.

If MPSK transmission is used with the application of multiplying type modulator and demodulator, a variant of Gray encoding and Gray decoding is implemented during the process of modulation and demodulation, which does not meet the requirements of the Gray codec denoted by G_1 and G_1^{-1} shown in Fig. 10.24a. Figure 10.25 shows the phase diagram of Gray encoding for $m=2$. A possible Gray encoding of the modulator should be compensated by G_2^{-1} decoding, and if the demodulator is simultaneously a Gray decoder, this should be compensated

---

* Obviously, the reduction of error multiplication by applying a Gray codec is also needed if the phase ambiguity is not eliminated by a differential codec but by other means. In these cases, the Gray codec function can be implemented separately, partially or totally, by the digital modem.

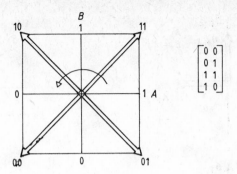

Fig. 10.25. Realization of Gray encoding by the utilization of an MPSK digital modulator.

by G_2 encoding (see Fig. 10.24b). It thus follows that encoding G_2 and decoding G_2^{-1} should be applied depending on the modulator type and demodulator type. It should be noted that multiplier type modems are generally realized in the i.f. band. In some cases, however, digital modulation is implemented at microwave frequencies by utilizing the phase shift principle, thus avoiding Gray encoding.

10.7 Phase ambiguity of QPSK transmission

An important property of QPSK transmission which is frequently applied in digital radio relay or satellite systems is the phase ambiguity at the receive side, *i.e.* the lack of recognition of the transmit side reference phase in the carrier recovery circuit.

Figure 10.26 shows the possible variants of vector phases due to this phase ambiguity at the input of the QPSK demodulator, introduced by either transmit side varying initial conditions, or by cycle slip resulting from the noisy transmission channel.

There are two basic ways to eliminate this phase ambiguity. According to the first method, symbol recognition is made possible independently of knowledge of the reference phase; this method is applied by differential encoding by which the differences between consecutive values rather than the symbol values themselves are transmitted. According to the second method, the reference phase information is transmitted separately to the receive side [7, 8].

There are two ways to implement the first method. Either differentially coherent detection is applied at the receive side, not requiring carrier recovery (MDPSK transmission), or differential encoding is used as discussed in Section 10.4.

The second method is based on the recognition of a periodically recurring unique word in the bit stream. This can be either the frame alignment signal in the case

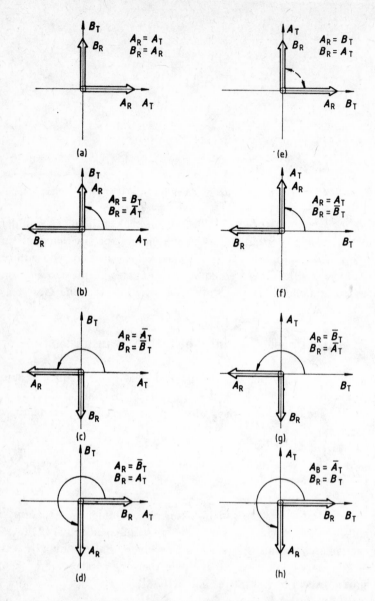

Fig. 10.26 (a)–(h). Phase ambiguity of QPSK demodulation. The random values of the reference phase result in the phasor diagrams within a column while the transmit side exchange of vectors **A, B** and the use of upper or lower conversion are shown by the diagrams within a row.

of PCM transmission according to the CCITT recommendations or, in the case of "burst mode" satellite transmission, the "unique word" (UW) following the synchronizing preamble and preceding the information bit stream [7, 8, 36]. In burst mode communication, this unique word has several functions: it is a verification of successful carrier and bit time recovery, it provides timing for the transmitted information signal and allows the determination of the reference phase and, if necessary, may provide phase correction.

Comparison of methods. Of the two variants of 4PSK transmission, the 4DPSK transmission, applying differentially coherent demodulation, shows a degradation of 2.3 dB with respect to 4CPSK transmission, applying coherent demodulation. This degradation is reduced to 0.5 dB by eliminating the phase ambiguity of the 4CPSK transmission by the application of differential codes. However, both transmissions will show a high degree of error correlation, *i.e.* there will be an uneven error distribution by exhibiting pairs of errors. This is extremely unfavourable if error recognizing or error correcting encoding is applied, and in this case, higher transmission capacity will be required.

In expensive systems, such as those of satellite communication, one of the most important design criteria is the effective utilization of the channel capacity. This is why in satellite systems utilizing 4CPSK transmission, the phase ambiguity is usually resolved by utilizing the unique word as explained earlier: in terrestrial systems, the frame alignment word is used for the same purpose. This method requires relatively complicated circuits, in contrast with the simple structure of the differential codec. The recognition of these words requires complicated hardware following special strategy, so the shift of the reference phase due to noise may result in frame loss, whereas in the case of a differential codec, only one or two errors may occur.

10.8 Examples for baseband circuit realization

This section presents a short survey of a few widely used baseband signal processing circuits used for digital modulation systems.

10.8.1 Biphase (WAL$_1$)–ASK transmission

For low transmission capacities, simple and economical digital communication is possible through the application of ASK (on–off type) modulation. The WAL$_1$ encoding is well suited to this kind of modulation. The WAL$_1$ encoded signal is simply generated by mod 2 addition of the binary information signal and the

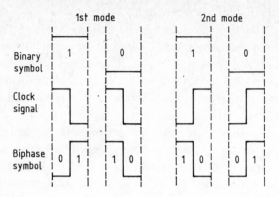

Fig. 10.27. Two modes of the WAL₁ type biphase encoding [37].

corresponding clock signal. As shown in Fig. 10.27, two different transmission modes may result following switch-on, depending on the relative phases of the clock signal and the information signal. However, at the receive side there is no information as to which of the two modes is transmitted, so there is a 180° phase ambiguity. This can be eliminated by applying the differentially binary codec shown in Fig. 10.21c. Biphase ASK transmission has the disadvantage of exhibiting error doubling but also has the following advantages:

— it is suitable for transmission without code restriction because at least every second transition of the biphase signal carries the clock information. The clock frequency component of the biphase signal is thus independent of the pattern of the information signal;
— automatic gain control can be applied at the receive side;
— continuous and accurate error probability measurement is made possible;
— there is the possibility of a signaling channel for carrying additional information (*e.g.*, by periodical violation of the biphase encoding law);
— the biphase signal has no DC component;
— the baseband circuits are simple.

In addition to the phase ambiguity mentioned earlier, another disadvantage is the two-fold increase in the biphase signal bit rate with respect to the original signal bit rate, and the restricted number of patterns which can be transmitted between the extremal limits of … 0101 … and … 00110011 … This results in a bandwidth requirement which is about 30 per cent higher compared with the bandwidth requirement of a binary signal. However, at low transmission speeds (60 to 200 kbit/s), this presents no difficulties because of the minimum i.f. bandwidth needed for other reasons.

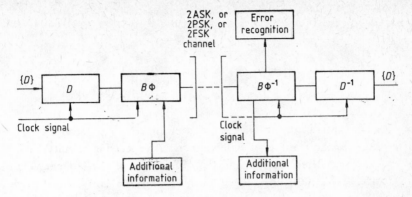

Fig. 10.28. Baseband building blocks of the biphase 2ASK (2PSK, 2FSK) system. Additional information can be transmitted by the periodical violation of the biphase encoding rule. The error probability can be measured by counting tribits of three equal components [37].

The block diagram of biphase transmission is shown in Fig. 10.28. The additional information capacity can be utilized for the transmission of signaling or frame alignment signals. Several error recognition schemes can be based on the redundancy of biphase transmission: errors can be indicated by recognizing the violation of the biphase rule or by monitoring three identical consecutive bit combinations. Biphase transmission is applied in the MIDAS high speed data transmission equipment [37]. A biphase 2PSK or 2FSK digital channel can be applied not only independently but also as a service channel of medium or high speed equipment.

10.8.2 M-PSK transmission

This is one of the most widely used digital transmission systems. This kind of transmission has the disadvantage of phase ambiguity at the receive side which can most easily be eliminated by using a differential codec (Section 10.4) so that in the following examples, the application of the differential codec for phase ambiguity elimination will be presented.

Figure 10.29 shows the baseband signal processing circuits applied with a phase shift type modulator and a coherent multiplying type demodulator. If a multiplying type modulator is used, then a G_2^{-1} type Gray decoder is also needed. In the case of a differentially coherent demodulator, the differential decoder D^{-1} is not necessary because decoding is performed by the demodulator itself. In all cases, the operation of the differential codec should be understood in the modulo M sense. In Table 10.10, the structure of the baseband signal processing circuits corresponding to the four basic variants of the M-PSK modem are summarized.

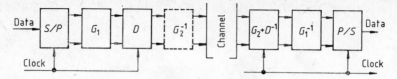

Fig. 10.29. Baseband signal processing applied with an *M*-PSK digital modem.

This table, together with the subject matter presented in Sections 10.4, 10.5 and 10.6, covers the features of baseband signal processing circuits used in the frequently applied 2PSK modem and the 4PSK modem, the latter having better performance. Evidently, the block generation and the Gray encoding is only meaningful for $M > 2$.

Table 10.10. Parts of baseband signal processing in the case of *M*-PSK transmission (excluding regenerator and clock recovery circuits)

Type of demodulator	Type of modulator	
	Multiplying	Phase shift
Coherent detection	$S/P—G_1—D—G_2^{-1}————G_2—$ $—D^{-1}—P/S$	$S/P—G_1—D————G_2—$ $—D^{-1}G_1^{-1}—P/S$
Differentially coherent detection	$S/P—G_1—D—G_2^{-1}————P/S$	$S/P—G_1—D————P/S$

G = Gray encoding; G^{-1} = Gray decoding; D = differential encoding; D^{-1} = differential decoding (in mod M sense); S/P = series/parallel conversion $(1 \rightarrow \log_2 M)$; P/S = parallel/ /series conversion $(\log_2 M \rightarrow 1)$; $————$ = M-PSK channel (modem).

In many cases, 8PSK modems are applied. However, these require further considerations to obtain a two-dimensional signal set resulting in an even distribution of the symbol vectors on the decision plane. According to Ref. [1], the 8PSK vector distribution, as shown in Fig. 10.30b, can be obtained by the quadrature modulation of two four-level signals *a* and *b*, as shown in Fig. 10.30a. These signal levels should be proportional to the levels of cos 22.5°, sin 22.5°, −sin 22.5° and −cos 22.5° where 22.5° is the even radial distribution of the vectors. The signal $\{D\}$ to be transmitted is transformed into a tribit signal; this is followed by Gray encoding and differential encoding in mod 8 sense, and finally, the digital drive of analog multiplexers, as shown in Fig. 10.30a, results in the four-level signals *a* and *b*.

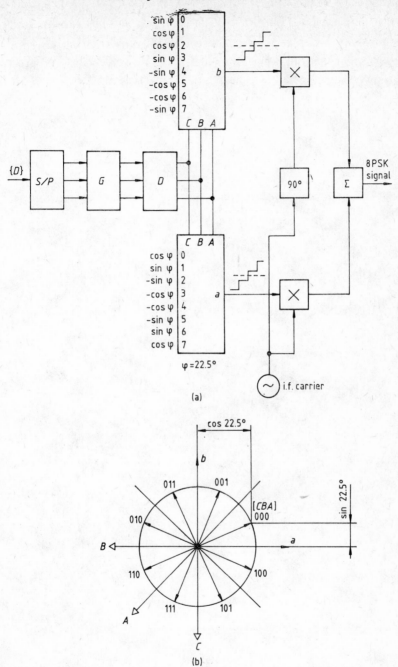

Fig. 10.30. 8PSK modulator and signal processing according to Ref. [1]: (a) block diagram; (b) vector diagram.

Another method of obtaining 8PSK modulation is given in Ref. [3]. According to this method, two 4PSK multiplying type modulators are used, driven by two carriers which are 45° out of phase. In addition to the preceding signal processing circuits, a conversion circuit T is also needed which has the function of allowing only 8 of the 16 vectors obtainable from the two modulators, according to the truth table shown in Fig. 10.31a.

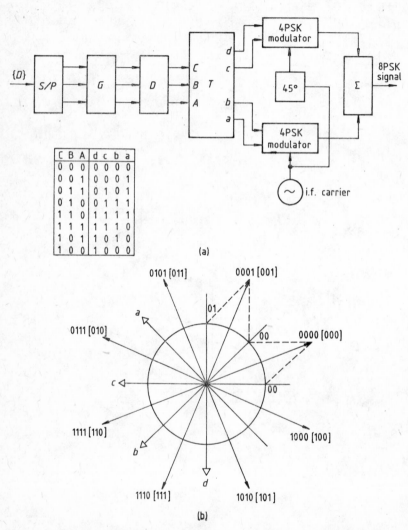

C B A	d c b a
0 0 0	0 0 0 0
0 0 1	0 0 0 1
0 1 1	0 1 0 1
0 1 0	0 1 1 1
1 1 0	1 1 1 1
1 1 1	1 1 1 0
1 0 1	1 0 1 0
1 0 0	1 0 0 0

(a)

(b)

Fig. 10.31. 8PSK modulator and signal processing according to Ref. [3]: (a) block diagram and truth table; (b) vector diagram

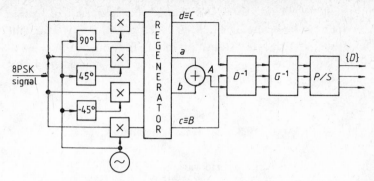

Fig. 10.32. 8PSK demodulator and signal processing.

The coherent demodulation scheme of an 8PSK signal is shown in Fig. 10.32. The A element of the tribit signal is obtained by the mod 2 addition of the a and b signals supplied by the regenerator; it can be shown further that $c = B$ and $d = C$. Other blocks in Fig. 10.32 provide operations which are inverted replica of the operations given in Figs 10.30 and 10.31.

10.8.3 O-QPSK transmission

Offset-QPSK has the effect of improving spectral conditions by concentrating the power spectral density in the vicinity of the carrier. It is usually applied for 4PSK modulation, and realized by introducing a one bit offset between the two binary sequences of the dibit signal obtained at the output of the differential encoder. At the receive side, the inverse operation is realized following the regeneration of the dibit signal. This additional signal processing has the effect of eliminating the possibility of having a 180° phase jump of the symbol vector as shown, for instance, in Fig. 10.25. In contrast with 4PSK allowing phase jumps of 0, ±90°

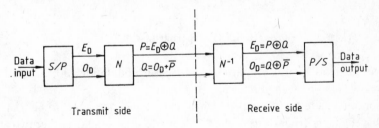

Fig. 10.33. O-4PSK signal processing. E_D and O_D designate the even and odd bits of the transmitted bit stream [39].

and 180° at half the bit rate to be transmitted, the frequency of transitions for O-4PSK is equal to the bit frequency itself, but only 0 or ±90° transitions take place. This means that the 180° transition of 4PSK is divided into two 90° transitions of O-4PSK. A simple signal processing method associated with O-4PSK modulation is shown in Fig. 10.33 [39].

10.8.4 MSK transmission

Minimum shift keying is the hybrid variant of 2PSK and 2FSK transmissions, and has several advantages as discussed in Section 10.8.2. Considering now baseband signal processing, a further advantage is the possibility of eliminating the phase ambiguity without differential encoding and without utilizing the synchronizing word or unique word discussed in Section 10.7. This possibility is given by the structure of the carrier recovery of the MSK demodulator, as detailed in Chapter 9. The carrier recovery comprises a frequency discriminator which supplies the transmitted information $\{d\}$ with correct polarity, though with degraded noise performance. However, this signal is suitable for deciding, by means of the comparator shown in Fig. 10.34, whether or not inversion due to phase ambiguity has taken place during transmission. In the case of a detected inversion, the phase ambiguity is eliminated by an EXCLUSIVE-OR gate inserted into the transmission path [39].

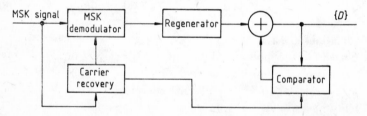

Fig. 10.34. Elimination of phase ambiguity for the case of MSK transmission [38].

10.8.5 QAM transmission*

QAM involves hybrid amplitude and phase modulation [30, 31] and is best realized by two 4PSK modulators.

Accordingly, a tetrabit signal is formed from the bit stream D to be transmitted, with two elements driving one of the 4PSK modulators, and the other two elements

---

* QAM is sometimes also designated QASK or 2A4PSK.

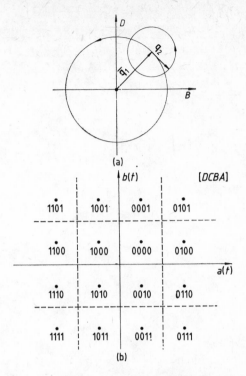

(a)

(b)

Fig. 10.35. QAM (QASK) modulation: (a) vector diagram; (b) suitable values of the grid network
according to Ref. [31].

driving the other 4PSK modulator. The first 4PSK modulator supplies vector
$\mathbf{q}_1$ defining the plane quadrant, while the other modulator supplies vector $\mathbf{q}_2$ which
defines one of the four symbols within the quadrant. In this manner, the grid
network shown in Fig. 10.35b is formed. It is seen that tetrabit elements D and B
define the plane quadrant, and tetrabit elements C and A define the location within
the quadrant. The condition for the even distribution shown in Fig. 10.35b is
given by $|\mathbf{q}_1|=2|\mathbf{q}_2|$ which is guaranteed by the 6 dB amplifier following the
first 4PSK modulator.

This modulation system, together with baseband signal processing, is shown in
Fig. 10.36a. The conventional demodulator shown in Fig. 10.36b serves for generating the $a(t)-b(t)$ quadrature system according to the grid network of Fig.
10.35b. If the regeneration system corresponds to that given in Chapter 8, then
the transmitted tetrabit signal DCBA can be recovered by comparisons as a result
of the regeneration. According to the modem arrangement shown in Fig. 10.36,
a phase ambiguity will be present in channels D, B and C, A. This is eliminated
by the Gray and differential codecs applied.

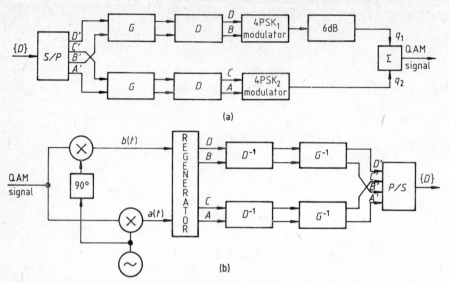

Fig. 10.36. QAM baseband signal processing: (a) transmit side; (b) receive side.

10.8.6 QPRS transmission

QPRS is a special combination of 4PSK and the precoded Partial Response System of class $C=1$ and order $n=2$. Favourable results can be obtained by a 9-state QPRS modem [8]. The QPRS modulator, together with the baseband signal processing circuits, is shown in Fig. 10.37a. The binary signal $\{D\}$ to be transmitted is transformed into a dibit signal, and following Gray encoding and differential encoding, the two binary signal components of the dibit signal are converted into three-level signals according to the considerations of Sections 10.1.3 and 10.1.4. This precoded partial response transformation corresponds to the precoded duo-binary transformation shown in Fig. 10.10, and can easily be implemented. The two three-level signals are used for quadrature modulation of two carriers. By denoting the range of signals $a(t)$ and $b(t)$ by symbols -1, and $+1$, the vector diagram of the multiplying type modulator will be given as shown in Fig. 10.37c. If the QPRS demodulation is realized by the arrangement shown in Fig. 10.37b, then following the regeneration of $a(t)$ and $b(t)$, the dibit value set in the vector diagram of Fig. 10.37c will be obvious. According to Section 10.1.4, the decoding of the precoded partial response signal may be implemented by a simple even–odd detection: logical zeros are assigned to even levels, and logical ones to odd levels. The dibit [11] will always be in the origin of the vector diagram as the elements -1 and $+1$ of the symbol set "-1", "0", "$+1$" are even (-1 is the zeroth level, 0 is the first level and $+1$ is the second level), and the "0" is odd.

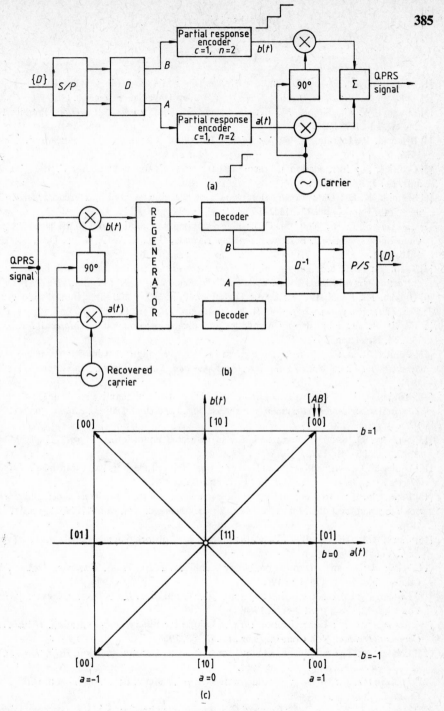

Fig. 10.37. Realization of a QPRS digital modem: (a) transmit side signal processing; (b) receive side signal processing: (c) vector diagram showing dibit values.

10.9 References

[1] Bennet, W. R., Davey, J. R.: *Data Transmission*. McGraw-Hill, London 1965.

[2] Colomb, S. W.: *Shift Register Sequences*. Holden-Day, San Francisco 1967.

[3] Lindsey, W. C., Simon, M. K.: *Telecommunication Systems Engineering*. Prentice-Hall, Englewood-Cliffs 1973.

[4] Bylanski, P., Ingram, D. G. W.: *Digital Transmission Systems*. Peter Peregrinus, London 1976.

[5] Croisier, A.: Introduction to pseudoternary transmission codes. *IBM J. Res. Develop.*, July, pp. 354–367, 1980.

[6] Schmidt, K. H.: Data transmission using controlled intersymbol interference. *Electr. Commun.*, No. 1 and 2, pp. 121–133, 1973.

[7] Cacciamani, E. R., Wolejsza, C. J.: Phase-ambiguity resolution in a four-phase PSK communication system. *IEEE Trans. Commun. Technol.*, Vol. Com-19, No. 6, Dec., pp. 1200–1210, 1971.

[8] Tsuji, Y.: Phase ambiguity resolution in a 4-phase modulation system with forward-error-correcting convolutional codes. *Comsat. Tech. Rev.*, Vol. 6, No. 2, pp. 357–377, 1976.

[9] Ohtake, K., Kasai, H., Taka, M.: Digital multiplexer for 400 Mb/s transmission system. *Rev. Electr. Commun. Lab.*, Vol. 24, No. 9–10, pp. 725–736, 1976.

[10] Rolls, R. G.: High density line codes and p.c.m. instrumentation. *Marconi Instrument*, Vol. 14, No. 1, pp. 16–21, 1973.

[11] Boulter, R. A.: A 60 kb/s data modem for use over physical pairs. In: *Proc. Int. Zürich Seminar Integrated Systems for Speech, Video and Data Communications*. pp. H3(1)–(3), 1974.

[12] Robinson, A. E. C.: An operational 2048 kb/s multi-level adaptive modem. In: *Proc. Int. Zürich Seminar Integrated Systems for Speech, Video and Data Communications*. pp.H4(1)–(6), 1975.

[13] Sachdev, D. K.: Digital transmission in restricted bandwidth. *Electro-Technol.*, March–April, pp. 103–110, 1972.

[14] Iwahashi, E., Fukinuki, H.: Analysis of probabilities of mark in PCM transmission system. *Electron. Commun. Jpn*, Vol. 51-A, No. 1, pp. 10–19, 1968.

[15] Benedetto, S.: Error probability in the presence of intersymbol interference and additive noise for multilevel digital signals. *IREE Trans. Commun.*, Vol. Com-21, No. 3, pp. 181–190, 1973.

[16] Clark, A. P., Harvey, J. D.: Detection processes for distorted binary signals. *Radio Electron. Eng.*, Vol. 45, No. 11, pp. 553–542, 1976.

[17] Sekey, A.: An analysis of the duobinary technique. *IEEE Trans. Commun. Technol.*, Vol. Com-14, No. 2, pp. 126–130, 1966.

[18] Lender, A.: Correlative digital communication techniques. *IEEE Trans. Commun. Technol.*, Vol. Com-12, Dec., pp. 128–135, 1964.

[19] Ketzmer, E. R.: Generalization of a technique for binary communication. *IEEE Trans. Commun. Technol.*, Vol. Com-16, Febr., pp. 67–68, 1968.

[20] Lender, A.: The duobinary technique for high speed data transmission. *IEEE Trans. Commun. Electron.*, May, pp. 214–218, 1963.

[21] Sullivan, W. A.: High-capacity microwave sytem for digital data transmission. *IEEE Trans. Commun.*, June, pp. 466–470, 1972.

[22] Houts, R. C., Green, T. A.: Comparing bandwidth requirements for binary baseband signals. *IEEE Trans. Commun.*, June, pp. 776–781, 1973.

[23] Han, B.: Analysis of a modified data test set by the use of generating functions. *IEEE Trans. Commun.*, Oct., pp. 1706–1710, 1974.

[24] MacWilliams, F. J., Sloane, N. J. A.: Pseudo-random sequences and arrays. *Proc. IEEE*, Vol. 64, No. 12, pp. 1715–1729, 1976.

[25] Meyer, F.: Gigabit/s *m*-sequence generation. *Electron. Lett.*, Vol. 12, No. 14, 8th July, p. 353, 1976.

[26] Savage, J. E.: Some simple self-synchronizing digital data scramblers. *Bell Syst. Tech. J.*, Febr., pp. 449–487, 1967.

[27] Leeper, D. G.: A universal digital data scrambler. *Bell Syst. Tech. J.*, Dec., pp. 1851–1865, 1973.

[28] Utlaut, W. F.: Spread spectrum principles and possible application to spectrum utilization and allocation. *IEEE Commun. Soc. Mag.*, Sept., pp. 21–31, 1978.

[29] Kurematsu, H.: The QAM2G–10R digital radio equipment using a partial response system. *FUJITSU Sci. Tech. J.*, June, pp. 27–48, 1966.

[30] Salz, J., Sheehan, J. R., Paris, D. J.: Data transmission by combined AM and PM. *Bell Syst. Tech. J.*, Sept., pp. 2399–2419, 1971.

[31] Weber, W. J., Stanton, P. H., Sumida, J. T.: A bandwidth compressive modulation system using multi-amplitude minimum shift keying (MAMSK). *IEEE Trans. Commun.*, Vol. Com-26, No. 5, pp. 543–551, 1978.

[32] Nave, P.: Spectrum des biternär codierten PCM Signals, *AEÜ*, Vol. 4, 1969.

[33] Roth, D.: Mittleres Leistungsdichtespectrum einer durch den duobinären Code frequenzmodulierten Funktion bei steiger Phase, *AEÜ*, Vol. 12, 1974.

[34] Gill, A.: *Linear Sequential Circuits*. McGraw-Hill, New York 1966.

[35] Ványai, P.: Baseband signal processing and bit synchronization of microwave digital systems using 4PSK modulation. In: *Proc. 6th Colloq. Microwave Commun. Budapest*, Vol. I, pp. 1–5/28, 1–4, 1978.

[36] Schrempp, W., Sekimoto, T.: Unique word detection in digital burst communications. *IEEE Trans. Commun.*, Vol. Com-16, Aug., pp. 497–605, 1966.

[37] Ványai, P.: Digital microwave radio relay system for high speed data transmission. *Budavox Telecommun. Rev.*, No. 3, 1977.

[38] Rudi de Buda, K.: Coherent demodulation of frequency shift keying with low deviation ratio. *IEEE Trans. Commun.*, Vol. Com-20, No. 3, June, pp. 429–435, 1972.

[39] C. Feher, K.: *Digital Communication, Satellite/Earth Station Engineering*. Prentice-Hall, Englewood-Cliffs, Ch. 4, pp. 168–170, 1983.

CHAPTER 11

AUXILIARY EQUIPMENT

11.1 Introduction

In the preceding chapters a rather detailed presentation has been given for equipment components of a digital microwave transmission system which are needed for the main function, *i.e.* the transmission of specified signals with given quality parameters. However, this main function requires the fulfilment of several auxiliary tasks, and this chapter is devoted to components used for this purpose. Primarily terrestrial radio relay systems will be investigated from this point of view, though some of the solutions presented are equally applicable in satellite systems.

In order to decide on the most important auxiliary functions, we recall the most important performance parameter of a digital transmission system, *i.e.* the error probability. The criterion for the operating condition of an item of equipment, system or transmission path is given by the error probability which should not exceed a specified limit.

Assume now that the status of an item of equipment or transmission path becomes non-operating in the above sense, owing to either propagation conditions or actual equipment failure. In this case, switch-over to a stand-by channel should be performed, requiring typically the exchange of information between the switching main stations. In radio relay or satellite systems, the exchange of further information such as service telephone calls between station personnel, and the transmission of telecontrol and telecommand data may frequently be required too.

From the above considerations, the three following most important auxiliary functions can be established (see also Fig. 1.4):

— the transmission performance should be continuously monitored, and an alarm signal should be generated when it is established with adequate confidence that the allowed error probability has been exceeded;
— the transmission of service information should be provided;
— the switch-over to a stand-by channel should be provided in the case of a transmission interruption.

Several implementations of these auxiliary functions are possible, and in the following, typical solutions, together with critical comparisons, will be presented. In some cases, a detailed analysis will be given while in others, only the basic principles will be outlined.

11.2 Continuous monitoring of transmission performance

It has been explained that the transmission performance monitoring of a digital connection is necessary to initiate several procedures such as interruption of the failed path, switch-over to a stand-by path, notification of the source and sink of the failure, *etc.* According to a CCIR Recommendation [1], a maintenance entity should be regarded as non-operating, *i.e.* as interrupted, when the error probability reaches the value of 10^{-3}. Such a maintenance entity may be a transmission path, a switching exchange, a multiplex equipment, *etc.*

At first sight, continuous error monitoring may seem paradoxical as an unknown signal sequence has to be classified as being correct or uncorrect. The solution is obvious: either a measurable quantity has to be found which has a unique relation to the error probability, or some kind or redundancy has to be found (or generated) in the signal sequence allowing correct/incorrect evaluation. Most of the following procedures fit into this latter category.

11.2.1 Monitoring of the receiver input signal

The receiver input signal can be determined relatively easily by measuring for example the AGC voltage, the signal-to-noise ratio of the recovered carrier, *etc.* On the other hand, there is a unique relation between the receiver input signal and the error probability in a given system, allowing calculation of the error probability from the input signal.

A disadvantage of this method is the fact that the error probability is an extremely rapidly changing function of the signal-to-noise ratio (and thus also of the received signal). This means that an extremely accurate measurement of the signal level is needed (within a few tenth of a decibel) even for a rough determination of the error probability, and circuit instability does not normally allow such an accurate measurement.

11.2.2 Monitoring by redundant codes

A typical example for redundant codes serving for error monitoring is the error detection encoding. This can also be applied in microwave channels, but in spite of this, it is by no means used universally. The reason is that this kind of encoding requires an increase in transmission speed (at least if signal space encoding is excluded), and this requires complex circuitry. In low speed systems, much simpler solutions are possible (a few of which will be presented in the following), while in high speed systems, the realization of the codes is a rather difficult task, even for present-day technology; however, it is sometimes applied in high speed systems. For error detection encoding, the transmission speed is higher than required by the information signal. The excess bits include parity checking bits and frequently also bits transmitting the service information (after analog-to-digital conversion). The increase in transmission speed is normally of the per cent order of magnitude, *e.g.* 1 to 2 Mbit/s at 140 Mbit/s speed.

A simpler method of redundant encoding, applied primarily in low speed systems, is rendered possible by a suitable redundancy of the line code. Redundant line codes are primarily applied in wirebound transmission. A typical example is the so-called bipolar code according to which zero levels correspond to binary zeros, and alternately positive and negative pulses correspond to binary ones. This results in two consecutive bits of the same polarity whenever an erroneous bit is received, and the number of errors is obtained by counting these impulse pairs (see Fig. 11.1: the shaded bits are in error). A similar redundancy is contained in the HDB3 line code described in Chapter 10 [2] which is similarly applied in wirebound transmission, and allows the measurement of the error ratio as explained above [3]. In microwave transmission, similar bipolar line codes are rarely applied (see Chapter 6); in these cases, the corresponding error counting method can be used.

The biphase code is another line code having redundancy; it is used extensively in wirebound transmission but is also applied in microwave transmission. The

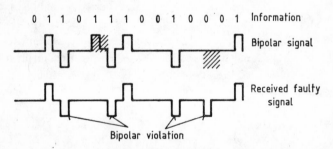

Fig. 11.1. Error monitoring by noting the bipolar violation.

biphase signal set has already been treated in Section 2.1.3. A somewhat modified variant of this type of code is the CMI code (Coded Mark Inversion), applied primarily in high speed transmission systems [2]. A common property of these two line codes is the fact that one or two transitions are contained in each bit time slot. The number of errors is thus obtained by counting the events during which three consecutive half-bits have the same polarity. (A somewhat more detailed investigation shows that this method yields the average number of errors, but in practice, this has no disadvantage.)

It can thus be concluded that the biphase encoding is preferable from the error monitoring point of view. It also has further advantages, *e.g.* its spectrum disappears at zero frequency as ω^2 (see Table 2.3), its power is independent of the modulation, *etc*. A substantial disadvantage, however, is the large occupied bandwidth which is approximately twice the bandwidth required by NRZ encoding. This is why it is used only exceptionally; we have presented an application example in Section 3.4.1. Some partial response signals also have redundancy which can similarly be utilized for error counting.

11.2.3 Monitoring by redundant modulation

Some signal sets obtained by some modulation and demodulation methods have redundancies which are similar to those of the codes discussed in the previous section. One of these is the phase modulation if a relaxation demodulator is applied (see Section 2.1.4). It is seen from Fig. 11.2 that a bipolar pulse train similar to that shown in Fig. 11.1 appears at the output of the discriminator. This means that at this terminal, the number of bipolar violations can be counted similarly to the method used for bipolar transmission.

Another modulation which is in itself redundant is the so-called pseudo-octonary or *B*-type modulation [4]. This is essentially a 4PSK system in which the carrier phase is shifted by 45° at the beginning of each dibit. This rather complicated

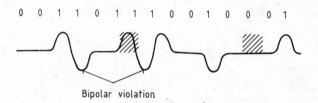

Fig. 11.2. Bipolar rule and its violation in a 2PSK system with relaxation demodulator.

modulation method has the aim of marking the beginnings of all dibits by phase changes, thus facilitating clock recovery. As a by-product, this modulation scheme allows error recognition by ascertaining whether the phases of consecutive dibits differ by odd multiples of $\pi/4$, as required for error-free reception.

11.2.4 Monitoring with non-redundant codes or modulations

The methods presented in the preceding two sections cannot be applied generally because a redundancy is required from either the code or the modulation. Realization of these methods is rather expensive; the bipolar system requires an increased signal-to-noise ratio, the biphase system has a higher occupied bandwidth, and the "B-type" modulation needs a circuit realization which is much more complicated than that of ordinary QPSK. Methods are therefore desirable which allow the performance monitoring of transmission systems not applying redundant codes, by utilizing an "internal" redundancy which is *a priori* possessed by the transmitted signal sequence.

An example of such an "internal" redundancy is the partial knowledge of the transmitted information. A known part of the information is the framing word which is always required when the transmitted bit stream is generated by time division multiplexing of several sources (see Section 5.3 and also Ref. [3]). The time slot of this framing word and the bits contained in the word are, of course, known. This allows an error count of the consecutive framing words. The error probability thus obtained gives the sampled mean value of the number of relative errors, and this is equal to the expected value, *i.e.* the error probability is correctly obtained. The main disadvantage of this method is the restricted application field: it can only be used for signal sequences comprising framing words. This means that the transmission system applying this specific signal sequence cannot be regarded as "transparent", and this transparency is frequently required. Another disadvantage of the system based on the monitoring of the framing word is the relatively low speed. The framing word information constitutes at most a few per thousand of the useful (and redundancy less) information. This means that the time interval required for the recognition of a given error probability is longer than the interval which would be required by monitoring every bit; the ratio of these intervals is inversely proportional to the time fraction of the framing word. Finally, a third disadvantage is the fact that the frame synchronization process has to be carried out at all repeater stations at which monitoring is wanted, and this requires extra circuitry.

The first disadvantage may be eliminated by adding the known part of the signal sequence information by the terminal station itself, thus rendering the system

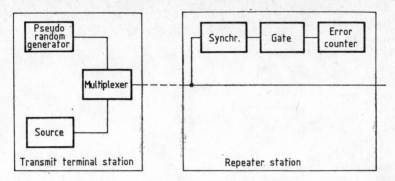

Fig. 11.3. Error monitoring by the application of speed transformation and the multiplexing of a short pseudo random bit sequence.

transparent. This method, applied by a few high capacity radio relay systems, is shown in Fig. 11.3. A pseudo random signal is added — *e.g.,* by pulse stuffing — to the source signal having speed $1/T$, by means of a multiplexer, the pseudo random signal speed being much lower. The transmitted signal sequence speed will then be $1/T'>1/T$. At the receiving terminal station, not shown in the figure, the inverse, demultiplication operation is carried out. Of course, this method requires even more circuitry than the previous one.

The method based on the monitoring of the redundancy of individual signals can be considered to be the best error monitoring method. This is the so-called pseudo error counting method [6, 7, 8], which has the following operation principle. Two decision circuits are connected to the output of the demodulator or i.f. amplifier. One of these is the "good" detector while the other is intentionally made "poor", and requires for the detection of a given error probability, a signal-to-noise ratio which is a few decibels higher than that of the "good" detector (see Fig. 11.4a). The comparison of the two signal sequences yields a "pseudo error count" which is a monotonically increasing function of the actual error count, though much higher. There are several orders of magnitude difference between the two error probabilities, corresponding to the few decibels quality difference. The output of the "good" decision circuit can thus be regarded as being errorless with respect to that of the "poor" decision circuit, *i.e.* the number of "pseudo errors" is, with good approximation, equal to the number of not coinciding bits of the two signal sequences. We thus have

$$P_{\mathrm{E}} = C(P_{\mathrm{E}})D$$

where P_{E} is the actual error probability, D is the relative number of the counted pseudo errors, and $C(P_{\mathrm{E}})$ is a factor which can be determined experimentally.

Figures 11.4b and 11.4c show two "poor" decision circuits: the first can be gen-

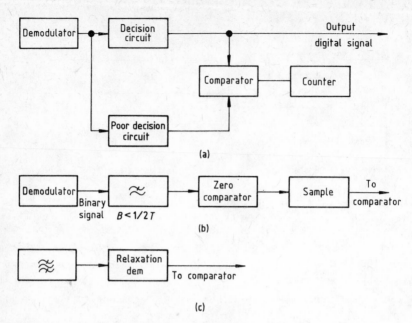

Fig. 11.4. Pseudo error detection: (a) operation principle; (b) and (c) realization of two poor demodulators.

erally applied while the second is intended for coherent or differentially coherent 2PSK systems.

The pseudo error counter can be fairly generally applied but also has certain difficulties. This is explained by the fact that the information from the error counter is needed when the error probability is rather high, between 10^{-3} and 10^{-6}. D is two orders of magnitudes higher, between 10^{-1} and 10^{-4}, and in the vicinity of 10^{-1}, the operation of the "poor" circuit is somewhat uncertain, making the calibration stability questionable after a long time.

According to a variant of the "pseudo error" detector [9], a redundant property of the signal sequence, *i.e.* the time instants of transitions, is monitored. When synchronous signals are transmitted, the transitions of a given signal sequence have to coincide with the edges of the bit time slots. Transitions detected in the central part of the bit time slots are, with high probability, due to errors. It is thus sufficient to detect the number of transitions within an interval τ around the central part of the bit time slots where $\tau = T/m$, $m = 5...10$. The actual number of errors can again be determined by calibration, but the "pseudo error" detector is now operating under less critical circumstances. This method can be applied to advantage in low and medium capacity systems.

11.3 Error counting methods

In Sections 11.2.2, 11.2.3 and 11.2.4 we have seen several possibilities of generating events having occurrence probabilities from which the error probabilities can be deduced. In this section, methods will be presented for counting these events with a satisfactory confidence of estimation. Our investigations are based on the methods of mathematical statistics and will therefore not be detailed; readers interested in mathematical details are referred to the literature (*e.g.*, [10], or any book on mathematical statistics). Furthermore, "error counts" and "bit intervals" will be used in the following investigations. However, the results obtained in the following can also be applied for cases in which pseudo errors are counted, or the counting does not include all bits.

According to the methods presented in Sections 11.2.2, 11.2.3 and 11.2.4, the occurrence number of given events has to be counted, and the probability of occurrence has to be deduced from this number. Furthermore, in most cases, it is not the actual value of the error probability which has to be estimated but rather the time instant, at which the error probability has exceeded a permitted limit, has to be recognized (this limit was denoted by P_s in Chapter 4).

11.3.1 Channels of constant performance

Let N be the number of errors counted during a time interval t_M. We then have

$$\hat{P} = N \frac{t_M}{T} \tag{11.1}$$

where T is the bit time and $\hat{P}$ is the estimated value of the error probability. The confidence of this estimation is obviously dependent on N. For instance, if no errors are detected during a measuring period this will mean $P_E = 0$ with very bad confidence. The known methods of mathematical statistics allow the determination of the confidence of measurement. We are only interested in the range of error probabilities below 10^{-3}, and it can thus be assumed that the individual bit errors are statistically independent events which thus follow the binomial distribution. This means that for measuring the error probability, the parameter of a binomial distribution has to be determined. It is also known [10] that the length of the measurement interval should be chosen to comprise 10 errors in order to achieve a 90 per cent confidence level and a ± 50 per cent confidence interval. (This means that if the number of errors counted is 10, the probability of obtaining an error ratio in the range of $10T/t_M \pm 50\%$ is 90 per cent. In the following, the upper and lower limit of this range will be denoted by P^+ and P^-, respectively.)

Based on the above considerations, our error counting device can now be designed: the counter should be reset at time intervals of t_M but if 10 errors occur within this interval, an alarm signal should be generated. In the given case, this signal should be generated when the error ratio is P_s, i.e. $t_M = 10/P_s$ according to Eq. (11.1). In a somewhat more accurate variant of this method [14], an up–down counter is used instead of a simple counter. According to this variant, the counter is stepped forward at every error by a step of t_M/T, and stepped backward by one at every consecutive clock signal. In this case, the alarm signal is generated when the counter reaches a count of $10t_M/T$. This results in a quasi-continuous error counting, in contrast to the preceding, quantized method.

11.3.2 Channels of rapidly degrading performance

The error counting procedures discussed in the previous section are suitable if no quick intervention is required, and the (tacit) neglection committed previously has no significance: we have not investigated the process during which the performance is being degraded, but the high error probability has been regarded as given. Actually, however, the performance degradation has some speed, allowing a stand-by switch-over before the performance has become unacceptable. This can be the case for example if the performance degradation is due to multipath fading, resulting in a finite speed of the performance degradation. If the performance degradation is recognized rapidly enough and another satisfactory channel is available (this has a high probability in the frequency diversity system discussed in Section 4.4.5), then the switch-over to that channel can be accomplished in due course thus eliminating the performance degradation of the connection.

In view of the preceding considerations, the effective application of for example frequency diversity requires the fast recognition of the onset of degradation. In the following, methods for measuring rapidly degrading error probabilities will be investigated. Utilizing the model described in Section 4.3.3, the worst case speed of the signal-to-noise degradation can be ascertained if the speed of the fading loss increase, measurable at a single frequency, is known. In the following, the method for estimating the error probability from this knowledge will be investigated.

The number of errors can now be written through generalization of expression (11.1) as

$$N = \frac{1}{T} \int_{-t_M}^{0} P_E(t) \, dt. \tag{11.2}$$

Equation (4.10) shows that P_E depends exponentially on the signal-to-noise ratio R. The Laplace method for asymptotic integration can be applied as R is high enough even in the worst case (P_E is low enough, less than 10^{-3}). Thus, for the evaluation of the integral (11.2), it is sufficient to know the derivative of the exponent in expression (4.10), in the vicinity of its maximum value. This will be at $t=0$ as the channel is investigated during the degradation period. Thus we have, after simple re-arrangement and substitution,

$$N = \frac{-\hat{P}(0)}{rT \ln \hat{P}(0)} \{1 - \exp [rt_M \ln \hat{P}(0)]\}. \tag{11.3}$$

On the other hand, by solving (11.3) for $P(0)$ knowing N, we get an estimate for the error probability at $t=0$; this can easily be carried out by knowing r and t_M. It can be seen without numerical investigations that in this case, arbitrarily low error probabilities cannot be measured. Of course, the lowest mesurable error probability depends on r, i.e. the faster the performance degradation the higher the lowest measurable P_E.

Our next task is the determination of the optimal N and t_M value. High speed systems will primarily be investigated for which, according to Chapter 4, the performance degradation is mainly due to multi-path distortion. We shall arrive at a surprising result according to which the measurement time has an optimal value for a given confidence and r. Let us first of all investigate the confidence properties under the present circumstances. Theoretically, this is much more complicated than the case of P_E=constant because, owing to the distortion, the individual bits (or, using a statistical term, the experiments of a series), are not independent of each other. The problem is much simplified in practice by taking into account the orders of magnitude. The highest error probability to be investigated is 10^{-3}; this can only be measured by many experiments which yield only rarely "favourable" results. Thus by assuming that the results of the experiments are independent of each other the error thus committed will not be essential. This has the consequence that the error distribution can again be regarded as binomial (in the range of interest). On the other hand, however, the parameter of the distribution is not constant. This variation can, however, also be neglected in the investigation of the time interval order of magnitudes, because even in the worst case, rT is much less than unity. (The following considerations can be taken as an example. It has been mentioned in Section 4.3.3 that in 140 Mbit/s transmission, r is not higher than ten times the rate of carrier-to-noise ratio decrease as measured at a single frequency, even in a 64QAM system without equalization which is regarded as the worst system, and the highest value of this decrease is always less than 100 dB/s. This yields $rT=1.5 \times 10^{-7}$ as the worst value. This means that even the short periods of the degradation process during which the error probability can be assumed to be constant comprise a high number of bits, i.e. a high number of experiments.) It follows from the above considerations that the relation between the number of errors and the confidence interval in periods of constant performance are, with good approximation, also valid for periods of degrading performance.

The parameters of the error counting device should thus meet the following requirements:

(i) the confidence interval should be small enough, say ± 50 per cent;
(ii) however, this confidence interval should take into account both ambiguity factors: measurement uncertainty is caused both by the non-deterministic nature of the measured quantity and also by the fact that the parameter r is also non-deterministic, its magnitude being anywhere in the range of zero to its worst case magnitude.

Lacking more accurate data, let us assume that all r values within the interval $0 < r < r_{max}$ have equal probabilities. The number of errors N will be highest for $r=0$ and lowest for $r=r_{max}$. The ratio of these quantities is given by

$$Q = \frac{-rt_M \ln P_N}{1 - \exp[rt_m \ln P_N]} \tag{11.4}$$

where $r=r_{max}$ has to be substituted, and P_N is the error probability pertaining to N errors at maximum r. This means that for a given confidence level, the highest error probability for the number of errors N is again P^+, but the lowest error probability is P^-/Q. This is explained by the possibility of having few errors during rapid degradation or having many errors during slow degradation. It thus follows that P^+/P^- has to be smaller than the value following from the confidence limits, i.e. N should be chosen higher than required for the case of $P_E = $ constant.

It is seen from the above considerations that the measurement time has really an optimal value: if t_M is too long, Q will be large so that P^+/P^- should be close to unity; this requires a high N value which can only be obtained at high error probabilities. If t_M is too short, again only high error probabilities can be measured because only then will N be high enough, as required by the confidence interval. Figure 11.5 shows the lowest measurable error probability *versus* t_M for different r parameters. The optimal values are also given in Table 11.1

Note that the numerical values of this table have been compiled on the assumption of continuous error counting, as explained at the end of Section 11.3.1. Should the counter be reset at intervals of t_M the optimal values of t_M and N will not change, but the lowest measurable error probability will be higher by 10 to 20 per cent.

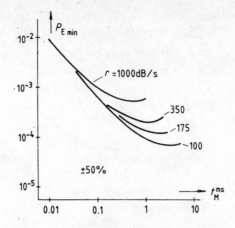

Fig. 11.5. Lowest measurable error probability as a function of measurement time for the case of an overall C/N decreasing at a speed of less than r dB/s.

Table 11.1 Optimal parameters of error counting for 140 Mbit/s transmission; ± 50 per cent confidence

$r^{\text{dB/s}}$	$t^{\text{ms}}_{\text{opt}}$	P_{min}	N_{opt}
100	3.15	6×10^{-5}	20
175	1.9	1×10^{-4}	20
350	1.0	2×10^{-4}	20

11.4 Transmission of service information

In the following, we will be concerned with the transmission methods of the service signals, and the sources of these signals (*e.g.,* voice signals, telecontrol/ telecommand signals, stand-by switching signals, *etc.*), and their multiplexing will not be dealt with (although, for completeness, the service multiplex equipment is also shown in Fig. 1.4, referred to earlier).

In wirebound systems, the service signal is normally transmitted over a separate pair. The analog solution in microwave transmission would be a separate transmission channel. This solution, though sometimes applied in earlier equipment, is no longer used because of its cost ineffectiveness.

In analog radio relay systems, the method of transmission depends on the main information transmitted, but in most cases, the service information is multiplexed by frequency division to the main information. This is possible since the lower limit of the FDM signal frequency range is always an accurately defined limit frequency [11]. Below this frequency, there is normally space for the transmission of the service signal (sub-baseband system). The spectrum of television signals has a defined upper limit frequency f_u, allowing the transmission of the service information by frequency modulation of a subcarrier of frequency $f_s > f_u$ (super-baseband system). A few methods applicable for digital transmission will be outlined in the following.

11.4.1 Exclusively sub-baseband transmission

First of all, it should be noted that there is no "free" frequency range in the spectrum of digital signals, at least in the spectrum of NRZ or M-ary NRZ signals (see Section 2.3.1). On the other hand, the spectral density of certain codes is zero at $f=0$, so that the range of $0 < f < f_0$ can be regarded as free if $f_0 \ll 1/T$. (An example of this case is the biphase code.)

Let us now investigate the effect on signal transmission of the band limiting of NRZ signals by a high-pass filter with a cut-off frequency $f_c \ll 1/T$. Obviously, the waveform of the pulses will be tilted, and this results in some kind of inter-symbol interference (see Fig. 11.6). The effect of this filter can be neglected if there are less than k identical consecutive bits in the sequence and $1/f_0 \ll kT$. This means that the error probability of a transmitted random sequence is not significantly increased if the signal is band limited by a high-pass filter which has a cut-off frequency of, say, 1 to 2 per cent of the symbol rate $1/nT$.

These considerations apply to systems which can be regarded as "linear". For instance, if digital signals are transmitted over an analog system, the sub-baseband service signal transmission is applicable if biphase line encoding is used. With some acceptable restrictions, the same is true for NRZ line encoding too. The transmitter part of this system is shown in Fig. 11.7.

However, most transmission systems are not linear. In most cases, modulators are digital, and the demodulation process involves essential nonlinear operations. For these systems, other methods have to be applied for the transmission of the service signals.

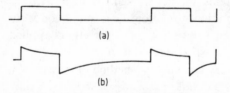

(a)

(b)

Fig. 11.6. Waveforms of a binary NRZ signal: (a) original waveform; (b) waveform at the output of a high-pass filter.

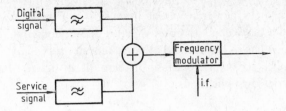

Fig. 11.7. Sub-baseband service signal transmission in an FM radio relay system.

11.4.2 Transmission within the multiplex frame

In all hierarchical levels of PCM transmission as recommended by the CCITT, there are a few bits in the multiplex time-frame which can be freely utilized [13]. These bits can be directly applied for the transmission of the service signals. For instance, in a 30 channel primary PCM multiplex system, there are 7 free bits in the zero time slot of two consecutive frames, realizing a 28 kbit/s channel which is suitable for the transmission of a suitably encoded voice signal (*e.g.,* by adaptive delta modulation). In addition, 4 free bits/multi-frame are available, with a total speed of 2 kbit/s; this can be used for the transmission of telecommand and tele-control signals. The speed of this channel can be 2 kbit/s, or less if these signals are homochronous or heterochronous with the multiplex signal, respectively. Figure 11.8 shows this kind of system.

In the case of transmission of higher-order digital multiplex signal sequences, this kind of service channel of the tributaries, and further an additional bit stream of the order of 10 kbit/s, may be available.

The above utilization of free bits corresponds to the sub-baseband system utilizing a free frequency range. This system seems to be universally applicable, but has a disadvantage: the transmission of the service signal, which is in connection with the transmission system, is combined with the multiplex system. The transmission

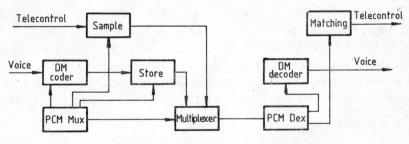

Fig. 11.8. Service signal transmission in the PCM frame.

system thus becomes non-transparent because it will be suitable only for the transmission of a particular multiplex signal. Further, some multiplex equipment at least simplified has to be installed at every repeater station at which the service channel should be accessible. Because of these disadvantages, this solution may have only restricted applications.

11.4.3 Transmission by increased bit rate

While the method presented in the preceding section was the time division analog version of the sub-baseband system applied in FDM systems, the method we describe in this section corresponds to the super-baseband system. This method is a further development of the error counting procedure explained with reference to Fig. 11.3: in addition to (or in place of) the pseudo noise signal shown in this figure, the encoded service signal is also multiplexed to the digital main information. The structure of the multiplex and demultiplex equipment corresponds to that shown in Fig. 11.3, with suitable extensions and modifications. This method is used in TDMA satellite systems in which no repeater stations are present, and the frame structure easily allows the insertion of the service signal. It is also used in high speed terrestrial systems if the high amount of service information cannot be transmitted by other means.

11.4.4 Transmission by additional modulation

According to this method, the service signal is transmitted independently of the main information, by an additional modulation of the carrier which differs from the modulation due to the main information. In an angle modulation (*i.e.* PSK or FSK) system, the carrier amplitude can be modulated by a narrow-band service signal. In PSK or QAM systems, the carrier frequency can also be modulated by the narrow-band service signal.

Transmission by amplitude modulation will not be treated because of its smaller significance. Using this method, the service signal frequency range has to be very narrow, less than 10^{-4} times the bit frequency, which is not sufficient in most applications. On the other hand, a service signal transmitted by frequency modulation is advantageous and is frequently applied. In the following, the design of this system will be dealt with in somewhat more detail, based on Ref [15, 16, 17].

11.4.5 Voice Under Data (VUD) system

This is a system suitable for the transmission of hybrid (*i.e.*, combined analog and digital) information in which the digital information content is much higher than the analog information content. This system can thus be applied for the transmission of a service signal available in analog form. It is characterized by transmission of the digital signal by either phase modulation (PSK) or combined phase and amplitude modulation (QAM), while the carrier frequency is modulated by the analog service signal. If the main modulation is QPSK, then the frequency range of the analog signal can be 0.2 to 0.3 per cent or even 1 to 1.5 per cent of the digital signal frequency range (with acceptable performance). If the analog signal is made up of FDM telephone channels, the number of FDM channels may reach 3 to 4 per cent, or even 10 to 12 per cent of the number of PCM channels, essentially without additional investment. (The designation of this system is a suitably modified form of the known Data Under Voice designation.) In the following investigation, QPSK digital modulation is assumed but the calculation method can also be applied to QAM modulation by suitable modification.

Figure 11.9 shows the block diagram of the VUD system. The frequency of the transmitter oscillator is directly modulated by the analog signal. The differential encoder of the digital path is also shown because it now has high significance. In the receiver, the carrier recovery loop also has the function of frequency demodulator. (The frequency modulation cannot be recovered from the suppressed carrier signal containing digital phase modulation.)

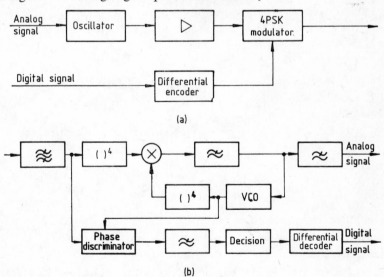

(a)

(b)

Fig. 11.9. Voice Under Data system: (a) transmitter; (b) receiver.

The digital modulation signal is a random NRZ (or in exceptional cases, a biphase) pulse train. The analog modulation signal is band limited and random. Its statistical properties are strongly dependent on the specific case but can be assumed to have a normal distribution if it is made up of FDM voice channels. Our investigations will be restricted to this case, and we shall even assume that the modulation process is white and has unit power. With these assumptions, the spectral density of the analog modulation process is given by

$$S_h(\omega) = \pi/a; \quad |\omega| < a = 2\pi f_a. \tag{11.5}$$

The time function of the phase due to the analog modulation is given by

$$m(t) = \Delta\omega \int^t h(t)\, dt \tag{11.6}$$

where $\Delta\omega = 2\pi\Delta f$ is the effective frequency deviation.

Figure 11.9 gives the impression that the digital and analog information paths are completely independent, and this is also true in the ideal case. Practically, however, the loop has a finite bandwidth and is thus unable to follow perfectly the analog frequency modulation, and a phase error will thus be present. When the instantaneous value of the phase error reaches π, then a cycle slip, or more precisely a phase jump, will occur causing a single error after differential decoding (see Section 3.4.3). This results in a residual error using the terminology of Chapter 4. On the other hand, intermodulation noise (also called "pattern jitter") is generated by the carrier recovery circuit because of the band limiting effect of the preceding filter. This appears as additive noise in the received analog signal (see Section 9.6) which is independent of the received signal level.

The system design thus requires determination of following parameters: the deviation $\Delta\omega$ yielding the satisfactory signal-to-intermodulation noise in the analog path; and the loop parameters so that the allowable residual error should not be exceeded. Finally, it should be checked whether the system meets the deep fade requirements.

For a given i.f. bandwidth, the signal-to-intermodulation-noise ratio can be calculated using the method given in Section 9.6: following FM demodulation, the signal-to-noise ratio in the analog channel is calculated to be

$$S/N = 48R_i(\Delta f/f_a)^2 \tag{11.7}$$

by taking into account that after frequency quadrupling, the frequency deviation will also be four times as large. From this equation, the frequency deviation required for a given signal-to-noise ratio can be calculated.

The phase error of the loop results from the additive noise and from the imperfect tracking of the frequency modulation. The additive noise itself has also two components, the thermal and the intermodulation noise. In the zeroth approximation, the variance of the phase error φ can be determined from the linear theory. The variance is utilized in higher-order approximations also taking into account nonlinear effects, so this will be calculated first.

It can be assumed that the three components of φ are independent of each other, allowing their variances to be summed:

$$\sigma^2 = \sigma_m^2 + \sigma_i^2 + \sigma_t^2, \tag{11.8}$$

where the terms correspond to modulation, intermodulation and thermal noise. By taking into account (11.5), we have

$$\sigma_m^2 = \int_0^a \frac{16\Delta\omega^2}{a} \frac{d\omega}{|j\omega + 4KAF(j\omega)|^2} \tag{11.9}$$

which, for a relatively wideband second-order loop $(a/W_L < 0.1, r=2)$ can be expressed as

$$\sigma_m^2 \cong (2\Delta f/f_a)^2 (3\pi f_a/W_L)^4 \tag{11.10}$$

and further

$$\sigma_i^2 + \sigma_t^2 = N_0' W_L/2 \tag{11.11}$$

where N_0' is the spectral density of the additive noise which is assumed to be white, by taking into account the quadrupling effect (see Section 3.4).

The preceding notation corresponds to those used in Section 3.2: W_L is the double sided bandwidth of the loop, $F(j\omega)$ is the loop transfer function, r is a loop parameter, and A is the input signal r.m.s. value.

In the following, the nonlinear effects will be taken into account by first approximation. According to Ref. [18], the effective input signal-to-noise ratio of the loop is given by

$$\alpha = \varrho \frac{r+1}{r}\left(1 - \frac{r+1}{r}\frac{2\Delta\omega}{W_L}\frac{X}{\sigma_{\sin\varphi}^2}\right) - \frac{1}{r\sigma_{\sin\varphi}^2} \tag{11.12}$$

where $\varrho = 2A^2/W_L N_0'$, the actual loop signal-to-noise ratio, $\sigma_{\sin\varphi}$ is the standard deviation of $\sin\varphi$, and X is the expected value of the product of $h(t)$ and $\sin\varphi(t)$. The latter two factors can be determined, for example, by the spectral method published in Ref. [19]. Applying this method results in the following final expression:

$$\alpha = \varrho\frac{r+1}{r}\left[1 - \eta\sigma^2\frac{4\sigma^2}{1-\exp(-2\sigma')^2} - \frac{2/r}{1-\exp(-2\sigma')^2}\right] \tag{11.13}$$

where

$$\sigma'^2 = \sigma^2(W_L\gamma^{1/2}) + 1/\varrho\gamma, \tag{11.14}$$

$$\gamma^2 = \frac{1}{2\sigma^2}[1-\exp(-2\sigma^2)]; \quad \eta = \exp(-\sigma^2/2) \tag{11.15}$$

where the first term in (11.14) means that $W_L \gamma^{1/2}$ should be substituted in place of W_L.

Knowledge of these quantities allows determination of the loop bandwidth for which the residual error P_{E0} will be minimal — this will almost be equal to the bandwidth for which σ^2 is minimal. It has been shown in Section 3.4.3 that $P_{E0} = T/T_0$, and the average time T_0 between cycle slips can also be determined according to this section. All these considerations allow the design of the VUD transmission if the number of service channels to be transmitted is known.

In the VUD system described above, the service signals are available in analog form and are multiplexed on an FDM basis. However, in many cases, even in those investigated in the previous sections, the service signals are available in digital form. A similar method can then also be applied, *i.e.* the service signal can be used to modulate the carrier frequency while the main information is transmitted by PSK or QAM. The design can then follow similar principles (see Ref. [20]), but details will not be given because of lack of space. A typical example is the following: along with 34 Mbit/s QPSK transmission, an FSK signal of 100 to 150 kbit/s speed can be transmitted with practically no performance degradation of the main information, and with acceptable quality of the additional signal. Along with 140 Mbit/s 16QAM transmission, an FSK signal of 100 to 200 kbit/s speed can be transmitted.

11.5 Stand-by channel systems

11.5.1 General considerations

The reliability of the transmission path required by the communication networks cannot be fully met by the equipment applied no matter how up-to-date the production technology used is, nor how reliable components applied are. This has several reasons. First of all, not even is the channel itself absolutely reliable because in a finite fraction of the time it does not provide the specified quality, *i.e.* in this time fraction it is not available. We have seen in Chapter 4 that only a small part of the allowable outage time percentage is allotted for equipment failure. Thus even in the case of high MTBF (Mean-Time-Between-Failures), the expected availability can only be guaranteed if the MTTR (Mean-Time-To-Repair) is very small. In the case of unattended repeater stations, even the traveling time to reach the failed station is in itself longer, which can be accepted as MTTR. Further, the reliable operation of some subsystems, *e.g.* the primary power supply, may be outside the competence of the equipment manufacturer or operator. For these and other reasons, a stand-by path is always required. If this is provided, the

MTTR is reduced to the time required for switch-over which is of the order of seconds or milliseconds or, in some cases, even zero.

Normally, communication networks include several nodal stations which can be switching or computer centres or other equipment, and these are connected by transmission paths. It is beyond the scope of this book to investigate the conditions for meeting the reliability objectives of the complete network or the realization of stand-by systems by roundabout paths. In the following, only the stand-by methods applied for a given microwave transmission path will be investigated.

There are several stand-by methods used for radio relay systems. According to the so-called station stand-by system, all main equipment parts such as transmitters, receivers, modulators, *etc.* are redundant at each station, an operating and a stand-by equipment part being available. In the case of a failure in the operating station, ascertained on the spot, a switch-over to the stand-by station takes place. However, as this significant stand-by equipment is normally not required by the reliability of the station, the so-called channel stand-by system is generally pre-

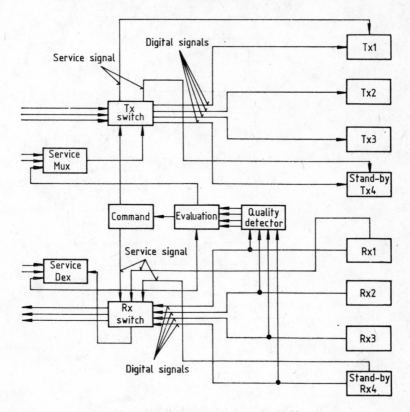

Fig. 11.10. 3/1 stand-by channel system.

ferred. According to this system, k stand-by channels are applied in addition to the m operating channels along the same route; this is the so-called m/k system, and in most cases, $k=1$. The complete transmission path is divided into switched sections. In the case of a failure at any station of a channel, the signal driving the transmitter of the failed channel at the beginning of the switched section is

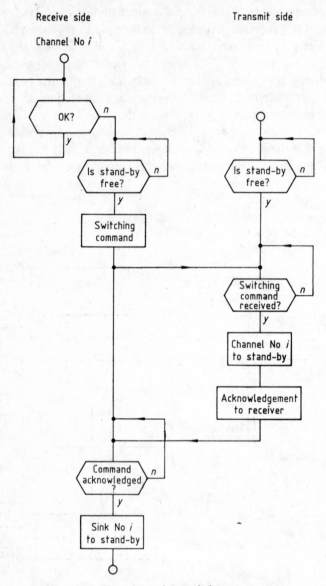

Fig. 11.11. Flow-chart of the switch-over process.

switched over to the stand-by transmitter, and at the end of the switched section, the signal coming from the stand-by receiver is passed on to the next section. The implementation of this switching process normally requires a dialog between transmitter and receiver which is carried, as mentioned earlier, by the service channel.

As an example, Fig. 11.10 shows a 3/1 stand-by system, and the flow-chart of the complete switch-over process is shown in Fig. 11.11. The main criterion of channel failure is the error probability exceeding an allowed limit. An additional criterion may be the failure of some selected circuits which does not result in an increased error probability, but in an interruption of the output signal. The two criteria are then combined in a logical OR gate.

There are two kinds of switching systems. According to the first, the process shown in Fig. 11.11 is realized without imposing severe restrictions on the time needed by the process. In this case, it is sufficient to count the errors according to the method given in Section 11.3.1, and the connection will then certainly be interrupted for a time interval determined by several factors. According to the second switching system concept, no interruption at all is endeavoured, requiring a switch-over process in zero time. This kind of switching, called hitless switching, can be realized in certain cases.

11.5.2 Subsystems included in the stand-by operation

The subsystems included in the stand-by operation depend on the error counting method applied (see Section 11.2).

If error detecting encoding is used, the decoder should be at the end point of the switched section. In this case, a failure at any point along the path can be detected, *i.e.* the complete system between the encoder and decoder can be included within the stand-by system. This means that in practice, only the error detecting codec and the circuits outside the codec section are excluded from the stand-by system. (Generally, only the HDB or CMI codec falls into the latter category — see Chapter 10). This system can thus be qualified to be a good stand-by switching system if circuit difficulties can be tolerated.

If the framing word errors are counted, then the complete transmission system is included within the stand-by system. However, it is not an advantageous method because of its low speed and the non-transparent nature of the transmission system.

If the pseudo errors are counted the error indication is present only at the failed station. In this case, an error counter is needed at each repeater station, and a separate channel is needed over which the failure is reported to the end point of the switched section. The transmission channel itself can also be used for this reporting if a non-hitless switching is satisfactory. For hitless switching, an omnibus data channel, accessible at every station, is required for this purpose. In spite of all these difficulties, this last method frequently proves to be the most feasible.

11.5.3 Non-hitless switching

This method is preferred because it is much more simple than the hitless switching system, but one must first decide whether or not the resulting interruption time can be tolerated.

If non-hitless switching is applied then interruption will certainly occur for the following reason. We have seen that the frequency of the stand-by channel differs from the frequency of the operating channel; the two signal paths are also different because of the different filters, cable and transmission line lengths, *etc.* This means that the delay of the two paths will also be different, *i.e.* the arrival time of a given bit transmitted over the operating and stand-by channel will be different at the output ports of these channels. This has the effect in high speed systems that in the switching instant, a few bits will be lost (if the delay of the stand-by channel is lower), or a few bits will be repeated (in the opposite case). This, in turn, will have the consequence of frame synchronization loss of the transmitted signal (which certainly has some kind of block or frame structure). The recognition and restoration of this synchronization requires time. In fact, if the transmitted signal has a multilevel multiplex structure (*e.g.* it is made up of primary, secondary, third-order multiplex components), all may loose frame synchronization, and re-synchronization can only follow successively. Table 11.2

Table 11.2. Outage time due to
framing loss

Multiplex order	Outage time, ms
1st	1
2nd	1.8
3rd	2.2
4th	2.3

summarizes the highest outage times for transmitted signals according to the CCITT European hierarchy and framing strategy.

In the case of voice transmission, an outage time even an order of magnitude higher will have no perceivable effect. This means that if only voice transmission is required a higher outage time than the inherent time given in table may be permitted, thus allowing the transmission of the failure information over the main transmission channel. Note, however, that this information is not automatically available: if the error probability has become too high at the Lth repeater station then, owing to regeneration, no pseudo errors will be experienced at the $(L+1)$th station. However, pseudo errors can be artificially generated, *e.g.* by intentionally modifying the timing of the outgoing signals upon receiving the error information. It also follows from Table 11.2 that during a non-hitless switch-over procedure, there is no need to carry out the switch-over process rapidly, *e.g.* over a high speed data channel.

11.5.4 Hitless switching

It was shown in Chapter 4 that frequency diversity is an effective multipath countermeasure even if a simple adaptive equalizer is used. We have also seen that a single diversity path for several operating channels can also be effective. It has finally be shown that the diversity effect is higher for a smaller frequency difference between channels. From the viewpoint of our investigations in this chapter, this means that an $m/1$ stand-by channel system can, in principle, be used as a multipath countermeasure, provided that the following conditions are met:

(i) the switch-over procedure commences as soon as the performance degradation of the channel is recognized, and the process is finished by the time the performance has become unacceptable (*i.e.,* the error probability has reached 10^{-3});

(ii) the transmission is not interrupted during the switch-over process, *i.e.* hitless switching is applied.

It follows from these considerations that a hitless switch is essential if frequency diversity is applied as a multipath countermeasure. However, the hitless property of the switch can be used to avoid equipment failure as well, notably in the case where the time constant of the degradation is not lower than that of the fading rate of the degradation. In most cases, power supply failures, and possibly others, may fall into this category.

The main criterion for designing this system is the knowledge of the available switch-over time. This can be determined by taking into account the considerations in Section 11.3.2, in which the lowest error probability to be measured by an optimally designed error counter has been determined. The available time, t_s, is that time during which the error probability is increased from this lowest measurable value to 10^{-3}. In Table 11.3, this time is given for a transmission speed of 140 Mbit/s.

Table 11.3. Time interval available for the switching process
(±50 per cent confidence)

$r^{dB/s}$	100	175	350
t_s^{ms}	16	8.5	3.4

Our next task is to determine the operations to be performed during this extremely short time. According to Fig. 11.11, we have

$$t_s = t_1 + t_2 + t_3 + t_4 \tag{11.16}$$

where the following notation is used:

t_1 time needed for alarm transmission,
t_2 time needed for status investigation of the stand-by channel,
t_3 time needed for transmission of telecommand and acknowledgment signals,
t_4 time interval of logic operations needed for switching.

t_1 and t_3 can further be subdivided into the propagation time, data transmission time and circuit delay time (t_{k1}, t_{k2} and t_{k3}, respectively). These will now be investigated individually.

The alarm signal and main information signal (being in the status of degradation) are propagated over the same path, thus $t_{11} = 0$; if the alarm is transmitted by b_1 bits and the data channel speed is r_1 bit/s, then the time needed is given by

$$t_{12} = \frac{b_1}{r_1} \tag{11.17}$$

and by estimation,

$$t_{13} = \frac{M}{r_1} \tag{11.18}$$

where M is the number of repeater stations.
The following estimations can be given:

$$t_{31} = \frac{2MD}{c}; \quad t_{32} = \frac{b_2}{r_2}; \quad t_{33} = \frac{1}{r_2} \tag{11.19}$$

where D is the distance between repeater stations and c is the speed of light.
In some cases, the circuit operation time can be neglected (this is certainly not true for microprocessor implementation).

The preceding considerations and relations allow the design of the complete system. It is seen that the propagation time cannot be neglected, so the switched sections should not be too long, and several switching main stations are required along a long connection. It is also seen that the data channel speed has high significance. In several practical cases, extremely high data transmission speeds and short switched sections may be required which cannot be realized by economical considerations. The error counting speed has then to be increased, and the only means of doing this is the counting of pseudo errors rather than real errors.

We now come to the design problems of the switch. We repeat that the switch has to equalize the delay difference between the two channels and the switch-over process has to be completed before the channel performance has become unacceptable. The switch-over process is thus not recognized by the information sink. In the complete switch-over system, transmit side and receive side switches are obviously included. The transmit switch-over process does not need special consideration. In accordance with Figs 11.10 and 11.11, the task here is simply to connect the signal driving the channel under degradation also to the input of the stand-by channel, as soon as the instruction for doing so has arrived from the receive side. The receive side switch is responsible for the delay time equalization and thus needs more consideration.

The delay equalization has to start as soon as the stand-by channel carries the required signal which appears without error at the output of this channel, and the clock recovery process has been completed. The delay equalization process involves the two following operations:

(i) the signal given by the operating receiver has to be sampled by the stand-by receiver clock signal, *i.e.* the clock signal has to be switched over. The two signals transmitted over separate paths are then isochronous, their transitions coincide, but they may be shifted with respect to each other by an integer number of bits;

(ii) following this operation, the signal transmitted over the stand-by channel has to be switched to the output, after insertion or exclusion of a suitable number of bits.

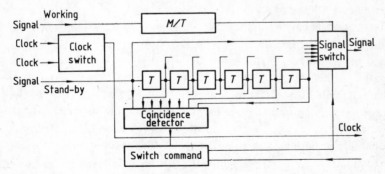

Fig. 11.12. Principle of a hitless switching.

An arrangement suitable for this task is shown in Fig. 11.12. It is seen that the integer part of the delay difference is adjusted by the tapped line, and the fractional part of the delay difference is adjusted by the clock synchronizing circuit. A few explanatory comments are given in the following.

A simple switch-over of the clock signal would not result in optimal sampling time instants, and the sampling edges of the stand-by channel clock signal could possibly even coincide with the transitions of the signal to be sampled, resulting in an unusable signal. This is avoided by writing the signal by the operating channel clock into a buffer store, and reading it by the stand-by channel clock. The buffer store may be quite simple.

Following the recovery of the operating channel, switch-back should be initiated as soon as possible so that the stand-by channel should be available for another possibly degrading operating channel. This requires the inverse of the preceding operations. However, the switch-back should be carried out with hysteresis in order to prevent several switch-over and switch-back operations: the error probability for switch-back should be at least one order of magnitude lower than that for switch-over. Here, the speed has no high significance, and the change of the error probability need not be taken into account.

Finally, the operation principle and design of the coincidence detector in Fig. 11.12 will be reviewed, noting that other principles may also be applied. The obvious task of the coincidence detector is to find the suitable tap which results in the correct delay. This is similar to the known frame synchronization task, with some simplification. As the two bit sequences are precisely equal, a bit-by-bit comparison can be carried out. This results in a relatively long time available for recognizing the coincidence. (This seems to contrast with our statement, repeated several times, according to which the switch-over has to be carried out quickly. However, the fading time constant is always longer than 1 ms. For instance, during 140 Mbit/s transmission, the number of bits in only 1 per cent of this time constant is 1400.)

By taking into account this characteristic, the following strategy can be applied. A complete coincidence of K consecutive bits in the two channels is required, and this investigation is carried out L times. If among these LK bits, K consecutive bits coincide, then the investigation is rated as positive, and the tap in question is connected to the output. If this is not the case, the next tap is selected, and the investigation is repeated.

It should be noted that this investigation may give a false result for two reasons: either false detection may take place because of unfavourable instantaneous properties of the bit sequence, or no coincidence is detected because of faulty bits. The probability of both errors can be determined by elementary but lengthy probability calculations, and these allow the design of the coincidence detector.

The starting point is the delay difference to be equalized, expressed by the corresponding number of bits, given by $\pm M$.

False detection is possible if the signal is periodic and the period is not longer than $M+B$ where B is the actual delay difference between the two transmission paths, expressed with the number of bits. The probability of false detection is then given by

$$Pr[fd] = 2^{-K} \frac{M(3M+2)}{2M+1}. \tag{11.20}$$

The actual coincidence will not be detected if at least one bit from K bits is detected erroneously; the probability of this event is denoted by $Pr(nd)$. If the investigation is repeated L times then the probability of missed detection will be $[Pr(nd)]^L$. A few calculated results are shown in Table 11.4, by assuming that the

Table 11.4. Parameters of the coincidence detector $Pr(nd)^L = 10^{-6}$

M	$Pr(fd)$	K	L	LK
3	10^{-8}	32	4	128
	10^{-6}	25	4	100
	10^{-4}	18	3	54
5	10^{-8}	35	4	140
	10^{-6}	28	4	112
	10^{-4}	22	4	88
10	10^{-8}	44	4	176
	10^{-6}	38	4	152
	10^{-4}	31	4	124

probability of missed detection is 10^{-6}.

In conclusion, the required time will be estimated, by assuming $M=5$ which seems to be plausible, an error probability of 10^{-3} and a false detection probability of 10^{-6}. The expected time of the coincidence detection is 1.1μs, and its worst case value is 9 μs.

11.6 References

[1] *CCITT Com. XVIII*, No–1E, 1976.
[2] *CCITT Recommendation G. 703*, Orange Book, Geneva 1977.
[3] *Handbook for Telecommunication*. Budavox, Budapest 1975.
[4] *CCITT Recommendation V. 26 bis*, Orange Book, Geneva 1977.
[5] Juichia, Y., Yosikava, M., Morita, H.: 20 G–400 M digital radio relay system. *Rev. Electr. Commun. Lab.*, Vol. 24, July–Aug., pp. 52–540, 1976.

[6] *CCIR Report 613-1*, XIII-th Plenary Assembly, Geneva 1975.

[7] Sant' Agostino, E.: Performance monitoring devices for digital radio-relay links. *CSELT Rapp. Techn.*, Dec., pp. 13–20, 1975.

[8] *CCIR Doc. 9/1069,* Interconnection of digital radio-relay systems, 1974–78.

[9] Almási, P., Frigyes, I., Ottó, E.: Method for the continuous monitoring of bit error rates of digital transmission systems. *Hung. Pat. Spec. No. 170878, 1974.*

[10] Burington, R., May, D.: *Handbook of Probability and Statistics.* Handbook Publishers. Sandusky, Ohio 1953.

[11] *CCIR Recommendation 380-3,* XIII-th Plenary Assembly, Geneva, 1975.

[12] Feher, K., Goulet, R., Morisette, S.: Order-wire transmission in digital microwave systems. *IEEE Trans. Commun.*, Vol. Com-22, May, pp. 676–681, 1974.

[13] *CCITT Recommendations G. 734, G. 744, G. 751,* Orange Book, Geneva 1977.

[14] McNicol, J., Barber, S., Rivest, F.: Design and application of the RD–4A and RD–6A 64QAM digital radio systems. *IEEE ICC'84*, paper No 22.4, Amsterdam, The Netherlands 1984.

[15] Frigyes, I.: Voice Under Data: a method for the transmission of hybrid information over microwave communication links. *Proc. 8-th European Microwave Conference*, pp. 615–619, Paris 1978.

[16] Frigyes, I.: Voice under data system transmitting hybrid information. *Budavox Telecommun. Rev.*, No. 1, pp. 12–21, 1981.

[17] Frigyes, I.: Comments on *M*-ary CPSK detection. *IEEE Trans. Aerosp. Electron. Syst.*, Vol. AES-18, No. 3, pp. 345–347, May 1982.

[18] Lindsey, W., *Synchronization Systems in Communication and Control.* Prentice-Hall. Englewood-Cliffs 1972.

[19] Tausworth, J.: A method for calculating PLL performance near threshold. *IEEE Trans. Commun. Techn.*, Vol. Com-15, pp. 502–507, 1967.

[20] Bi, Ph.: Simultaneous transmission of FSK and QPSK over radio channels (in Hungarian) Ph. D. Thesis, Hungarian Academy of Sciences, 1985.

SUBJECT INDEX